The Pests of Protected Cultivation

The Biology and Control of Glasshouse and Mushroom Pests

The Pests of Protected Cultivation

The Biology and Control of Glasshouse and Mushroom Pests

N. W. HUSSEY, B.Sc., Ph.D.

W. H. READ, M.Sc., F.R.I.C.

and

J. J. HESLING, B.Sc., A.R.C.S.

Edward Arnold (Publishers) Ltd. London

First published, 1969

SBN: 7131 2231 5

Printed in Great Britain by
Spottiswoode, Ballantyne and Co. Ltd.
London and Colchester

Preface

It is twenty years since the publication of that admirable book by Professor and Mrs. H. W. Miles, *Insect Pests of Glasshouse Crops*. The present volume seeks to bring the subject up to date and has been enlarged to include the pest problems of the expanding mushroom industry.

The last two decades have seen a rapid increase in the number of available pesticides though their intensive use has brought its own problems. The most serious is the development of resistance in mites, aphids and leafminers but the choice of pesticide is also limited by the phytotoxicity induced in many species and cultivars of plants grown under glass. The time has come to re-evaluate alternative non-chemical methods either for direct pest control or to lessen the pesticide selection pressures that lead to the evolution of resistant strains.

Although protected environments present many unique pest problems they also permit the exploitation of techniques which would be uneconomic elsewhere. These methods result in less pesticide usage which, in turn, may allow some pests which have largely disappeared in recent years to again assume importance. We have, therefore, included many species which are of only minor importance at the present time.

This book provides current information on the chemical control and biology of the pests of protected crops. The sectional bibliographies are not complete but the references cited provide the most useful sources of biological information. While the chemical controls recommended are applicable only to the United Kingdom the extensive biological data should be of interest to those concerned with glasshouse and mushroom culture anywhere in the world. Little space has been devoted to basic zoology as this is covered adequately elsewhere.

We are greatly indebted to Mr. B. D. Moreton who painstakingly criticized the manuscript and suggested improvements. Thanks are also due to Mr. W. J. Parr who is largely responsible for the section on thrips. Professor N. Jefferson, Dr. P. Osborne, Mr. P. W. Miles and the Shell Chemical Company contributed photographs and their help is gratefully acknowledged in the text. Miss M. Gratwick and Mr. Morrison of the Plant Pathology Laboratory also provided invaluable assistance by making available official photographs. Miss B. Gurney kindly provided the illustrations of Lycoriid genitalia.

Glasshouse Crops Research Institute, Littlehampton, 1969

N.W.H.
W.H.R.
J.J.H.

Contents

Chapter 1

The Pest Problem in Closed Environments

The term 'protected cultivation' could include almost all crops where some type of shelter is used. Such a wide definition would embrace all forms of temporary cloche culture, glasshouses, and the recently developed plastic 'balloon' coverings of water-cress beds as well as mushroom and rhubarb forcing houses. The vast assemblage of pests found in such situations would include many which are dealt with adequately in other books on horticultural and agricultural pests. The present volume is, therefore, restricted to those pests likely to be encountered in glasshouses or mushroom houses.

A number of special features govern the rates of increase and the limitation of pest populations in these protected environments. These features must be appreciated if the pest problems are to be fully understood. For convenience, the glasshouse and mushroom cultural environments are considered separately since the importance of problems caused by resistance to, and application of, insecticides are quite different in the two situations.

THE GLASSHOUSE ENVIRONMENT

Under glass plants grow in predominantly warm conditions; temperatures, even at night, rarely falling below 50 °F (10 °C). Air movement is very gentle, and only in sunny weather, when the ventilators are opened to prevent excessive air temperatures, is there even the slightest breeze. Relative humidity usually varies between 60–90%. Drier, or more humid, conditions outside this range are either unsuitable for plant growth or liable to encourage disease, and are therefore avoided. Within this equable environment the plants are regularly watered and fed so that growth is rapid and luxuriant. Often, a glasshouse is devoted to a single crop and so provides a large uniform environment in which an introduced pest finds ideal conditions for rapid multiplication. Abundant food is available, the relatively still and warm atmosphere facilitates insect flight, and hence spread, within the crop, while the temperature, between 60–70 °F (16–21 °C) for long periods, favours rapid breeding. To appreciate the advantages conferred on an introduced insect by such environments it is pertinent to consider the natural restraints acting on an insect population outdoors. For convenience, the factors tending to limit increase can be classified as physical and biotic. The physical factors are the summation of the environment (temperature, humidity, wind and rainfall) while the biotic factors include competitors, predators, parasites and disease.

Considering the environment first, how are outdoor populations affected by climatic variations? Temperature affects the rate of population increase by influencing the speed of development of immature stages as well as the longevity and fecundity of adults. Andrewartha & Birch (1954) introduced a statistic referred to as the *net reproductive rate* which, under conditions of unlimited food supply and protection from natural enemies, is a measure of the multiplication per generation. Nickel (1960), working on a species of spider mite, found that higher average temperatures caused a striking increase in the reproductive potential. At 86 °F (30 °C), the net reproductive rate was 128 compared with 111 at 77 °F (25 °C) and only 54 at 60 °F (16 °C). In 1 month, at 60 °F, a single female mite produced about 50 mites but, theoretically, at 77 °F a female could produce over 100 million. Many factors prevent this vast increase in numbers but the figures demonstrate the effect of higher temperatures on the biotic potential of glasshouse pests. Outdoors, excessive fluctuations in temperature within short periods cause much mortality, but severe weather, so long as it is seasonal, has little effect because the insects are normally in a stage of development adapted to withstand extreme seasonal conditions. Turning to humidity and precipitation, Nickel's data also showed that even a 70% change in relative humidity increased the net reproductive rate only fivefold. The most important influences of rainfall are more direct. For instance, insect flight is curtailed, thereby affecting both the spread of the species in a crop and, where these involve flight, its mating habits. Any insects washed to the ground may be killed. Interference with sexual flight rhythms is probably the most important effect of wind, although, in some circumstances, desiccation may cause death.

An insect introduced into a glasshouse finds almost all environmental variables in its favour and this, coupled with the luxuriance and uniformity of its host plants, leads to rapid population increase. Because maintenance of glasshouse space is expensive the grower tends to cultivate the greatest possible number of plants both by using all available space and by planting as densely as possible. The resultant mass of interwoven foliage is difficult to spray evenly with insecticides, especially as the sides of the structure prevent direct access to much of the cultivated area. This sheltering effect tends to protect the insect from chemicals to a greater degree than on outdoor crops.

A population of glasshouse pests is also largely unaffected by biotic factors for it is almost entirely protected from predators and parasites. Even the few enemies that may be accidentally introduced are usually killed by the regular programme of chemical sprays before they can become effective.

The role of pathogenic diseases is more difficult to evaluate, and few have been found associated with glasshouse pests. This may merely reflect the pests concerned for many are sucking insects which are not often attacked by bacteria or fungi. However, even where these have been studied (e.g. *Cephalosporium* on whiteflies and *Entomophthora* spp. on aphids) the glasshouse environment did not normally permit the establishment of a severe disease outbreak without predisposing the plants to other fungal diseases (Hussey, 1958).

Therefore fewer factors restrict the increase of pests within the favourable glasshouse environment than those breeding more slowly outside.

Development of resistance to chemicals

Despite the great increase, over the past decade, in the number of pesticides available to the grower, insect control in glasshouses has become progressively more difficult. The circumstances, already outlined, encourage pest populations and necessitate frequent treatments which can become excessive. It is not uncommon for more than twenty sprays to be applied within the span of a single cucumber crop. Such an intense spray programme, acting on all the developmental stages of successive generations, exerts a powerful *selection pressure* in favour of individuals with genes for resistance, originally present in very small numbers.

One of the most striking characteristics of insecticide resistance is the suddenness with which it appears as a barrier to efficient control. A chemical may give consistently satisfactory results against a particular pest for many months, or even years, before unexpectedly becoming practically ineffective in the space of as few as two or three additional applications. This can largely be attributed to the manner in which selection operates on the genetic pattern of the species. Assuming that a factor for resistance doubles the survival rate of its carrier and is dominant in the genetic sense though present, say, only once in every million individuals then, under continuous selection pressure, its frequency will increase to 1 in 300 individuals after fifteen generations. But, after only seven more generations half the population will possess the resistance character. Where the character is recessive in a genetic sense, or controlled by more than one gene, the time intervals may be much greater. In all cases, however, the general pattern of a long initial lag, followed by an almost precipitous increase, can be expected. It has been postulated that selective mechanisms determine many characteristics in natural populations of living organisms but ordinarily the evolutionary process is so slow that there is little evidence of it within a human life span. The rapid evolution of insecticide resistance is, however, an outstanding example of the effectiveness of this process.

Control problems accentuated by the development of resistance have increased steadily during the past decade. The first glasshouse pest to develop resistance was the red spider mite (*Tetranychus urticae* Koch), which became tolerant to azobenzene as long ago as 1949 on nurseries in the Lea Valley. It has since become resistant to many organophosphorus and some organochlorine acaricides. More recently the aphid *Myzus persicae* Sulz. has developed resistance to organophosphorus compounds on year-round (YR) chrysanthemums.

It may be thought that the grower has a large armoury of pesticide products with which to combat a given pest and that alternation of these within the spray programme will prevent or delay the selection processes. Unfortunately, as explained in Chapter 2, pesticides fall into a limited number of groups within any one of which all the compounds have the same toxic mechanism. Resistance to any single representative of a group may rapidly confer resistance to the rest.

Possible techniques for overcoming, or at least circumventing, resistance are outlined in Chapter 2, but within glasshouses the problem, besides being more acute, is complicated by the tendency for the most serious pests to develop distinct strains within a single house. This strain differentiation is most marked (a) when the species is relatively immobile, e.g. because it is incapable of flight (red spider mite) or, at most, only a weak flier (aphid) depending on air currents for dispersal, and (b) when the reproductive mechanisms restrict gene combinations, as in asexual aphids or arrhenotokous spider mites (the females can produce eggs without fertilization to produce males which are haploid and therefore homozygous). Glasshouse populations tend to develop in isolation with virtually no genetic dilution by 'wild' untreated individuals. The reproductive mechanisms mentioned tend to confer a genetic stability on the population. In these ways there may arise, even in individual houses, distinct 'strains' within a species between which the pattern and degree of resistance to a range of pesticides tend to differ. It then becomes almost impossible to recommend spray programmes designed to overcome specific cases of resistance. There are, therefore, many reasons to expect the development of complex patterns of insecticide resistance within glasshouses; both the speed and degree with which resistance is currently developing call for changes in emphasis in pest control procedures.

Outdoors, attention is being directed towards integrated control in which the pest population is reduced without disrupting the activity of natural enemies. This is attempted either by using selective pesticides which are not toxic to the natural enemies or by preserving untreated reservoirs of natural enemies which can subsequently reinvade the treated area. It has already been emphasized that natural enemies, unless artificially introduced, are rarely present within glasshouses. Introductions into geographical areas from which natural enemies have been absent have been made many times in the past fifty years. Normally, a new insect introduced to a new environment is expected to establish and subsequently perpetuate itself indefinitely. But one of the most important features of glasshouse hygiene is the complete emptying and cleaning out of the structure after one or two crops and in these circumstances neither host nor natural enemy can survive permanently. Any attempt to use biological control under glass must involve frequent reintroduction of the natural enemy, thereby necessitating some permanent source of supply.

A final complicating issue of pest control under glass is the high commercial value of the crop. The slightest blemish can reduce the market value, so that even dead insects on open blooms may have economic repercussions.

While the complicating factors outlined above tend to create special problems in pest control, the environment under glass confers advantages which, as explained later, may be exploited in the development of efficient techniques.

THE ENVIRONMENT WITHIN MUSHROOM HOUSES

Physical and biotic factors

The habitat afforded by a commercial mushroom crop is rendered unusually suitable for insects by the pasteurizing or 'peak heat' process which ensures that the spawn can be inoculated into a 'sterile' medium. This 'sterility' creates a biological vacuum within which any insect capable of feeding on the compost or mycelium finds ideal conditions for multiplication. Most crops are prone to attack because many mushroom pest-insects are positively attracted either to the compost or to developing mycelium; their parasites and predators are partially excluded by the ventilation systems and so occur even more rarely than those of glasshouse pests.

Within the mushroom house the environment is unusually equable and favourable for reproduction. During the first 2 weeks after spawning the air and compost temperatures are maintained at about 75 °F (24 °C) with humidity approaching saturation. Since fresh air is unnecessary during this phase of mycelial development air movement is negligible. As the spawn-running room is normally almost completely filled with boxes of compost the grower rarely enters during the first fortnight of mycelial growth except to ensure that the extra metabolic activity does not raise compost temperatures unduly. Any pests which find their way into the building do so largely unobserved and, in the relatively high temperatures, large populations may develop by the time the trays are removed for casing. Once covered with the layer of peat and chalk or sterilized subsoil, into which the mycelium grows before the hyphae coalesce to form the 'pinhead' primordia, the trays are maintained at about 60 °F (16 °C). The ventilation rate is increased but the actual air movement is still slow since the average rates of air change involve wind velocities of only about 1 ft/sec. The insect pest, therefore, remains in an equable, though cooler, environment.

The limited ventilation in the spawn-running stage and the need to maintain a temperature of 75 °F (24 °C) throughout the year calls for a well constructed and insulated building. It is therefore quite possible to make the structure fly-proof and so prevent infestation of the beds by even the most persistent fly seeking the source of the attractant odour emanating from rapidly growing mycelium. Once cased, the trays are placed in cropping houses where they become more vulnerable to infestation. As these houses must be entered frequently, both for cultural attention as well as for picking, and are ventilated at rates up to 20 volumes/hr, it is difficult to exclude flies, which often gain access quite readily.

A most important feature governing the degree of infestation is the extent of mycelial colonization of the compost at casing. Where the compost has been correctly prepared it will be completely colonized by the mycelium within 14 days at 75 °F (24 °C). After casing, the odour of growing mycelium is produced only while growth continues up to the surface of the casing. It is therefore exposed to attracted pests for a much shorter period than if the compost is incompletely colonized at casing. In the latter case the mycelium may continue to grow for about a fortnight after the trays have been returned to the cropping house. The difference in degree of pest infestation in these two situations can be striking and serves to illustrate the widely accepted belief that insect pests are less important in healthy crops.

The almost complete absence of predators and parasites of pests within mushroom houses is partly due to the fact that individual crops only remain in the cropping house for about 9 weeks. In order to eliminate all pests and diseases on completion of cropping the house is either 'cooked-out' for several hours at 160 °F (71 °C), fumigated with methyl bromide, or is at least emptied and washed down with some disinfectant. These hygiene treatments naturally eliminate any pest population established in the crop. Even if some naturally occurring parasite of a pest of wild mushrooms or other agarics became established in a commercial mushroom house it would scarcely be able to complete its development within this relatively short period.

Every pest which becomes established in newly spawned compost can therefore look forward to almost 2 months uninterrupted population increase. Once breeding in the compost, a pest population is immune from control since there is no way in which pesticides can be applied to reach the larvae.

Difficulties in use of pesticides

The invulnerable nature of the breeding site dominates the problem of mushroom pest control and increases the difficulties created by the long period of positive attraction by the mushroom for some insect pests. The crop can therefore be protected in only two ways: one is to render the air within both spawn-running rooms and cropping houses lethal to flies, the other to introduce into the compost, before the trays are filled, an insecticide that is lethal to the larvae.

Aerial treatment necessarily implies very frequent applications. Since most insecticides are introduced as aerosols, fogs or smokes their effect is ephemeral and they must be repeated almost daily to combat the continuous ingress of attracted flies. Indeed, the rate of decomposition and lack of persistence of some materials is such that twice daily use would be necessary during long, fine days. In economic terms it is theoretically more profitable to consider incorporating insecticides in compost. Then the first problem is the persistence of the chemical. As already outlined, a lethal amount of the material must persist for at least 5 weeks after filling and this very need involves the use of a class of compounds, the organochlorines, which, due to their very long persistence and subsequent deleterious effects on wildlife, are falling out of favour. Even where suitable compounds with the requisite short persistence are available there are problems involved in the actual incorporation in compost. It is not easy to mix small quantities of insecticides, in the order of 10 p.p.m., evenly through a stack of several tons of compost. As the manure is periodically turned by machine during composting it is logical to assume that the grower would wish to introduce an insecticide during this process. Normally the incorporation is made during the last turn of composting, this, and the subsequent filling operations, being relied upon to achieve an even mix. Such a procedure is not completely efficient and one must expect variable concentrations around the target mean. The insecticide is normally applied as an emulsion, at a rate of about 15 gal/ton, by watering the 'face' of the compost exposed as the stack is turned. Following this somewhat inexact incorporation the insecticide has to withstand the 'peak-heating' process. Temperatures up to 140 °F (60 °C) are usually maintained for several days, often in the presence of large quantities of ammonia. Naturally, some decomposition occurs, the amount depending on the conditions and duration of pasteurization, and due allowance must therefore be made in the amount originally applied. While practical experience has suggested that reasonable accuracy and uniformity is possible, it is nevertheless preferable to mix the insecticide in smaller lots of compost after peak-heating. Where spawning machines, using rotating tines to 'through spawn' the compost, are available it is a simple matter to meter on the exact quantity of insecticide to every tray. This method ensures that no portion of the crop escapes treatment and very efficient results have been obtained. Practical difficulties are (i) the discomfort to workers who often object to wearing rubber gloves for long periods while moving the heavy trays (the use of gloves is desirable in view of the long duration of contact with treated material—often a working day), and (ii) the materials often give off unpleasant odours which preclude the use of the technique in closed or confined buildings.

Use of aerial insecticides within mushroom houses is also attended by problems caused by the treatment of a crop which is normally present in all stages of development on the surface of the beds throughout the 6- to 12-week-picking period. An individual sporophore develops from a small pinhead to a 'button' mushroom in 4 days and to a 'cup' in 5 days. The major mushroom pest, the phorid, is usually controlled only to prevent its nuisance value to the pickers who resent the flies crawling on, or about, their person, and an infested crop may need almost daily treatment. An individual mushroom may therefore be treated several times before harvesting and then, as picked mushrooms are subjected to rapid deterioration, it may be marketed within 24 hr. The choice of insecticides is therefore restricted to those with a very low mammalian toxicity and a rapid breakdown.

Site and locality hygiene

Where, as in several important mushroom pests, flies are attracted to the crop, the location and size of wild populations has a considerable effect on the problem posed at the commercial holding. Differences in the wild populations probably account for the very different scale of the pest problem sometimes existing between otherwise comparable farms.

Mention has already been made of the fact that houses may be 'cooked-out' at the end of cropping. This highly desirable practice is only possible where the structure of the building will withstand prolonged high temperatures, usually attained by injecting live steam. On many small farms, where the cropping houses are merely modified buildings, this process is either impossible or too costly. Pests are therefore afforded an opportunity to 'carry-over' to the next crop which is brought in, within a few days, in its vulnerable pre-pinhead stage. Infestations of mites and cecid (Diptera; Cecidomyiidae) larvae are very difficult to eradicate in these conditions.

Whether 'cooked-out' or not, the disposal of spent mushroom compost presents problems if large accumulations close to the farm site are to be avoided. While those pest species whose life cycle is integrated with mycelial development rarely persist in important numbers in such material, others may continue breeding for a long time. Any accumulations of spent compost must be sited as far as possible from the farm, preferably out of line with the prevailing wind. This precaution is necessary as phorids are attracted upwind by the wind-borne scent of growing mycelium at the farm.

Because many of the smaller pests (cecid larvae, mites and eelworms) are easily transported on tools, or clothes, special efforts are desirable to avoid widespread pest dispersal by restricting equipment and protective clothing, etc., to specific houses.

REFERENCES

ANDREWARTHA, H. G. & BIRCH, L. C. (1954). *The distribution and abundance of animals*. University of Chicago Press, Chicago. 782 pp.

HUSSEY, N. W. (1958). Notes on a fungus parasite on Greenhouse Whitefly. *Pl. Path.*, **7**, 71–2.

— (1966). Aspects of the development of resistance to chemicals by British insect and acarine pests. *Proc. 3rd Insecticide and Fungicide Conference* (1965), 28–37.

KEIDING, J. (1967). Persistence of resistant populations after relaxation of selection pressure. *Wld Rev. Pest Control.*, **6**, 115–30.

NICKEL, J. L. (1960). Temperature and humidity relationships of *Tetranychus desertorum* Banks with special reference to distribution. *Hilgardia*, **30**, 41–100.

Chapter 2

Pest Control: Materials and Methods

Insects and other organisms are called pests when they cause economic loss to growers or constitute a nuisance to workers. To the cucumber grower even bees may be a serious pest for, if they pollinate the cucumber flowers, the resulting swollen fruits are unacceptable to consumers.

Pest status varies with both the species and the crop attacked. Pests which feed on, or spoil, flowers must be eliminated to prevent loss of marketable blooms but considerable foliage damage to tomatoes and cucumbers can be tolerated if there is no loss of yield. In contrast, a tomato plant may be killed by a single wireworm and other plants by only a few woodlice or millepedes.

The high value of glasshouse crops permits expenditure on pest control which would be uneconomic on outdoor crops. However, as there is frequently a conflict between the labour needed for pest control and that for cultural operations, the 'economics' are more often influenced by shortage of man hours than by the cost of pesticides. The dictum that 'prevention is better than a cure' cannot be too strongly emphasized with pests of glasshouse crops for their eradication is made difficult by modern cultural practices, such as close planting, and by the susceptibility of many plants to injury by chemicals.

Some pests, such as aphids and thrips are common outdoors and readily, although accidentally, enter glasshouses. Others, such as mushroom flies, are attracted by odours from the actively growing mycelium while noctuid moths may be attracted to the lights above artifically illuminated chrysanthemums and thence to the crop. Other outdoor pests, including whitefly, carnation fly, red spider mite, vine weevil, and tomato moth enter greenhouses less frequently. Less mobile pests such as scale insects, mealy bugs and eelworms are taken into glasshouses on plants and tools or in soil.

Pests that gain entry into greenhouses may survive and overwinter, even when the glasshouse is unheated and unoccupied for several months. Some mushroom pests also may survive for long periods in the absence of mycelium or compost. Woodlice, springtails, symphilids and millepedes hibernate either as adults or juveniles, while leafminers and moths overwinter as pupae. The red spider mite *Tetranychus urticae*, and females of some *Thrips* species, hibernate within crevices in the glasshouse structure or in debris on the soil surface.

Clearly, it is important to exclude, or minimize, the entry of pests into glasshouses and to prevent their establishment if entry is gained. If establishment cannot be prevented succeeding crops may be safeguarded by modifications of the cultural programme.

When deciding on the type and timing of control measures that are most likely to be successful a knowledge of the bionomics of a pest is essential. If pesticides are to be used it is necessary to decide the best chemical, type of formulation, and the method, as well as the time and frequency of application in relation to the life history of the pest. Consideration must also be given to the presence of other pests which must be simultaneously controlled, while the safety of workers and consumers must not be forgotten. With a few pests the occurrence of chemical-resistant strains adds to the difficulty of selecting effective control measures.

PEST INTRODUCTION

The risk of introducing pests, especially those transmitted by human agency, is very dependent upon the care of the persons involved. For example, potato-root eelworm (a pest of tomatoes) is so common that any container, soil or tool that has been in contact with potatoes should be sterilized.

Pests are introduced most frequently on vegetatively propagated perennials. Amateurs' glasshouses are particularly vulnerable owing to the exchange of many kinds of plants and cuttings. It is unfortunate that chemical-tolerant pests may be widely distributed from propagating nurseries where the frequent and routine pesticide applications used to obtain pest-free plants encourage their selection. Salesmen's returnable containers, often used for fruit collecting, have in the past spread red spider mite and tomato moth; such containers are largely superseded by non-returnable types.

In the last decade increased public appreciation of flowers and pot-plants has led to changes in both growing methods and plant species or varieties; the latter are often obtained from countries from which new pests can be imported.

To obtain suitable cutting material throughout the year several of the specialist producers of chrysanthemums and carnation cuttings grow the stock plants in countries with a Mediterranean-type climate, e.g. in Malta, the Azores, Cyprus and South Africa. Unrooted cuttings, transported by air to Britain, are usually rooted by cutting producers and then sent to nurserymen, in Britain, other European countries, and America. There is also trade in unrooted cuttings which may be rooted and grown by persons less competent to detect new pests than the specialist propagators. The environment into which infested plants are introduced may favour the survival of pests carried with them. In 1963 caterpillars of the Mediterranean climbing cutworm (*Spodoptera littoralis*) were found on chrysanthemums in nurseries in Great Britain. The caterpillars had been introduced, probably as eggs, on imported cuttings. In a similar way the South African tortrix *Epichorista ionephela* was introduced into several European countries, probably as eggs or small larvae on carnation cuttings.

In the past pests such as glasshouse whitefly (*Trialeurodes vaporarorium*), scale insects (*Pseudococcus* spp.), and several species of *Thrips*, including glasshouse thrips (*Thrips haemorrhoidalis*) and palm thrips (*Parthenothrips dracaenae*), were introduced into Britain and became established pests. The discovery in Britain of *Encarsia formosa*, a parasite of the glasshouse whitefly, shows that predators and parasites of pests may sometimes be imported with them.

The ease with which a pest is detected depends on its size, number, and developmental stage as well as damage symptoms. A plant should be regarded with suspicion if it has pitted or mottled foliage; distorted stems, foliage or flowers; or galled roots.

QUARANTINE AND PLANT HEALTH

To prevent the entry of pests and diseases on plant material from overseas the Statutory Instrument 1965 No. 1426 imposes restrictions on the importation into England, Scotland and Wales of plants and certain seeds, raw fruit and vegetables under the provisions of 'The Importation of Plants and Plant Products (Health) (Great Britain) Order 1965'. This order prohibits, *inter alia*, the landing of certain genera and species of plants and plant products from specified countries or areas except under licence. Importation of all other plants is permitted from these countries or areas and from all other countries provided they are accompanied by a phytosanitary certificate and have not been grown on land where non-European strains of potato wart disease have occurred at any time.

Additional conditions apply to some plants and stipulate the manner in which tests are to be made and what pests and diseases must have been absent for specified periods before export. In the case of cuttings, the mother plants must also have been examined. Cacti, ferns and orchids collected in the wild are exempt from these additional conditions. Similar regulations apply in most countries throughout the world. The requirements are less stringent for the movement of plants between Great Britain, Northern Ireland, Eire, the Channel Islands and the Isle of Man.

Although the regulations regarding the importation of plants from other countries are stringent, pests often escape detection for it is impossible to examine thoroughly many thousands of plants or cuttings. To prevent an introduced pest from becoming established, the Ministry of Agriculture in Britain, and its counterparts elsewhere, are empowered to issue Orders for the control of pests, including the destruction of their host plants. Occasionally the measures taken are not effective, sometimes because a few growers take a short-term view and risk the establishment of a new insect as a permanent pest rather than suffer some immediate financial loss. Orders are difficult to enforce and the co-operation of growers is essential if the required results are to be achieved.

The chrysanthemum midge (*Diarthronomyia chrysanthemi*), introduced into England from America in 1924, was found on several nurseries in the autumn of 1927. Methods for controlling this pest were devised and an Order issued compelling growers to spray infested plants. Speyer (1929) reported that it had not reappeared on any nursery in the Lea Valley and attributed this to the energetic co-operation of chrysanthemum growers. Later reintroductions of this midge were not dealt with so effectively and when it became established in Britain the Order was rescinded. To eradicate an attack of this pest, when it first occurred in Jersey in 1964, action was taken under the Destructive Insect and Pests (Jersey) Law 1960; this required the removal and destruction of chrysanthemums from an infested glasshouse.

It is essential that plant material for export is pest free for, if it is not, its entry into another country may be refused, or the plant material may be liable to damage by a control treatment, e.g. orchids may be damaged by fumigation with methyl bromide.

REFERENCE

SPEYER, E. R. (1929). Chrysanthemum Midge (*Diarthronomyia hypogoea* F. Low.). *Rep. exp. Res. Stn, Cheshunt*, **1928**, 100.

RESISTANT VARIETIES

The horticulturist regards a plant variety (i.e. a cultivar) or species as 'resistant' if the damage done to it by a pest is less than that caused to most other varieties by the same pest under similar circumstances.

Basically, resistance can be chemical or physical. In some plants resistance is due to a low level of attractiveness to a potential pest; attractiveness or 'host preference' may be a function of characters such as aroma, colour, texture or taste. In other cases pest reproduction is reduced by specific nutrient deficiencies or the presence of toxic substances. Yet again the micro- or macro-structures of the plant (e.g. thick plant cuticle or woody tissue) may hinder feeding, or interfere with pest movement (e.g. by dense hairs or sticky secretions).

Economic loss may also be reduced if the growth vigour of a variety enables it to withstand the loss of root or leaf; this faculty is termed 'tolerance'.

Few varieties of the more important glasshouse plants are sufficiently resistant to, or tolerant of, attack by their common pests to have practical commercial use.

However, some varieties of tomato with resistance to the root-knot eelworms *Meloidogyne javanica*, *M. incognita acrita* and other *Meloidogyne* spp., have been bred in the U.S.A. European workers have also produced root-knot eelworm-resistant varieties suitable for commercial use and so avoid the tedious practice of grafting susceptible varieties on to resistant rootstocks. Useful resistance to *Meloidogyne* arose from a single cross made in 1943 between *Lycopersicum esculentum* 'Michigan State Forcing' and *L. peruvianum*. It must not be assumed that such resistance is permanent, for strains of eelworm may evolve which will be able to break the resistance.

Many workers have studied the complex factors determining the preferences of insects for particular varieties or species of plants. Hussey & Gurney (1962) investigated differences in the susceptibility of some chrysanthemum varieties to chrysanthemum leafminer, and Wyatt (1965) compared the fecundity and behaviour of the aphid *Myzus persicae* on some commercial varieties of chrysanthemum. Most experienced specialist growers are aware of the relative susceptibilities of varieties to attack by the common pests and are continuously seeking substitutes for the most susceptible varieties.

Resistance to pests in plants seems less pronounced than resistance to disease and this is reflected in the greater effort spent on breeding disease-resistant plants. Often there are difficulties in breeding resistant varieties because many commercially important characters must be incorporated in the resistant plant. For instance, a mutant of the tomato variety 'Potentate' with abnormally hairy leaves was very resistant to the glasshouse red spider mite but it was unacceptable commercially because the fruits were downy. The value of host resistance or tolerance depends on the efficiency and economics of control measures. Resistance to one insect may be accompanied by greater susceptibility to another insect or to disease. Insects, like fungi and eelworms, may produce variants able to overcome resistance. Attempts to breed pest-resistant varieties of plants, other than varieties resistant to eelworms, do not offer prospects of early and permanent success.

An interesting means of overcoming the problem of pollination of cucumber flowers by bees is based on a breeding line of cucumber which produces no male flowers under normal glasshouse and field environment. A commercial variety having this character has been produced in Holland. In England, Darby (1965) is breeding this and other desirable characters into the popular variety 'Butcher's Disease Resister'.

REFERENCES

DARBY, L. A. (1965). Plant breeding—cucumber. *Rep. Glasshouse Crops Res. Inst.*, **1964**, 55.

GRAHAM, C. W. (1966). Potato root eelworm and tomato rootstocks. *Pl. Path.*, **15**, 76–85.

HUSSEY, N. W. & GURNEY, B. (1962). Host selection by the polyphagous species *Phytomyza atricornis* Meigen (Dipt. Agromyzidae). *Entomologist's mon. Mag.*, **98**, 42–7.

PETERSON, C. E. (1960). A gynoecious inbred line of cucumber. *Q. Bull. Mich. St. Univ. agric. Exp. Stn*, **43**, 40–2.

WYATT, I. J. (1965). The distribution of *Myzus persicae* Sulz. on year-round chrysanthemums. 1. Summer season. *Ann. appl. Biol.*, **56**, 439–59.

CONTROL BY CULTURAL AND SANITARY PROCEDURES

Healthy, well-grown plants are usually able to withstand pest attack better than those that are badly grown, though the rate of pest reproduction is often greater on the vigorous and rapidly growing plants than on those that are poorly nourished and slow growing. Aphids, in particular, prefer young growth on well-fed plants, especially when nitrogenous fertilizers have been used. With glasshouse plants there is no evidence that 'natural' manures or fertilizers differ in any way from 'artificials' in their effect on plant susceptibility to pests; indeed, natural manures and composted plant material can introduce pests such as potato-root eelworm, woodlice, springtails and millepedes. There is abundant evidence that the nutrition of plants effects both the fecundity of their pests and their suitability as hosts but, in general, nutrient modifications that control pests are themselves phytotoxic.

Specialization in the cultivation of a few glasshouse crops may make it inconvenient and uneconomic to change cropping programmes as a pest-control measure. Much can, however, be done to minimize pest outbreaks (with the subsequent need to use costly chemical sprays) by adopting good cultural and hygienic practices. As soon as cropping is finished plants and weeds should be removed and destroyed. There is then a much smaller chance that pests will reach a developmental stage in which they can survive in the absence of the host. For example, plant senescence in late summer and autumn induces red spider mite to enter diapause and aphids to become winged (alate) and so to disperse to other plants. If prompt plant removal is impossible the stems of some, such as tomato, cucumber and bean, can be severed near soil level, followed by formaldehyde fumigation (p. 376) or by a spray with commercial formalin (40% formaldehyde) at a dilution of 1 in 50.

Because weeds harbour pests, they should be cleared from close proximity to glasshouses. Only non-volatile and non-leaching herbicides should be used for their control, for many glasshouse plants (especially tomatoes) are extremely sensitive to most herbicides.

REFERENCE

ANON. (1965). Phytosanitary problems in glasshouses and under plastics. *European and Mediterranean Plant Protection Organisation.* 38 pp.

BIOLOGICAL CONTROL

The favourable environment which the glasshouse affords to an introduced pest has already been stressed (Chapter 1). Its advantages are also conferred on parasites and predators whose searching efficiency is greatly increased by the equable temperature and lack of wind and rain. Since these same factors permit mating at almost any time there are few influences limiting the full expression of the efficiency of natural enemies. Outdoors, parasites are often attacked by hyperparasites (their own parasites) and predators but the absence of these under glass increases the efficiency of introduced natural enemies still further.

The first attempt to develop biological control in commercial glasshouses was pioneered by Speyer, at the Cheshunt Experimental Station, who used the parasite *Encarsia formosa* as a control for the whitefly *Trialeurodes vaporariorum* (p. 90). Within a few years this small wasp was being distributed to over 800 nurseries every year. There were two major reasons for the decision to discontinue this distribution scheme after the war. Firstly, the development of DDT afforded, for the first time, a really efficient and safe chemical control, and secondly, the rearing technique, based on large-scale glasshouse production, had fallen under some criticism as other pests such as spider mites and thrips were frequently, inadvertently, distributed to growers on the leaves carrying the parasitized whitefly scales.

No further developments took place until 1951 when Doutt demonstrated that, on commercially grown gardenias, the mealy bug *Planococcus citri* could be successfully controlled by a combination of two encyrtid parasites *Leptomastix dactylopii* Howard and *Leptomastidea abnormis* Girault and the predatory ladybird *Cryptolaemus montrouzieri* Mulsant. The only other practical programme for the biological control of a glasshouse pest was that proposed by Bravenboer (1955) for *Tetranychus urticae* on peaches. He demonstrated that control by the ladybird *Stethorus punctillum* Weise and the predatory mite *Typhlodromus longipilus* Nesbitt could be successfully integrated with mite control based on chlorbenzilate and tetradifon.

Other natural enemies have been tested from time to time (Hussey, 1965) but no attempt has been made to develop commercial control programmes. However, the ever-increasing problems caused by the development of pesticide resistance in strains of red spider mites and the aphid *Myzus persicae* has highlighted the need to reduce the selection pressure resulting from chemical control programmes. Even a partially successful biological control programme may drastically reduce the number of pesticide applications which will, in turn, extend the life of the materials used.

To date most attention has been directed to insect natural enemies but the use of fungi or viruses is possible and on p. 313 reference is made to the potential use of entomogenous eelworms. Bacterial diseases may also be exploited in the future. Already *Bacillus thuringiensis* has been used to control several kinds of caterpillars on outdoor crops. Experimental work has shown that the mushroom phorid *Megaselia halterata* is susceptible to this disease though the dosage required is uneconomic. Viruses are, theoretically, the most attractive agents for biological control since they can kill their hosts without any particular environmental requirements other than the effective dispersal of the virus particles. Unfortunately none have yet been found attacking glasshouse pests. Fungi normally require high humidities for the germination of the spores and these near-saturated conditions may, if maintained too long, encourage the development of plant pathogenic fungi. Other limitations prevent the effective use of the fungus *Entomophthora coronata*, a pathogen of the aphid *Myzus persicae*. The germinating fungal conidia seem able to penetrate the cuticle of aphids only when these are in dense colonies; individuals in dispersed populations are unaffected even when covered with conidia.

The strategy of biological control may vary in different cultural situations. One approach has been suggested for the control of *M. persicae* on year-round chrysanthemums. The greater proportion of the cuttings required by the flower producing industry are grown by a handful of large-scale producers who, at present, apply pesticides frequently in an effort to despatch clean plants to their customers. Once the development of pesticide-tolerant strains was recognized it became apparent that survivors of these routine treatments were distributed to other growers. This distribution accounted for the rapidity with which resistance to organophosphorus sprays developed throughout the country in 1964. Such undesirable dispersion could be prevented by an effective biological control programme based on the mass release of natural enemies on the cutting beds 2 or 3 weeks before despatch.

Chemical control of less important pests may have to be integrated with biological control of aphids and mites. On the dense foliage of cucumbers, carnations and chrysanthemums, any method of pesticide application achieves only indifferent coverage. Even where as much as 500 gal/acre of diluted insecticide is applied 25% of the lower leaf surface may receive no deposit at all. Experience suggests that, if the interaction between host and natural enemy has reached a point where the latter is widely distributed through the house, any hosts surviving chemical treatment will be effectively destroyed by surviving natural enemies. A chemical having a toxic vapour would, however, tend to be lethal to such enemies under nearly all conditions. For instance, the vaporization of sulphur, at $\frac{1}{2}$ g/1000 ft^3, for the control of powdery mildews, is almost always lethal to the red spider mite predator *Phytoseiulus*.

The parasite, *Aphidius matricariae*, could be used in an integrated control programme for resistant *Myzus persicae* on the flowering chrysanthemum crop. Early introduction of the parasite leads to the synchronous development of parasite pupae within the mummified body of the host aphid. At this stage, the parasite is protected from pesticides so that a carefully timed programme would permit chemical control of other pests without affecting biological control of the resistant aphid.

It is widely believed that systemic insecticides, applied to the roots and translocated to the leaves to kill sucking insects, are relatively non-toxic to natural enemies. Again, however, experience with *Phytoseiulus* has shown that such chemicals may have complicated side effects. After demeton-methyl drenches, *Tetranychus urticae* is often able to recommence breeding within 2 weeks for individually they imbibe sublethal doses. But the predator, by feeding on 4 or 5 host mites daily, may acquire a cumulative lethal dose.

The short-term control programmes required under glass could naturally be jeopardized by these side effects of chemical applications and a strategy based on a separation of chemical and biological controls in time or space seems the most practicable course. For instance, 'spot' chemical treatment of localized outbreaks of less serious pests might prove practicable while soil treatment to eliminate pupating *Thrips tabaci* descending from cucumbers could be applied generally without deleterious effects to either *Phytoseiulus* or *Encarsia*.

The most practical method to avoid the complications of integration with chemicals would be to control the principal pests on a crop by the simultaneous use of several natural enemies.

The success of the first attempts at biological control on a commercial scale suggests that the technique could be extended. However, the commercial future depends on the development of economic and reliable rearing techniques to provide the required number of natural enemies. To facilitate dispersal on the plants, any practical scheme must be able to provide several natural enemies simultaneously, in large numbers, suitably packaged. Reliable and large-scale production requires a single, adequately staffed and equipped unit. A venture on this scale would require much capital. This could only be provided by the Government, who should require evidence that the industry was willing to contribute its share. The initiative to launch industrial biological control must therefore come from the industry. If present developments are successful, the far-reaching advantages inherent in 'non-chemical' control should encourage the industry to contribute collectively to a partnership with an official organization. Such a partnership should be economically viable and must not be afraid of making modest profits. Non-profit-making organizations all too often cost the taxpayer money. Chemical control of mites on cucumbers costs over £200/acre/annum so that any biological control scheme could cost up to £100/acre and still be attractive to the grower. Success could mean both cheaper, and more lasting, pest control, as well as larger crops, since any pesticide programme may involve some phytotoxicity.

REFERENCES

BRAVENBOER, L. (1955). Chemical and biological control of red spider (*Tetranychus urticae* Koch). *Meded. TuinbVoorlDienst*, **18**, 672–80.

DOUTT, R. L. (1951). Biological control of mealybugs infesting greenhouse gardenias. *J. econ. Ent.*, **44**, 37–40.

HUSSEY, N. W. (1965). Possibilities for integrated control of some glasshouse pests. *Ann. appl. Biol.*, **56**, 347–50.

PHYSICAL CONTROL

Heat

The disinfestation or 'sterilization' of glasshouse soils by steam heating has long been practised for eelworm control, and, until the advent of effective and cheaper chemical methods, was sometimes used to destroy pupae of tomato leafminer in the soil of houses to be used for tomato propagation.

With the exception of symphilids and, in open soils, eelworms (both of which may be at depths below which the soil can be heated economically), all soil-inhabiting insects are killed by the standard practice of passing steam into the soil until a temperature of 200 °F (94 °C) is attained throughout its upper 10–12 in. Pests are also killed at the lower temperatures obtained by passing mixtures of steam and air into the soil, an experimental procedure designed to overcome certain disadvantages of orthodox steaming.

Nowadays the destruction of root-invading fungi is the main reason for disinfesting glasshouse soils by 'steaming', for it is very rarely economic—at a cost of £350 to £700/acre—to 'steam' for the sole purpose of controlling eelworms or insects except where only a very short interval is permissible between disinfestation and planting.

When disinfesting glasshouse soils, the usual technique is to lay perforated pipes in soil trenches 10 in. deep and, after covering the pipes with soil from adjacent trenches, pass steam through them until the required temperature is reached and maintained for 10 min. Alternatively, perforated spikes about 9 in. apart are inserted (either singly or fabricated into grids) vertically in deeply cultivated soil, and steam is passed through them. Both of these techniques enable the soil to be heated sufficiently to kill pests to a depth of about 15 in. An innovation, termed 'sheet', or 'balloon', steaming, involves the liberation of steam from pipes on the soil surface beneath a plastic sheet cover. Usually this method does not sufficiently heat the soil deeper than 6 in.; this is inadequate to control eelworms but is effective against woodlice and millepedes. By passing a mixture of steam and air through soil it can be uniformly heated to 150 °F (65 °C) which is lethal to all insects and nematodes. All steaming methods can be used in addition to the injection of nematicides.

Soils and composts intended for propagating or potting should, if possible, be disinfested by heating to ensure freedom from both insect and fungal pests.

The soil is usually steamed over perforated grids; other methods such as baking may be employed if care is taken to avoid overheating. Some plants and plant material can be heated sufficiently at certain stages of growth to kill pests. There is only a small margin, however, between the intensity and duration of treatment required to kill pests and injurious over-treatment. The heat treatment of bulbs and other plants to control nematodes is an example (p. 293).

Low temperature treatment

Chilling of plant material controls the Mediterranean climbing cutworm (*Spodoptera littoralis*). Storage of imported cuttings at between 32 and 35 °F (0–2 °C) for 10 days kills the eggs of this moth and so ensures that pest-free material is available for the trade in cuttings.

Light

Some insects are attracted and others repelled by light, but these responses rarely offer practical prospects for pest control in protected cultivation. The daily light duration, however, influences the hibernation of both insects and mites. It has been found that the red spider mite (*Tetranychus urticae*) does not hibernate in Britain if artificial illumination is used to extend the day length to 12½ hr in the autumn and the temperature is kept high enough to maintain normal growth of its host plant (p. 195).

Irradiation

Because many important glasshouse pests are parthenogenetic it is unlikely that effective use could be made of liberating males sterilized by atomic or other radiation. The enclosed environment provides the main prerequisite for taking advantage of sterile male release, but only minor pests (which are easily controlled by pesticides) have a suitable mode of sexual reproduction which could be nullified by the use of sterile males.

Screening

Insects may be excluded from glasshouses by inserting wire or fabric mesh screens at doors and ventilators. Screens which are sufficiently fine to exclude minute pests such as thrips, midges and mites would, however, seriously interfere with the atmospheric conditions within the glasshouse by impeding the passage of air through the ventilators. Relatively large mesh screens are, however, occasionally used to prevent the entry of bees into cucumber houses.

Special glasshouses used for the propagation of virus-free plants usually combine mechanical ventilation with screens, and rely on the rapid outflow of air through screened vents to prevent the entry of small insects. If necessary, the temperature within the glasshouse can be regulated by cooling the incoming, filtered air.

Unless screen apertures are large in relation to the size of the insect the latter usually alights and crawls through. Electrified grids which electrocute alighting insects have been devised, but their commercial value has yet to be demonstrated. So far no practical use has been made of lethal atomic or ultrasonic radiations.

REFERENCES

ANON. (1965). *Soil Sterilization*. Bull. 22. M.A.F.F. London, 49 pp.

HUSSEY, N. W. & GOSTICK, K. G. (1965). Effects of low temperature storage on the eggs of *Spodoptera littoralis* (Boisd). *Nature, Lond.*, **203**, 323–4.

CHEMICAL CONTROL

Terminology

The term insecticide has become accepted in common usage as denoting a substance which will kill, or adversely affect, not only true insects but also other arthropods such as mites, symphilids and millepedes. Sometimes it is extended to include even substances used to control eelworms, slugs and snails. The terms acaricide, nematicide and molluscicide are, however, more correct to describe substances that kill mites, eelworms, and slugs and snails respectively. When dealing with the control of important families or developmental stages of insects, terms such as aphicide, larvicide and ovicide may be used.

Chemicals: their development as pesticides

Modern insecticides are mainly the products of complicated organic chemical syntheses which have largely replaced both the naturally occurring insecticides such as pyrethrum, derris and nicotine, and the inorganic insecticides lead arsenate, sulphur and potassium and calcium polysulphides. Much research has gone into the development of modern insecticides which are highly toxic to insects, and also less hazardous to man than many of the earlier synthetic products.

While the number of insecticides listed in Martin's *Pesticide Manual* exceeds 200, only 59 of these were included in the M.A.F.F. List of Approved Products (1969) and only about half of this number are commonly used for the control of glasshouse insect pests.

The outstanding landmarks in the development of modern insecticides were the discovery of the potent insecticidal property of DDT by Lauger, Müller & Martin in the Basle laboratories of J. R. Geigy S.A., and the synthesis of some extremely insecticidal and acaricidal organophosphorus compounds by Schrader of Messrs. Farbenfabriken Bayer A.G., Leverkusen, Germany.

The discovery in 1939 of the insecticidal properties of DDT, a compound which was first described in 1874, resulted from a reasoned anticipation of its properties following work on moth-proofing chemicals for woollen textiles. From then onwards the effect of molecular structure on insecticidal action and, more recently, its effect on mammalian toxicity have been studied; in this way new compounds with improved characteristics have been synthesized. Even so, large numbers of related chemicals have to be made and 'screened' by laboratory tests on a selected range of insects and other organisms. Those showing promise have to be compared further with existing products. Problems such as the potential market, large scale manufacturing methods and costs, formulation, mammalian toxicity, and methods of determining residues have to be investigated. Finally, the value of the product has to be demonstrated by large scale trials conducted under conditions likely to occur in commercial practice.

This work is very costly and it is obvious that no manufacturer will contemplate vast expenditure to discover, or develop, insecticides having too specific, and consequently small-scale use. Similar economic considerations apply to the expenditure of time and funds by persons engaged in pest control investigations and it is inevitable that major pests of the more important crops receive most attention. When an established insecticide is known to be effective against a pest which occurs infrequently, and particularly if that pest is restricted to plants of minor commercial importance and does not constitute a hazard to more important crops, trials with newer insecticides are often unwarranted. Under such circumstances information on the efficiency and safety to host plants of new insecticides is obtained only slowly by making use of occasional natural infestations in *ad hoc* tests. It is much easier to predict the efficiency of a chemical against many such pests than to predict its phytotoxicity.

The properties required by an insecticide depend both on the pest and on the host plant to be protected, but some are common to all insecticides. Obviously, insecticides should be highly toxic to the pest and harmless to man and to the host plant. Unfortunately these ideals are difficult to achieve. Certain other properties, highly desirable in insecticides for outdoor use, are of less importance when the chemical is to be used under glass where it will have little, or no, impact on wild life and, in any case, will be used only in limited amounts. With some crops edible produce must be picked at frequent intervals over long periods and only chemicals of low mammalian toxicity, or very limited persistence, can be used once cropping has commenced. Residues may, however, persist longer on glasshouse crops because they are not removed by weathering.

It is often claimed that the ideal pesticide should be highly specific. This is almost certainly so with pesticides for outdoor use where the destruction of predators, parasites and other beneficial insects is undesirable, and it may be equally important in glasshouses if biological control of the major pests becomes possible. However, it is doubtful whether the breeding and distribution of parasites or predators for the specific control of minor pests will ever be economic and consequently it will still be necessary to control them chemically. Highly specific substances, active against minor pests and either harmless to the beneficial insect or capable of application in a highly specific manner, may still be desirable.

Although it is frequently stated that systemic insecticides do not interfere with control by predators this is often incorrect. Predators may be killed if their prey has ingested systemic insecticides, and larvae of parasites may be killed either by the insecticide in the host tissue, or as a result of the premature death of the host. The red spider predator *Phytoseiulus* (p. 196) may be killed by demeton-S-methyl up to 10 weeks after application through consuming poisoned prey.

At present, chemicals effective against a wide range of pests have considerable advantages. Joint acaricidal and aphicidal activity is extremely useful (e.g.

organophosphorus compounds), as is joint acaricidal and fungicidal action (e.g. dimethionate; dinobuton).

CLASSIFICATION OF INSECTICIDES

The older system of grouping insecticides according to the manner in which they enter the insect, *viz* stomach poisons, contact poisons and fumigants, is now unsatisfactory. Many modern insecticides can gain access to a vital part within the insect by more than one route. In some cases, it is difficult to differentiate between cuticular (contact) and tracheal (vapour) absorption.

A more satisfactory classification of insecticides may be based on their chemical structure, but even this also leads to anomalies. In view of these difficulties a classification based both on the chemical nature and specificity of biological action has been adopted here. Certain chemicals of diverse chemical and biological properties, which cannot be classified, are considered separately. Brief descriptions of the properties of the chemicals used in Britain to control insect pests are given in Appendix I, p. 371.

Organochlorine compounds

This group comprises the relatively complex, highly chlorinated compounds having a broad spectrum of insecticidal activity. Excluded are those compounds containing chlorine which are highly specific acaricides (e.g. dicofol), simple chlorinated compounds such as dichloropropene, and compounds dependent on phosphoric or carbamate esters for their activity.

DDT was the first member of the organochlorine group of insecticides and is probably the best known from its high success in controlling the insect vectors of typhus and malaria.

Many of the organochlorine insecticides are highly persistent, particularly in soils. Traces have also been found in the fatty tissues of humans, other mammals, birds and fishes. Some pesticides have been detected in animals from areas in which no pesticides have been used. In view of this, and to allay the possibly exaggerated alarm from certain quarters, a committee under the chairmanship of Sir James Cook (now termed the Cook Committee) was set up by the Minister of Agriculture, Fisheries and Food in 1963 to review the use of organochlorine pesticides. In the report of this committee certain recommendations were made which were subsequently accepted by the Government after consultation with Industry and resulted in agreed restriction of certain uses of aldrin, dieldrin and heptachlor. A supplementary report of July 1964 recommended minor restrictions on the use of endosulfan.

These recommendations were designed to reduce possible long-term hazards to human beings and wild life. Although such hazards arising from the use of these pesticides for soil treatment and the occasional foliar treatment of ornamentals in the restricted environment of glasshouses may appear negligible, aldrin, dieldrin, endosulfan, endrin and heptachlor are no longer accepted for use against most glasshouse pests.

BHC was considered much less likely to have undesirable effects than DDT, the use of which should be minimized. Methoxychlor, a compound related to DDT, and not at present used to control glasshouse pests, is not stored in fatty tissues as is DDT.

Organophosphorus compounds

These constitute the largest group of insecticides and are mostly esters of phosphoric, thio- and dithio-phosphoric acids. The earliest discoveries, TEPP (HETP) and sulfotep, however, are esters of bisphosphoric anhydride and bis-thiophosphoric anhydrides respectively, while schradan is a bis-dimethylamine phosphoric anhydride, its common name commemorating Schrader, its discoverer. This group can be divided into two sub-groups—systemic and non-systemic organophosphorus compounds. The former are so named because they can be absorbed through the leaves or roots into the conducting systems and transported to other parts of the plants. Susceptible insects feeding on the sap or tissues of treated plants are killed.

Although the use of systemic compounds may appear to provide a simple means of controlling insect pests, many factors influence their practical efficiency. Not all systemic compounds are absorbed and translocated with equal ease. Uptake of a chemical through the roots is influenced by climatic conditions, plant species and stage of growth. The nature of the soil and placement of the compound therein also affects its uptake. The rate of transpiration is another important factor affecting the rate of toxicant accumulation in the aerial parts of the plant, for substances absorbed through the roots are carried in the xylem by the transpiration stream. The maximum amounts of toxicant tend to be found in the young, fully expanded leaves.

The poor control of aphids and red spider with many soil-applied systemic insecticides on the buds and opening flowers of chrysanthemums compared with that obtained on the foliage is probably partly attributable to the lower transpiration/weight ratio of the former. Worthing (1965) found that systemic compounds differ in the extent to which they are transported to chrysanthemum flowers.

Due to guttation and subsequent evaporation, systemic compounds may accumulate at the edges and lobe-tips of leaves, resulting in a typical scorching of these areas on susceptible plants. The stability of the systemic compound, or its insecticidal metabolites, may be an important factor, particularly when the compound is applied to the soil when the transpiration is at its minimum.

The entry of a systemic compound through leaves and its subsequent distribution within the plant is more complex than if entry occurs through the roots. Movement from a leaf to other parts of the plant probably occurs through the phloem and is more restricted than movement in the xylem. Downward transport of systemic compounds to the lower leaves or roots is much less pronounced than the counter flow; the roots are very seldom rendered toxic to insects or nematodes by foliar applications. The action of foliar sprays of thionazin on eelworms and bulb scale mites within narcissus bulbs may be due to downward transport. Systemic activity must not be used to justify bad spray application although it undoubtedly removes the necessity for thoroughly covering the undersides of leaves.

The nature of the leaf surface and the relative solubilities of the chemical in water and waxy or fatty substances (lipoids) influence the penetration of compounds into sprayed leaves. Insecticides which are water-, but not lipoid-, soluble may not easily penetrate waxy foliage, whereas lipoid-soluble compounds can do so readily.

Systemic organophosphorus compounds are used extensively to control pests, particularly aphids and red spider, on a wide range of glasshouse plants, but their use on tomatoes, cucumbers and lettuces must obviously be restricted to ensure that toxic residues are absent at harvesting. The period between the last application and harvesting depends on the particular systemic compound and, in some cases, on the time of year when it is applied. The application of systemic compounds as high volume sprays requires less active ingredient than when applied as a drench to the soil. Most systemic chemicals are somewhat phytotoxic to many plant species and varieties and this is more apparent when a compound is applied as a spray than when absorbed via the roots.

Non-systemic organophosphorus compounds are equally important for controlling many glasshouse pests. They have the advantage of being suitable for application as aerosols (pp. 26, 31). Parathion, one of the first members of this group to be developed, is extremely toxic to mammals and much work has been undertaken to develop less toxic compounds, of which malathion and diazinon are available to amateurs. Parathion, however, has the advantage of very low toxicity to plants, and is recommended for some purposes where other insecticides are likely to cause damage.

Most organophosphorus insecticides are acaricidal. Possibly their greatest practical disadvantage is the facility with which resistant strains of red spider mite and the peach-potato aphid (*Myzus persicae*) have been selected by their use. This selection appears to be more pronounced with non-systemic than with systemic organophosphorus compounds.

Specific acaricides

Compounds which are highly toxic to mites but have slight, or negligible, effect on true insects are usually referred to as 'specific' acaricides. The alkylated dinitrophenolic esters, dinobuton, binapacryl and dinocap, and the compound dimethionate, all of which combine acaricidal and fungicidal activities, are included in the miscellaneous group.

Most of the 'specific' acaricides have a bridged-diphenyl structure and contain chlorine in one or both phenyl groups. The first compound of this type found to have acaricidal action was azobenzene. Following this discovery the effects of different bridging groups and substituents in the phenyl groups were investigated by workers in Britain, America, Holland and elsewhere and led to the development of a series of very active acaricides.

Although compounds of this type are not absorbed and transported in plants as are systemic compounds, most of them, when applied to the upper (dorsal) surface of a leaf, are able to exert acaricidal action on mites on the under (ventral) leaf surface. This property, which is termed translaminar activity, is advantageous since it reduces, to some extent, the necessity for complete spray coverage of the ventral leaf surface and enables the compounds to be applied as smokes and aerosols. Particles of the pesticide produced by these techniques are deposited almost entirely upon the dorsal leaf surfaces. Nevertheless, the possession of translaminar action should not encourage poor application for, when dicofol was applied in a spray to the dorsal surface of bean leaves, its toxicity to mites feeding on the ventral surfaces was very much lower than when that surface was sprayed with the same fluid (Cooke, 1963).

These bridged-diphenyl compounds are more toxic to the eggs and immature mites than to adults but are sufficiently persistent to achieve almost complete control of non-resistant strains. Dicofol is the most active against adult mites. An interesting feature of tetradifon is its ability to reduce the fertility of the females. Mixtures of specific acaricides and organophosphorus compounds effective against adult mites are used when rapidity of control is necessary.

The mammalian toxicity of most of the 'specific' acaricides is low. With the exception of azobenzene, which is volatile, they are persistent and in some structural characteristics resemble DDT. Consequently, the phytotoxicity of many to cucumbers (a notoriously DDT-intolerant subject) is not unexpected. It is, however, strange that dicofol, which most closely resembles DDT, is, like tetradifon, safe on cucumbers.

Certain other compounds, including Aramite, (a complex tert. butylphenoxy sulphite) and 'Pentac' [bis(pentachloro-2,4-cyclopentadien-1-yl)] are specifically acaricidal, and are used in the U.S.A. for the control of tetranychid mites on ornamental crops, particularly against mites resistant to organophosphorus and most 'specific' acarides.

Carbamates

The number of established insecticides in this group is small and up to 1966 their only practical use under glass has been for the control of earwigs on ornamental crops. Chemists and biologists are actively investigating further compounds of this group, and products of value for glasshouse pest control are likely to be discovered (p. 119). Most experimental products so far have been more toxic to plants and mammals than organophosphorus compounds of similar insecticidal efficiency.

Nematicidal soil fumigants

Although most of the chemicals used as soil disinfestants against nematodes are also insecticidal, their *raison d'être* is almost entirely as nematicides, combined with some fungus control.

The actual fumigants are simple chemicals such as halogenated alkyl compounds and methyl isothiocyanate. The latter is usually derived from the more complex compounds metham-sodium and dazomet by decomposition in the soil. Proprietary mixtures of nematicidal and fungicidal fumigants are available.

Miscellaneous

To avoid a multiplicity of categories, the naturally occurring insecticides nicotine, derris and pyrethrum, which have very different chemical structures, are placed in this group, as are sundry unrelated compounds such as copper aceto-arsenite (Paris green); dinitrophenolic esters (binapacryl, dinobuton), etc.

REFERENCES

ANON. (1969). Agricultural Chemicals Approved Scheme, List of Approved Products. *M.A.F.F. Lond.*, p. 150.

COOKE, V. A. (1963). The translaminar effect of Kelthane on three strains of glasshouse red spider mite *Tetranychus urticae* Koch (*T. telarius* L.) on cucumbers. *Ann. appl. Biol.*, **51**, 485–8.

FINLAYSON, D. G. & MACCARTHY, H. R. (1965). The movement and persistence of insecticides in plant tissue. *Residue Reviews*, **9**, 114–52.

MARTIN, H. (1968). *Pesticide Manual*. 1st Ed. Brit. Crop Prot. Council. 464 pp.

WORTHING, C. R. (1965). The uptake, distribution and metabolism of demeton-S-methyl and oxydemeton-methyl in chrysanthemums. *Rep. Glasshouse Crops Res. Inst.*, **1965**, 106–7.

CHOICE OF INSECTICIDE

Serious practical problems are created for the grower by the development of resistance either to a single compound or, by cross-resistance, to a number of related materials with the same mode of toxic action. So far, resistance among glasshouse pests has been restricted to the red spider mites (*Tetranychus* spp.), the aphid *Myzus persicae*, and the leafminer *Phytomyza atricornis*. However, the intense selection pressure provided by the insecticidal programmes necessary to maintain control, will inevitably lead to induced resistance in other glasshouse insects should the genes for tolerance exist.

It is widely believed that a control programme based on 'ringing the changes' will delay, or even prevent, the selection of resistant strains. The success of such rotations depends firstly on the absence of cross-resistance between the chemicals employed and secondly on the necessity for any resistance induced by one material to have reverted to susceptibility within the space of one complete rotation.

There is considerable conflict in the scientific evidence derived from experiments to establish the merits of rotational control programmes. However, while success cannot be guaranteed, it is probably wise to alternate materials from the different groups of compounds outlined in earlier sections. As a rough guide the use of materials within a group should be continued for a full growing season before switching to other compounds with a different mode of action unless, of course, control efficiency declines earlier. It is also vital to ensure that insecticidal applications are restricted to situations in which they are economically essential.

REFERENCE

HELLE, W. (1965). Expectations of delay of resistance in *Tetranychus urticae* by rotational sprays. *Meded. LandbHoogesch. OpzoeekStns Gent*, **30**, 1397–9.

THE APPLICATION OF PESTICIDES

Formulations

Most insecticidal chemicals are insoluble in water and in order to distribute very small amounts over the soil or the surfaces of plants they must be formulated into products which can be used by growers. A few can, however, be used without being formulated. The soil fumigants DD, methyl bromide and methylisothiocyanate are injected into the soil without dilution while it is possible to vaporize some chemicals without formulating them with other chemicals in smoke generators, e.g. nicotine, azobenzene and naled. Nicotine is also readily soluble in water and is usually purchased by commercial growers as the technically pure chemical from which sprays are prepared by adding the required amount of nicotine and wetting agent to the water.

With other compounds the complexity of the formulation depends both on the properties of the chemical and the method of application.

Aerosol (atomizing) concentrates

To enable a compound to be applied as an aerosol, i.e. by dispersion (atomization) as an extremely fine mist (particle size 10–100 μ), it is usual to dissolve it in a volatile organic liquid. Acetone is usually chosen as the solvent on account of the very low toxicity of the vapour to man and plants, and the absence of any risks of forming an explosive mixture with air, within the glasshouse, at concentrations arising from its use in atomizing concentrates. The concentrates are, however, highly inflammable, and unless the containers are tightly closed may rapidly change in composition through evaporation of the solvent.

Aerosol concentrates of poisonous chemicals are particularly hazardous to operators as the very rapid evaporation of the solvent may result in contamination of the hands and clothing being overlooked.

The *modus operandi* of insecticidal aerosols has not been investigated fully. Not all pesticides are effective when applied in this way, as the deposits are restricted to upward-facing surfaces. The vapour pressure of the compound, its ability to exert insecticidal or acaricidal activity through the laminae of leaves (translaminar action) and the manner of the insect's initial response are important factors.

Baits

Some chemicals, including Paris green or metaldehyde, if purchased in powder form, are easily mixed with attractant materials such as bran or dried blood to form baits, though most growers find it more convenient to obtain ready mixed baits from manufacturers. If baits are mixed 'on-site' immediately before use, precautions should be taken to avoid inhaling the chemical, a dust mask being worn if necessary.

Dusts

Dusts require expensive manufacturing machinery, either to comminute and mix the chemical and an inert carrier material such as kaolin, talc, silicates or other absorbent mineral sufficiently finely for efficient application, or to impregnate the finely powdered carrier with the chemical. Dusts having a particle size less than 150 μ and which produce a dust cloud suitable for applying a chemical to plants are termed 'fine' dusts; those which are intended for application to the soil, seedlings or low plants, and by virtue of their larger particle size (150–250 μ) and often greater density of the carrier material, rapidly fall to the ground are termed 'coarse' dusts. As different inert carrier materials and other additives affect the stability of the pesticide as well as the 'flowability' of the dust in the dusting appliance, considerable expertise is necessary in producing satisfactory products.

Spraying concentrates

Many insecticides are applied to glasshouse plants as high volume (H.V.) sprays and consequently insoluble compounds have to be formulated into products which can be mixed with water to give a suspension or emulsion of the chemical of sufficient stability to permit its practical application. This is achieved in several ways.

The active chemical can be mixed with, or absorbed by, an inert material such as kaolin, and combined with wetting and dispersing chemicals to yield a product which disperses readily on stirring into water to give a stable suspension. Products of this type are termed wettable powders (w.p.). Alternatively the chemical can be dissolved, together with synthetic wetting and emulsifying agents, in a suitable solvent to yield a fluid product termed an 'emulsifiable concentrate' (e.c.) or, occasionally, a 'miscible concentrate'. The solvents for such concentrates are usually insoluble in water, e.g. xylene, solvent naphtha or a petroleum fraction, and on addition to water this type of formulation spontaneously produces minute emulsified droplets of the solution of the chemical. Occasionally the pesticide together with wetting agents is dissolved in a water-soluble organic solvent to give a formulation which, when added to water, similarly yields a suspension of minute droplets or particles of the pesticide. Provided care is taken to close the container securely to prevent evaporation of the solvent, emulsifiable concentrates are normally stable on storage. Under very cold conditions the active material may crystallize out, but it redissolves when the temperature becomes normal.

Petroleum oil, used to control red spider, is formulated as a stock emulsion: a thick, creamy dispersion of oil droplets in water containing emulsifying agents. Such emulsions are less stable than emulsifiable concentrates as the oil tends to separate slowly on storage. If frozen, the separation of the oil may be complete and the product become unusable.

Manufacturers strive to formulate compounds into spray concentrates of maximum insecticidal efficiency and safety to plants when diluted with waters of different degrees of hardness. As the solvents, wetting and dispersing agents in different products may be incompatible, mixtures of two or more products of the same, or different, manufacturer should not be used in combination unless the maker confirms that it is safe to do so.

Smokes

Some pesticides are sufficiently stable at high temperatures to permit their volatilization as smokes, by the heat produced with suitable pyrotechnic mixtures. Much skill is required in the selection of suitable chemicals to ensure that ignition can be readily effected and yet burn at a rate sufficiently low for their safe use. Some pesticides are mixed with the heating chemicals and compressed into pellets; others require more complicated mixtures and ignition systems, and so are made up into smoke canisters. Where effective smokes are available they provide the most convenient means of controlling pests but the cost exceeds that for control by using aerosols, particularly where canister-packed smoke-producing mixtures are necessary.

Granules

For application to the soil or to low plants, insecticidal chemicals, especially those which are systemic, are sometimes mixed with a carrier material and formulated into granules having a particle size approximating to 1–2 mm.

One advantage of this type of formulation is that the large particle-size reduces the hazard to operators handling and applying the insecticide—the proportion of dust-size particles in granules containing scheduled poisons is restricted by the Safety Precautions regulations. Another is that the most effective rate of toxicant release can be obtained by using different carrier materials.

Content of active ingredient

Information concerning the content of active ingredient in proprietary products, where given by the maker, is usually expressed as a percentage, either weight in weight (% w/w) or weight in volume (% w/v); a concentrate having 1 lb of active material per gallon thus being expressed as a 10% w/v concentrate. Recommended concentrations of chemicals for the control of specific pests are therefore given as percentages in order to simplify calculations of the required dilution. Tables which will be of assistance in preparing sprays of the required strength are given on p. 391.

Relative efficiencies

The following generalizations can be made regarding the phytotoxicities and insecticidal efficiencies of different types of formulations. Dusts are the least phytotoxic followed by wettable powder and smoke formulations. Dusts are, however, difficult to apply uniformly and, as there is no 'run-off' as with sprays, unsightly deposits and undesirable toxic residues may arise from excessive application. While dusts are of little value against sucking insects which remain on the undersides of leaves, they are suitable for the control of soil insects (followed by cultivations where necessary) and for those insects which move actively over dusted surfaces or bite sections from treated leaves.

Aerosols (p. 31) are cheap, quickly applied, of medium phytotoxicity but effective with only a limited number of insecticides. At equal toxicant rates smokes are often rather more effective but less phytotoxic than aerosols although the products of combustion of the pyrotechnic mixture may be injurious.

Emulsifiable concentrates are more phytotoxic but usually more insecticidally potent than wettable powder formulations, particularly when leaf penetration is an advantage, as against leafminer larvae, or where translaminar action is desirable. They are also more effective against insects having waxy cuticles or coverings, e.g. scale insects and mealy bugs. The residues from wettable powders are usually more noticeable than those from emulsifiable concentrates. The dermal toxicity of aerosol (atomizing) concentrates is higher than that of wettable powders of the same chemical.

Wetting agents

The leaves of many plants can only be wetted with difficulty by water and, to reduce the surface tension of the diluted material and so promote wetting of foliage sprayed at high volume, wetting agents are incorporated in proprietary spraying products. As an excess of wetting agent may lower the efficiency of some sprays by reducing the retention of the particles or droplets on sprayed surfaces, the manufacturer may incorporate only sufficient wetting agent for foliage of average wettability. Additional wetting agent may therefore be required to give the best results on 'difficult' leaves such as those of carnations and cucumbers.

Wetting agents are classified as anionic, cationic and non-ionic according to their chemical structure. As the first two are incompatible it is essential that any additional wetting agent is compatible with those used by the manufacturer in formulating the insecticide. Suitable wetting agents are usually recommended on the labels or instructions accompanying a product.

SPRAYING EQUIPMENT AND ITS USE

High volume (H.V.) spraying

Although, as a result of the introduction of alternative methods of pesticide application, the need for H.V. spraying machines has declined, there are relatively few nurseries without such sprayers. Development in the fields of biological control, and automated application of systemic insecticides, will further diminish the need for H.V. spraying to control insects.

A motorized spraying machine is necessary on all but the smallest nurseries where the somewhat laborious use of a knapsack sprayer may suffice. The choice of sprayer and motive power must be decided by the intending user who should take into account the likely amount of use, ease of maintenance and manoeuvrability. Where only a small sprayer is required and the nursery is supplied with electricity the convenience of electric power outweighs its higher cost. Electrically driven sprayers can be taken into glasshouses to reduce the length of hose required and so prevent excessive fluid pressure loss when using a medium pressure pump and long, small-bore hose. Elsewhere the advantages of machines which can be mounted on a tractor and driven by its power 'take-off' may be decisive. The main requirement of a spraying machine is that it shall provide the required rate of fluid output at the pressure necessary for efficient application to the crops which will be grown, the spraying time being included in the assessment of efficiency.

Unless the person spraying is very skilled and dextrous it is rarely possible to spray more than 60 gal/hr to plants in glasshouses. On well-trimmed tomatoes and cucumbers a skilled operator can apply 80 gal/hr without undue waste. To allow for lowered efficiency due to wear, and flow restriction in long hoses, an output of at least 160 gal/hr at 200 lb/in^2 should be stipulated for tomato and cucumber nurseries. On a small nursery a machine having an output of 100 gal/hr at 80 lb/in^2 will usually suffice. The stiffness of hoses required for high outputs and high pressures can, under some circumstances, be a serious disadvantage and high pressures with large nozzle orifices are undesirable.

To provide the higher pressures and outputs, displacement pumps of the plunger (ram) or roller-vane types are preferable. Ordinary vane and gear pumps are subject to a high wear rate under nursery conditions. The lighter and cheaper centrifugal pumps are very suitable where moderate pressures are satisfactory, and, being largely free from wear, require little maintenance. The spray tank can either be attached to the pump or be a separate item, but where a non-priming centrifugal pump is used it is convenient to have the tank arranged so as to feed the pump by gravity. An easily cleaned filter should always be inserted between the tank and pump, and contents of the tank should be agitated by mechanical stirring or 'feed-back' via the pressure release valve.

The most important factor in efficient H.V. spraying is the skill of the man with the lance. To get adequate coverage of the foliage when spraying most glasshouse crops, the spray lance must be moved upwards, downwards and sideways, and, in order to avoid the shading of one leaf by another, it may be necessary to insert the nozzle between the leaves of plants such as cucumbers.

The volume of spray (H.V.) needed per unit area of crop depends on so many factors that precise figures are impossible. Cucumbers at the height of their growth require from 500–800 gal/acre, and tomatoes, when carrying the maximum amount of foliage, may require about the same volume. Approximately $1\frac{1}{2}$ gal/100 ft^2 of bed will be needed to thoroughly spray contact insecticides on to YR chrysanthemums, when in bud, or 2-year-old carnations because the physical difficulty of adequately covering both lower and upper leaf surfaces leads to considerable wastage.

The design of the spray lance is also important. A lance about 27 in. long with the nozzle set at an angle of about 120° to the shaft, and with a comfortable handle, *cum* lever-tap, is most suitable. The nozzle assembly should be as free as possible from projections which can catch in supporting wires or plant stems. Single jet lances are preferable to twin jet assemblies, and where eddy chamber and disc nozzles are used, disc apertures of $\frac{3}{64}$ or $\frac{4}{64}$ in. are suitable for pressures of 80–150 lb/in^2. The hoses used must withstand the fluid pressure applied with ample safety margin and also resist the solvent properties of insecticidal emulsions.

To promote exchangeability of equipment on a nursery the sizes and screw threads of unions on hoses, lances, etc., should be standardized as much as possible.

Low volume (L.V.) spraying

In low volume spraying the insecticide is diluted with only a small volume of water and is conveyed to the plants as small droplets in a stream of air with the intention of covering the leaves with small, non-coalescing droplets by direct impact. The method is seldom satisfactory.

The machines used—knapsack or wheeled machines of the spinning-disc air blast type—are difficult to manoeuvre in the pathways between rows or beds of plants; damage may result at short range from the air blast and even with open-spaced plants the cover is inadequate. The poor cover obtained by L.V. spraying was demonstrated by Staniland who incorporated a fluorescent dye in the spray and examined the leaves in ultra-violet light when dry.

Somewhat better results can be obtained with paint-spray guns or small, hand carried machines of the spinning-disc type, on crops such as tomatoes and roses. Even so it is difficult to get adequate cover of both surfaces of the leaves with any worthwhile economy of time compared with H.V. spraying. Furthermore, the amount of chemical deposited is not regulated by 'run-off' and this may lead to phytotoxicity or hazardous residues on edible produce. It is inadvisable to apply, at low volume, pesticides which have little safety margin at high volume. Further, precautions should be taken to avoid the inhalation of mists of poisonous substances. L.V. spraying is most successful where persistent insecticides, e.g. DDT, are used against mobile insects such as caterpillars. It is therefore useful, for example, for the control of pests on chrysanthemum flowers during the period November–January, when the use of H.V. sprays may encourage flower diseases such as grey mould (*Botrytis cinerea*) and petal blight (*Itersonilia perplexans*). Even so, either dusts or H.V. sprays in which a fungicide is incorporated are usually more satisfactory.

'Aerosol' generation (atomization)

Although both these terms exaggerate the degree of dispersion obtained, they have become adopted by the glasshouse industry.

Canister-type aerosol generators in which the insecticide is dissolved in a solvent which is gaseous at normal pressure, e.g. methyl chloride or Freon, are too expensive for most commercial uses, although large cylinders containing such solutions are used in America and, to some extent, elsewhere. Devices for the automatic dispersion of pyrethrum aerosols from domestic 'aerosol' canisters are available and have shown promise for the control of flies entering mushroom houses.

The methods employed for producing aerosols from solutions (aerosol or atomizing concentrates) are (a) with a compressed air paint-spray gun; (b) with electrically driven devices in which the fluid is fed to the centre of a disc rotating at very high speed. The minute droplets developed by the centrifugal forces are then carried forward by a peripheral air stream from a fan; (c) with a small hand device from which the insecticide is forced through a jet by pressure produced by a gas charge of the 'Sparklet' type.

Spray guns have several advantages for, under some conditions, they are much easier to handle and suction-fed types can easily be fitted with glass fluid containers which can be graduated, thereby enabling the operator to measure the quantity of concentrate required and so guide him in obtaining uniform distribution. With a motorized spinning disc, or similar type apparatus, the accuracy of the discharge rate regulator must be checked and the time required to treat a glasshouse calculated so that the rate of the operator's movement can be regulated to provide uniform distribution.

Where time saving is important and 'production' type guns are used an air compressor having an output of 8–10 ft^3/min at 80 lb/in^2 is required for each gun in large commercial glasshouses. For small glasshouses and mushroom rooms smaller 'light-production' guns are satisfactory. These function satisfactorily with only 4 ft^3/min of air at 40–50 lb/in^2.

The air pressure at the gun must be adequate, otherwise the droplet size will be too large for effective distribution. Application should commence several yards from the far end of the glasshouse, the operator walking backwards whilst directing the discharge over the plants or below them in the case of arch-grown cucumbers, or tomatoes when trimmed to a height of 4–5 ft towards the end of the season. The spray must not be directed at the plants for, if foliage is wetted, local damage and excessively high residues may occur. In wide houses it may be desirable to apply the aerosol from two pathways, preferably by two operatives working in parallel.

To restrict losses and provide as much vapour action as possible no ventilation should be given for several hours following aerosol treatment. Moderately high temperatures are usually required (70–80 °F; 21–26 °C) and in bright, warm weather it is usually best to treat glasshouses a short time before sunset in order to take advantage of the existing temperature and at the same time prevent risk of excessively high temperatures developing in the unventilated houses. *Caution.* The inhalation of poisonous chemicals is very dangerous—the required protective clothing must be worn (p. 39).

Smokes and fumigants

No special equipment is necessary when fumigating with insecticidal smokes, but protective clothing may be required (p. 38). The pellets or canisters should be distributed along the glasshouse and placed on a non-inflammable base. To obtain good smoke distribution, there should be at least four generators in each 100 ft length of narrow houses and at least three in wide, high houses. The same provisions regarding ventilation and glasshouse temperature apply as for aerosols. In interconnected blocks of houses a number of operators should ignite the smokes in each house simultaneously, otherwise a dust mask or respirator must be worn.

Occasionally, houses are fumigated with formaldehyde vapour by adding 1 lb of permanganate of potash to 3 pints of commercial formalin for each 1000 ft^3 to be treated. The reaction is violent and should take place in large vessels—the operator leaving the house directly the materials come into contact.

Thermal fogging

Thermal fogs are usually produced by injecting an oil-based solution of the insecticide into the exhaust from the pulsed, explosive firing of an air-fuel mixture. The shearing action of the exhaust flow across the jet introducing the insecticide, coupled with a further heating process in which fresh air is mixed with the exhaust, results in a fog of droplets ranging in size from 5–60 μ.

The short duration of heating permits the use of some substances which are normally thermally unstable. Some insecticides may require formulating specifically for use in 'fogging' machines.

The protective clothing regulations (Table 2.1, pp. 38–40) stipulated for the application of aerosols apply to the application of thermal fogs.

Dusting

Dusts can be applied either with hand or motorized dusting machines or by compressed air-dust guns operating in a manner similar to paint-spray guns. Except where small hand dusters are used it is generally best to direct the stream of dust over, rather than at, the plants to prevent localized excessive deposits.

Dipping

It is practically impossible to wet both upper and under surfaces of the leaves of some plants (e.g. saintpaulia) by spraying and it may be necessary to dip them in a vessel containing the diluted pesticide. Rubber gloves should always be worn unless other gloves are stipulated in mandatory protective-clothing requirements.

Application to the soil

The application of insecticides in soil drenches is laborious unless the liquid can be applied through a hose fitted with a coarse watering 'rose'. The latter provides a uniform distribution when large dosages are applied to unplanted soil. When the lance is replaced with a rose some spraying machines may have sufficient throughput for the purpose. By elevating a large tank having an outlet for a hose connection, or pumping the diluted pesticide from a tank standing on the ground, large areas can be drenched expeditiously. An alternative method is to dilute the pesticide in a venturi-type, liquid fertilizer applicator, if necessary, after preliminary partial dilution. This method is very satisfactory with water-soluble concentrates provided the solution in the applicator is of sufficient density (sp. gr. 1·05 or above). In many cases the density can be adjusted to this level by the addition of a soluble fertilizer such as potassium or ammonium nitrate, but advice should be obtained from the manufacturer of the pesticide before adjusting the density in this manner. A few growers have successfully used this method for the application of some emulsifiable concentrates of systemic insecticides. Preliminary dilution with a solution of a fertilizer salt (e.g. potassium nitrate at 1 lb/gal) is necessary. This procedure has been combined with trickle irrigation systems, though where this is done the protective clothing required when cleaning spray apparatus must be used when handling and cleaning the dilutor and watering equipment. Dilution of poisonous substances is not permitted by water supply authorities where there is direct connection between the supply and dilutor.

Several volatile, fluid nematicides (e.g. DD, ethylene dibromide and methyl isothiocyanate) can be injected into the soil without dilution, with an injector which incorporates a hand-plunger pump. The stroke of the plunger is adjustable so enabling the required amount of the fluid to be ejected through a number of peripheral holes near the end of a spike which is thrust into the soil to the required depth. The 'in-line' pump and spike assembly is surrounded by a cylindrical vessel containing the fluid.

Soils should be well cultivated before injecting and, as work proceeds, the hole made by the spike closed with the heel. The soil surface is usually trodden or rolled and then wetted with water or a diluted phenolic acid emulsion (p. 261) to retain the vapour. Motorized injectors have been devised, but pipe and purlin supports restrict their movement and great care is necessary to ensure that untreated areas do not result from orifices becoming blocked with soil.

Injectors with the container made from semi-transparent plastic, so enabling the operator to see the content of fluid, are now obtainable. The fitting of a spacing-distance gauge to the adjustable depth of the injection stop-plate will enable the operator to inject at required intervals along lines marked on the glasshouse border.

Liquid soil fumigants corrode the metals used in the construction of injectors, and it is important both to thoroughly wash out the apparatus after use, with paraffin or other solvent recommended by the manufacturer, and to oil it before storage. If metham-sodium is injected water should be used for washing-out.

Fumigation with methyl bromide (p. 262)

Methyl bromide is a gas at normal temperatures (b.p. 4° C; 39 °F) and penetrates cultivated soil (provided it is not wet) sufficiently to kill eelworms to a depth of 24 in. or more by downward diffusion from the surface. It is available liquified under pressure in canisters (usually containing 1 lb) or in gas cylinders. The gas is liberated under a polythene sheet, the edges of which are buried in the soil or weighted down to prevent the gas escaping. The required dosage is given by using the necessary number of canisters, by weighing the cylinder (a spring balance is convenient) or by using a 'dosimeter' such as that illustrated by Hague, Lubatti & Page (1964). If cylinders are used, either a vaporizer and perforated pipe attached to a 'T' piece, or discharge of the liquid into trays distributed over the area to be treated, is necessary to obtain good distribution over large areas.

The dosage required is preferably calculated on the concentration $\times$ time product (C.T.P.), which is usually expressed as oz hr/1000 ft^3 or mg hr/l. For all practical purposes these expressed dosages are equivalent. The underlying principle in relating dosages to the C.T.P. is that, within extreme limits, the exposure of organisms to a low concentration of a fumigant for a long period has the same effect as exposure to a higher concentration for a proportionately shorter time.

Precise regulation of the dosage requires expensive analytical apparatus, consequently it is usual to apply sufficient methyl bromide (or, indeed, other soil fumigants) to ensure that effective control is obtained under all normal conditions. The diffusion of vapours in soil is a highly complex process since it is dependent on the temperature, porosity, moisture content, and chemical nature of the soil. A brief resumé of these factors is given by Peachey & Chapman (1966). The recommended procedures for soil fumigation are based on the results of repeated trials and performance in practice, rather than on the results of fundamental research.

Avoidance of plant injury

When an effective and safe control has been found for a pest on a host plant of minor economic importance, the continued examination of other chemicals for that specific use may be unwarranted. There is, in consequence, a lack of information on the tolerances of many of the least important plant species and their cultivars to many insecticides which may potentially be applied to them.

The safety, or otherwise, of certain chemicals when applied at the normal concentrations to some of the more important plant species is included in Appendix II (p. 382). It must be remembered, however, that many factors influence the susceptibility of plants to damage, and in cases of doubt, information should be sought from the suppliers of the insecticide or the N.A.A.S. If information is unavailable, an initial, small-scale test should be made.

Injury by pesticides is usually most pronounced on tender, succulent growth—seedlings and young plants are particularly vulnerable, especially when grown under poor light conditions.

Sprays or soil drenches should never be applied to plants when the soil is dry, when the plants are wilting due to recent planting or, with tomatoes, at the onset of tomato mosaic disease. Excessively dry atmospheres may increase the likelihood of damage from soil-applied systemic compounds. The application of sprays in bright sunshine is risky. On the other hand spraying should not be done under conditions which may result in the plants remaining wet for long periods (e.g. overnight) for this encourages fungal attack in addition to chemical damage.

Some ornamental pot-plants, such as bromelids, certain orchids and palms, may retain considerable quantities of spray in hollow crowns or leaf crotches, and it is advisable to drain this off by inverting the plants or laying them on their sides as soon as the foliage has dried.

Aerosols, smokes and fumigants should not be applied when the foliage is wet or when there is risk that the temperature may become excessive.

Care must be taken to ensure that no residues of volatile nematicides are present when plants are taken into treated glasshouses. Their retention is greatest in soils of high organic content (e.g. peaty soils) and in wet soils. On occasions, the elimination of soil fumigants has been greatly retarded by a high water table. Severe damage to susceptible plants has been caused by the passage of vapours from treated houses through pipe ducting, partitions or even ventilators.

REFERENCES

GOULD, H. J. & KINGHAM, H. G. (1964). The efficiency of commercial high volume spraying with acaricides on cucumbers under glass. *Pl. Path.*, **13**, 60–4.

HAGUE, N. G. M., LUBATTI, O. F. & PAGE, A. B. P. (1964). Methyl bromide fumigation of nematodes in soil under gas-proof sheets. *Hort. Res.*, **3**, 84–101.

PEACHEY, J. E. & CHAPMAN, M. R. (1966). Chemical control of Plant Nematodes *Comm. No. 36. Commonwealth Bureau of Helminthology, Farnham Royal, Eng.* p. 119.

READ, W. H. (1966). Methods and problems of glasshouse spraying. *Proc. 3rd Brit. insect. fung. Conf.*, 1965, 377–83.

STANILAND, L. N. (1960). Field tests of spraying equipment by means of fluorescent tracer techniques. *J. agric. Engng Res.*, **5**, 42–82.

THE SAFE USE OF INSECTICIDES

It is to be expected that chemicals which are lethal to insects may also be toxic to mammals and so present hazards to persons applying them, handling treated plants or even consuming treated produce. When insecticides are applied within the restricted environment of glasshouses and mushroom houses hazards to wild life, including beneficial insects, are of little significance.

Pest control chemicals should be effective, economic and safe. Consumers are unlikely to accept a lower grade of produce in exchange for reduced pesticide usage. Safer and, if possible, cheaper pesticides must be continually sought. Non-chemical methods of control may be developed but it should not be overlooked that even sex attractants and sterilants are chemicals, as are the toxic, crystalline inclusions of the insect pathogen *Bacillus thuringiensis*.

The toxicities of chemicals to man, whether ingested through the mouth (oral) or absorbed through the skin (dermal), cannot be determined directly, and consequently their toxicities to standard laboratory animals, usually rats, are determined. Sometimes other animals are used, for it has been found that man has the doubtful honour of reacting to at least one chemical more like a domestic hen than a rat.

The acute oral and dermal toxicities of a chemical are the best indications of its potential hazard to human users. These are expressed as L.D.50's—the amounts, which as a single dose, will kill with one application, 50% of the test animals. As this amount depends on the body weight of the animal the L.D.50's are expressed in mg/kg body weight. The toxicities of insecticides, inhaled as mists or vapours, are probably roughly proportional to their acute oral toxicities. It should be realized that inhalation leads to very rapid absorption and is likely to constitute the greatest risk to persons applying insecticides in glasshouses. With some compounds toxic effects rapidly follow contamination of the skin, e.g. with nicotine. Figures for the acute oral, and dermal L.D.50 are given in Appendix I, p. 371.

Some chemicals, especially persistent organochlorine insecticides, are accumulated in the body so that a succession of small doses may ultimately produce poisoning symptoms. Toxic organophosphorus compounds may destroy an enzyme, cholinesterase, which is only slowly regenerated by the body. Absorption of organophosphorus compounds with low acute toxicity (e.g. malathion) makes the recipient much more susceptible to the action of more poisonous related compounds. This interaction is termed potentiation. Likewise persons medically advised not to use organophosphorus compounds may be very susceptible to organophosphorus insecticides normally considered safe.

Safety to workers

As a result of public concern following the deaths of agricultural workers who had been abnormally exposed to 3,5-dinitro-*o*-cresol (DNOC) when spraying in hot weather an official working party was set up by the Government in 1950. This working party was set three tasks, which, expressed briefly, were:

1. To make recommendations for the safety of workers applying agricultural pesticides.
2. To investigate risks to consumers from treated agricultural and stored products, and to make recommendations if protective measures were considered necessary.
3. To investigate possible risks to wild life and to make recommendations.

On the receipt of the report on the first of these, the Agriculture (Poisonous Substances) Act was passed in 1952. This authorized the Ministry of Agriculture, Fisheries and Food and the Secretary of State for Scotland, after consultation with the industries concerned, to make regulations to safeguard agricultural and horticultural workers. The regulations do not apply to self-employed persons or members of the general public, but they impose obligations both on employers and on employees of farmers, growers and contractors, both of whom can be charged with infringements. Similar Acts have been passed by the legislatures of Northern Ireland, the Channel Islands and the Isle of Man.

Inspectors, with the responsibility of enforcing the act and regulations made under it, have been appointed. They have the right of entry to holdings and can demand the production of certain documents or take samples for analysis. Inspectors are authorized to give advice and assistance in connection with the regulations, and can grant a certificate of exemption from some or all of the requirements of the regulations if they are satisfied that such requirements are unnecessary under the particular circumstances.

The general requirements of the Regulations, which are summarized in a leaflet entitled *The Safe Use of Poisonous Chemicals on the Farm*, concern the provision of protective clothing including listed, approved types of respirators and dust masks together with the supply of replacement filters for these; accommodation for workers' personal clothing; washing facilities; the supply of drinking water and the keeping of records of exposure to scheduled pesticides and of absences due to illness. The employer is also responsible for ensuring that protective clothing is washed, that spraying apparatus and the outsides of containers are kept free from contamination. He must also ensure that a register is kept containing particulars of the period of exposure of workers engaged in scheduled operations, as well as any real, or suspected, illness they may have had after using scheduled substances. This register need not be kept if the total area of greenhouses, including mushroom houses, is not more than 2 acres. No person under 18 is permitted to undertake scheduled operations; the employer must ensure that any worker is adequately supervised and warned in taking precautions.

The obligations on workers require that they use, but not misuse, the protective clothing provided and do not eat, drink or smoke when there is any risk of contamination. The detailed requirements of the regulations are stated in Statutory Instrument 1966 No. 1063 and are summarized in the leaflet

The Safe Use of Poisonous Chemicals on the Farm, issued free of charge by the Ministry of Agriculture, Fisheries and Food.

In the Agriculture (Poisonous Substances) Regulations 1966, the chemicals, classified according to their danger, are placed in four lists and referred to as 'Part I–IV Substances' according to the list in which they occur. None of the most hazardous Part I substances are used in glasshouses in Britain.

The regulations take into account the fact that the inherent danger of a compound is greatly influenced both by the manner in which it is applied and the purpose for which it is used. The use of poisonous chemicals in glasshouses and mushroom houses is generally more dangerous than in the open. When spraying crops such as tomatoes and cucumbers which have grown to the tops of the wires, it is difficult to avoid considerable contamination of the clothing from contact with the plants and by overhead drips. The special risks inherent in the use of aerosols have been mentioned already. A leaflet dealing with the risks involved in the use of insecticides in mushroom growing is published by the M.A.F.F. entitled *The use of chemicals by mushroom growers*. Operations involved in the control of glasshouse and mushroom pests, for which protective clothing is required when using specified substances (Scheduled Operations), are given in Table 2.1.

The classification ('Part' no.) of insecticides in the schedule of substances to which the Agriculture (Poisonous Substances) Regulations apply, is also given in Table 2.1.

TABLE 2.1. **Protective clothing and other precautionary requirements**

The Regulations made in 1966 under the Agriculture (Poisonous Substances) Act, although applying only to employers and employees of farmers, growers and contractors, should always be observed, insofar as the safety precautions are concerned, by self-employed persons and members of the general public. Proprietary products containing substances to which these regulations apply carry instructions for their safe use. It must not be overlooked that practically all insecticides are poisonous and the user should avoid contamination even with non-scheduled substances.

The protective-clothing requirements when using those substances recommended for glasshouse pest control and to which the Regulations apply (usually classified as Parts I, II, III and IV substances, see above) are as follows. No Part I substances (dimefox) should be used in glasshouses or mushroom houses.

	Operation requiring protective clothing	Clothing, etc., required to be worn
Part II Substances		
Including: Disulfoton, DNOC and its salts Endosulfan Endrin Parathion Phorate Schradan	Opening a container; transferring from one container to another; or mixing or diluting	Rubber gloves, rubber boots, face-shield and either (a) an overall and rubber apron; or (b) a mackintosh. (Where DNOC is used solely as an insecticide and the preparation contains less

TABLE 2.1 (*continued*)

	Operation requiring protective clothing	Clothing, etc., required to be worn
Sulfotep TEPP Thionazin		than 5% DNOC by weight, only rubber gloves and a face- or eye-shield)
	Opening a container containing granules or transferring the contents	Rubber gauntlet gloves and either an overall or mackintosh, with the sleeves worn over the cuffs of the gloves
	Washing or cleansing spraying or other apparatus	Rubber boots, face-shield and either (a) an overall and rubber apron; or, (b) a mackintosh
	Spraying a ground crop in the open	Overall, hood, rubber gloves, rubber boots and either a face-shield or dust-mask
	Spraying in a greenhouse*	Rubber gloves, rubber boots, hood, face-shield and either an overall or mackintosh
	Applying aerosols	Overall, hood, rubber gloves and approved respirator
	Granule placement by hand or with hand-operated apparatus	Rubber gauntlet gloves and either an overall or mackintosh, with the sleeves worn over the cuffs of the gloves
	Soil-application in the open	Overall, rubber gloves and rubber boots
	Soil application in a greenhouse	Overall, rubber apron, rubber gloves and rubber boots
	Entering, or being present in, a greenhouse in which a specified (Part II) substance has been used *except* where (a) at least 12 hr after its application; or (b) where the house has been thoroughly ventilated for at least 1 hr following a period of 6 hr after application	(a) Where aerosols or smokes are used—overall, hood, rubber gloves and respirator (b) Where sprays are used—rubber gloves, rubber boots, hood, face-shield and either an overall or mackintosh

TABLE 2.1 (*continued*)

	Operation requiring protective clothing	Clothing, etc., required to be worn
Part III Substances		
Including: Azinphos-methyl, Chlorfenvinphos, Demeton-*S*-methyl, Dichlorvos (see note 2), Nicotine (see note 3), Oxydemeton-methyl, Phenkapton, Phosphamidon, Vamidothion	Opening a container; transferring from one container to another; or mixing or diluting	Rubber gloves and face-shield
	Opening a container containing granules or transferring the contents	Rubber gloves
	Applying aerosols	Overall, hood, rubber gloves and approved respirator (see note 2)
	Entering or being present in a glasshouse in which a specified (Part III) substance has been used *except* where (a) at least 12 hr after its application; or (b) where the house has been thoroughly ventilated for at least 1 hr following a period of 6 hr after application	(a) Where aerosols or smokes are used—overall, hood, rubber gloves and respirator; (b) Where sprays are used—see note 1 below

* A 'greenhouse' is any building or erection in which plants, including mushrooms, are grown

Note 1. Although protective clothing is not compulsory when applying sprays containing Part III substances, it is obviously desirable to wear a hood or other head protection, and an overall or mackintosh to prevent contamination of ordinary clothing when spraying tall crops such as cucumbers, tomatoes or climbing French beans.

Note 2. Where an aerosol is produced from a preparation containing not more than 0·4% by weight of dichlorvos and no other scheduled substance, the wearing of protective clothing is not compulsory. Impregnated resin strips containing not more than 20% by weight of dichlorvos are also exempt.

Note 3. Products containing not more than 7·5% of nicotine, and no other scheduled substance, are exempt.

'Spraying' does not include 'soil application' when pesticides are applied to the soil in unbroken, liquid form, i.e. with a watering can or hose.

Although many insecticides used to control glasshouse pests are not scheduled in the Regulations as poisons this must not lead to neglect of simple precautions. A light dust filter should be worn when applying aerosols of non-scheduled substances except where unconnected houses are being treated and the operator can walk away from the mist so as to avoid exposure to mist drifting from previously treated, interconnected, houses. A mask should also be worn where there is danger of inhaling dusts or sprays of non-scheduled substances.

Regulations made under the Poisons and Pharmacy Act 1933 govern the sale of the more poisonous insecticides, many of which are not readily available to the amateur. Poisonous substances must be kept in a locked store, on which a warning has been fixed, preferably sited away from other materials. They must at all times be clearly labelled 'poison'. Measured amounts for preparing sprays must not be placed in bottles or other containers which have contained beverages or are similar to such containers—several fatalities have occurred through doing so. When empty, the containers should be washed out and burnt or smashed and safely buried. Unwanted pesticides and surplus liquid from spraying, or washing from apparatus must not be run where they can contaminate streams, ponds or drains.

Safety to consumers

The working party's report on the second of its tasks was published in 1953 and as a result of its recommendations a Pesticides Safety Precautions Scheme has been established. This embraces a Notification Scheme which, although officially voluntary, is in fact binding upon members of the various trade organizations, such as the Association of British Manufacturers of Agricultural Chemicals. Under this scheme manufacturers of pesticides (or their agents) notify the Ministry of Agriculture, Fisheries and Food when they have developed a new chemical, or a new use of an existing, accepted chemical, to the stage at which it may be used in trials involving agricultural workers or when treated produce from such trials may be marketed for human or animal consumption. The working party also recommended the setting up of a standing committee to advise it on all risks arising from the use of pesticides in food production and storage.

In the Notification Scheme the onus is placed on the notifier to provide information on the toxicity of the compound, the residues resulting from its use, risks to wild life, etc., all of which will enable a decision to be made on the stage of clearance granted. Commercial clearance, the final stage, is only given when no further information is required from the manufacturer. The other stages—trials clearance, limited clearance and provisional commercial clearance—give progressively increased freedom to the manufacturer in undertaking the large scale developmental trials required to ensure that the new chemical is a reliable economic pesticide. Where there is existing evidence that the proposed use of a chemical is safe, commercial clearance may be granted without the preliminary stages.

Clearance can be withdrawn if further information becomes available suggesting the use of a chemical is undesirable. Fluoracetamide is no longer permitted for use as an insecticide and the uses of aldrin, dieldrin and heptachlor have been restricted. The use of a product containing a new chemical or formulation, or a new use of an existing chemical having been approved, the recommendations for its safe use are summarized on the labels or directions accompanying the product. They are also published in full by the M.A.F.F. as 'Recommendation Sheets'. The recommendations cover matters such as the crops for which the product is approved, the rate of application, the number and timing of applications and the period which should elapse between their application and the harvesting of edible crops on which they may be used.

As a result of the decision of the Advisory Committee on Poisonous Substances used in Agriculture and Food Storage (p. 36) that there was undesirable contamination of some food stuffs and the environment with traces of some of the persistent organochlorine compounds, revised recommendations for certain of these were issued as a precautionary measure (p. 21). The terms of reference of this committee, now frequently referred to as the 'Cook Committee', have since been enlarged and it is now responsible to the Secretary of State for Education and Science. The problems associated with other persistent pesticides will be periodically reviewed.

The approval scheme

This, the oldest voluntary scheme arranged between organizations representing pesticide manufacturers and the Government, commenced in 1942 with the object of enabling growers to select efficient and suitable pest control products. Originally named the *Crop Protection Products Approval Scheme* it was changed to the *Agricultural Chemicals Approval Scheme* in 1960, and is still financed by fees paid by manufacturers on application for approval.

Manufacturers and their agents can obtain approval for specific uses of a product by submitting data in support of the claims made for the product. If these claims are accepted as correct and the chemical had been cleared under the Safety Precautions Scheme approval is granted. Approval extends to the directions, claims and precautions given on the label attached to the product, and the label must carry the approval mark—a bold capital 'A' supporting a crown. Very frequently, information derived from trials by independent investigators is also considered when deciding whether a product should be approved. Reapproval is required if the composition of the product is changed.

As the scheme is voluntary the fact that a product is not approved does not indicate that it is inferior; indeed, certain products of value for the control of some glasshouse pests are not approved.

It is uneconomic for manufacturers to test insecticides on the less important plants grown under glass or against minor pests and consequently, when products are submitted for official approval, few specific claims are made regarding their use on many plant species.

Nevertheless, growers can select approved proprietary formulations of many pesticides. Two lists of approved products are published, one being an annual *List of Approved Products for Farmers and Growers* and the other *Chemicals for the Gardener*. The former is free and lists proprietary names of products under the headings of the common names of the active ingredient.

Users of pesticides should always study carefully the 'directions for use' as well as the precautions to observe. Products should not be mixed unless it is known that the procedure is safe.

REFERENCES

ANON. (1960). Fumigation with methyl bromide. Precautionary measures. (Home Office) *H.M.S.O. Lond.* 8 pp.
— (1964). Review of the persistent organochlorine pesticides and Supplement. *Rept. Advis. Comm. on Poisonous Subst. used in Agric. and Fd. Storage. H.M.S.O. Lond.* 68 pp. Suppl. 8 pp.
— (1965). Chemicals for the Gardener. *H.M.S.O. Lond.* 35 pp.
— (1966). Pesticides Safety Precautions Scheme agreed between Government Departments and Industry. *M.A.F.F. Lond.* 9 pp. (and appendices).
— (1969). Agricultural Chemicals Approval Scheme. List of Approved Products *M.A.F.F. Lond.* 150 pp.
— (1968). Safe and efficient fumigation practice. *M.A.F.F. Lond.* 5 pp.
JONES, K. H., SANDERSON, D. M. & NOAKES, O. N. (1968). Acute toxicity data for pesticides (1968). *Wld Rev. Pest. Contr.*, **7**, 135–143.
RILEY, E. A. (1964). Government control of pesticides in Great Britain. *Proc. 2nd Brit. Insect. Fung. Conf.* 1963, 13–19.

Chapter 3

The Economics of Pest Control

With the ever present dangers of inducing resistance to crop-protection chemicals it is essential to ensure that unnecessary applications are avoided. An application of pesticide is naturally superfluous when the cost of an operation giving immediate control or preventing future damage does not result in a larger cash return.

Cost of treatment

The cost of a chemical application includes (i) the cost of the pesticide, (ii) labour charges involved and (iii) the repair and depreciation of equipment. The percentage depreciation is difficult to assess but Nix (1954) suggests that it should diminish as the equipment ages so that it is written off in 8 years, for obsolescence is normally more important than wear and tear. Interest on capital should be charged at compound interest (5%) over the 8-year period. With equal annual payments of interest plus capital the charge amounts to approximately 3% of the initial cost. Such expenditure should be regarded as a real cost to a grower having to borrow the money to buy a machine, or as 'opportunity cost', i.e. income from an alternative investment forfeited, in the case of a man drawing on existing capital. There is no difficulty in calculating these components of the cost and arriving at an accurate estimate.

Economic effectiveness

It is first necessary to appreciate how pests affect crops. There is usually some form of direct damage caused either to the roots, stems, or leaves, by organisms tunnelling within, or feeding superficially on, the tissues. Insects with sucking mouthparts, e.g. whiteflies and red spider mites, may merely cause chlorotic marks on the leaves, but occasionally necrotic brown lesions occur, e.g. capsids. In other cases distortion of the foliage is caused by salivary secretions injected into the tissues. With insects which produce honeydew, damage may result from the growth of sooty moulds which interfere with the respiratory function of the leaves and increase labour costs by necessitating cleaning of the fruit (Hussey *et al.*, 1959).

Under protected culture, one of the most important effects of a pest is to cause contamination or reduction in the quality of the crop, e.g. distorted and marked blooms caused by severe thrip damage. The presence of aphids in chrysanthemum flowers, and midge larvae on mushrooms, are examples of loss by contamination. A serious, indirect effect may result where the pest acts as a vector for plant diseases. The importance of aphids and leafhoppers in the dissemination of viruses is well known but the role of the tarsonemid mite, *Pediculopsis*, as a vector of the fungus *Fusarium* in carnation bud rot, and of the *Lycoriidae* (Diptera) as transmitters of several mushroom diseases, is less widely known. In these cases the presence of even a small pest population may cause disproportionate losses. In contrast, the mushroom phorid *Megaselia halterata* provides an instance where large numbers of flies constitute a hazard, not by reducing the crop, but merely by their nuisance value to the pickers.

When considering the effect of pests on crop plants, it must be appreciated that an initial increase in the pest population is not necessarily followed by a proportionate loss of yield. Further, since the harmful effect of individuals decreases as their numbers increase, the relation between the initial pest population and crop yield, or growth, is not linear (Justesen & Tammes, 1960). The effect on yield falls off as the population increases.

The exact relationship between injury and loss of yield is complex and can be divided into three phases which may occasionally be absent or modified. As the plant can compensate for the damage up to a certain threshold value (i.e. the economic threshold) there is no effect on yield during the initial phases of injury. There follows a phase showing direct correlation between decrease in yield and the degree of injury which may be followed by a third phase with a maximum degree of injury which, through competition within the pest population, is often self-limiting. This third phase may give zero yield or, where injury can only occur after a portion of the crop is harvested, a fixed minimum yield.

Another variable affecting the degree of damage caused by an insect population is the timing of the attack relative to the stage of plant growth. Lycoriid larvae causing damage to plant roots are a good example. Larval populations of 50, or less, in small pots severely retard growth of seedlings in the cotyledon stage but, when the plant has several leaves, almost 1000 larvae would be required to produce a similar check to growth.

Calculation of the potential financial loss is complex and often it can only be guessed. It is, however, desirable to establish the economic threshold for each pest yet it has been attempted in only a few cases under glass—e.g. with whiteflies on tomatoes (Hussey *et al.*, 1959), red spider mite on cucumbers (Hussey & Parr, 1963) and the Worthing phorid on mushrooms (Hussey, 1961). Judenko (1965a) has stressed that the problems involved in determining the success of a given treatment (i.e. comparison of treated and pest-free plots) are similar to those posed by assessing losses caused by pests (i.e. comparison of pest-attacked and pest-free plots). Outdoors, the study of unattacked plants is often difficult without recourse to some form of control (Judenko, 1965b). Such action may produce side effects other than simple control. The chemical may kill other less obvious pests, or affect parasites and predators so that further pest increase results in less yield increase than expected. There may, in addition, be direct interactions between the plant and chemical which may increase or decrease yields. However, in protected cropping it is often possible to avoid such complications by artificial infestation of experimental plots. There will frequently be no difficulties from emigration on to pest-free plots either from outside or from infested plots.

Ordish (1952) suggested that the ratio between the treatment cost and its potential financial benefit could be useful in deciding when, and how, control measures should be applied. These ratios must be determined experimentally both for different pest populations and degrees of damage and, to date, the basic data are available in only one or two cases.

A different approach to the determination of profitable pesticide application was suggested by Hillebrant (1960) who used the dosage/response curve, i.e. yield increase v. dosage rate of pesticide. The slope of this curve depends on: the effect of the pesticide on both pest and crop; the effect of the pest on the crop; on the effect of the application process on the crop; the interaction of the pesticide with other factors affecting crop production. Despite variation in these factors the dosage/response curve has the general form shown in Fig. 3.1. The optimal rate of pesticide use is determined, in

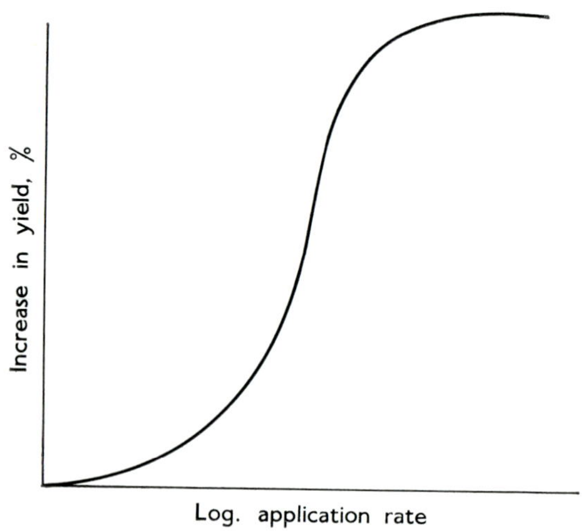

FIG. 3.1. Typical dosage–response curve.

any particular situation, by the relation between the cost of the treatment and the cash value of the crop on the one hand, and the slope of the dosage/response curve on the other.

There is, of course, one possible and curious by-product of successful pest control which should not be overlooked. A new technique, or particularly effective treatment, confers an immediate advantage to the first growers who exploit it but, if it is widely adopted, the greater overall production may result in a price reduction in the markets which more than counteracts the initial economic advantage.

Practical economics of pest control in protected environments

The theoretical arguments considered so far have been developed for problems encountered on outdoor crops where, in general, pest populations are relatively low. As a result, single pesticide applications usually cause spectacular reductions in pest populations with correspondingly large, clear-cut increases in crop yields. In protected cropping, although the kinds of pests affecting a single crop are relatively restricted, those that do occur tend to constitute a constant threat. This permanent threat derives partly from the high reproductive potential of many species and partly from the difficulties of achieving complete kills.

For various economic reasons the grower commonly adopts cultural methods involving very dense plantings. This situation makes complete spray coverage almost impossible, and multiple applications are necessary to contain the pest population below the economic threshold. In these circumstances efficiency is the main prerequisite with fumigant action conferring a distinct theoretical advantage.

The grower should always consider the degree and timing of control required to safeguard his crop. In recent years, there has been a tendency to strive for a higher degree of control than is justified. Such 'insurance' control may be dearly bought, both economically and, where applications are made over-frequently, from the development of resistance. These points can be illustrated by considering the control of red spider mites on cucumbers. This pest does not affect the cucumber fruit itself but, if the leaf damage (chlorosis caused by the minute feeding punctures) exceeds a certain critical level approximating to a reduction of the total photosynthetic area by about 30%, then there is a reduction in the number of fruits formed (Hussey & Parr, 1963). The degree of leaf damage may be estimated by a simple scoring system (ranging from 0–5) which permits a *mean leaf-damage index* to be calculated. The critical degree of damage below which no loss of crop can be detected, however long the damage persists, is 1·9. It has been shown that, where the damage index rises to 2·0, the number of cucumbers picked 5 weeks later (period of fruit maturation) is reduced by 8%; damage levels of 2·5 and 3·0 result in losses of 42% and 70% respectively. By calculating the weekly cash value of an average cucumber crop throughout its growth (weekly yield × average market price) it is possible to illustrate (Fig. 3.2) the differences in financial loss which would follow a leaf-damage index of 2·0 at various times in the cropping season.

Since the average cost of each H.V. acaricide application, including labour, depreciation and interest charges, is about £20/acre (smokes are rather less expensive at £14 15*s.*/acre) it is apparent that even twice-weekly applications are economically advantageous throughout most of the season. On the other hand, from considerations of the rates of change in leaf damage caused by the feeding of increasing and declining populations, and assuming that really efficient acaricides are available, a maximum of six sprays in a season should maintain the leaf damage below the economic threshold (Hussey, 1966). As the average number of acaricide applications normally used is between 10 and

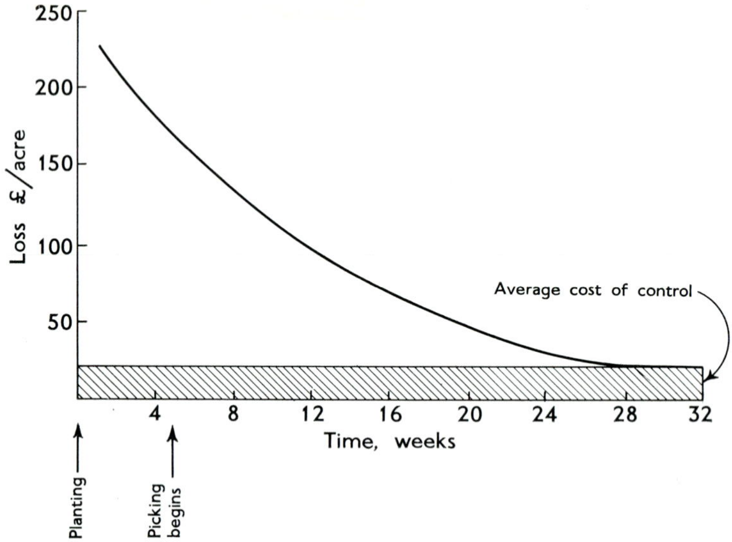

FIG. 3.2. Financial loss (£/acre) throughout a cucumber crop following mite damage equivalent to a mean index of 2·0.

20 there would be a valuable economy in timing spray applications by leaf damage relative to the economic threshold rather than adhering to a regular programme.

Similar considerations apply to any pest situation and growers should calculate the various financial parameters of spraying costs and potential benefits to judge the economic soundness of any pesticide application.

REFERENCES

COURSHEE, R. J. (1965). The economics of plant protection. *P.A.N.S. Sect. A.* **11,** 466.

GRAINGER, J. (1956). The economic effects of crop diseases. *Res. Bull. W. Scotl. Coll. Agric.,* **16,** 80 pp.

HILLEBRANT, P. M. (1960). The economic theory of the use of pesticides. Pt. I. *J. agric Econ.,* **23,** 464–72 and Pt. II. *J. agric. Econ.,* **24,** 52–61.

HUSSEY, N. W. (1961). Recent work on the control of the Worthing Phorid *M.G.A. Bull.,* **144,** 495–505.

— (1966). The role of leaf-damage assessment in the control of glasshouse red spider mite on cucumbers. *Proc. 3rd Brit. Insect. & Fung. Conf.,* **1965,** 317–22.

HUSSEY, N. W. & PARR, W. J. (1963). The effect of glasshouse red spider mite (*Tetranychus urticae* Koch) on the yield of cucumbers. *J. hort. Sci.,* **38,** 255–63.

HUSSEY, N. W., PARR, W. J. and GURNEY, B. (1959). The effect of whitefly populations on the cropping of tomatoes. *Rep. Glasshouse Crops Res. Inst.,* **1958,** 79–86.

JUDENKO, E. (1965a). The assessment of economic effectiveness of pest control in field experiments. *P.A.N.S.,* **11,** 359–68.

— (1965b). Some methods of assessing crop loss caused by pests. *Proc. XII. Congr. Ent.,* **1964,** 614–5.

JUSTESEN, S. H. & TAMMES, P. M. L. (1960). Studies on yield losses. I. The self-limiting effect of injurious or competitive organisms on crop yield. *Tidjschr. Plziekt,* **66,** 281–7.

NIX, J. S. (1954). The economics of crop spraying. *Farmers Bull. No. 16*. Farm Economics Branch School of Agric., Cambridge.

ORDISH, F. (1968). Some notes on the short-term and long-term economics of pest control. *Pestic. Abstr.*, **14**, 343–55.

ORDISH, G. (1952). *Untaken Harvest*. Constable, London. 171 pp.

SYLVE'N, E. (1968). Threshold values in the economics of insect pest control in agriculture. *Pestic. Abstr.*, **14**, 356–66.

TAMMES, P. M. L. (1961). Studies on yield losses, II. Injury as a limiting factor of yield. *Tijdschr. Plziekt.*, **67**, 257–63.

Chapter 4

Structure and Classification of Insects

The insects, or Insecta, consisting of almost a million described species, are the largest class of the phylum Arthropoda which includes the lobsters, crabs, millepedes, centipedes, scorpions, spiders and mites. All arthropods possess the following characters: bilateral symmetry, the body with an external skeleton composed of a series of segments each bearing a pair of jointed appendages, a dorsal heart and a ventral nervous system. The vast assemblage of insects, all similar anatomically yet each biologically distinct, has evolved through the possession of four important attributes. Firstly, their ability to fly, which enables them to spread geographically and to seek out mates and food as well as providing a ready means of escape from enemies. Secondly, their adaptation to unusual environments, which has enabled them to colonize every available habitat from snow fields to deserts, including even salt lakes and oil wells. The horny exoskeleton, conferring as it does the greatest possible strength with minimum materials, has afforded insects a third attribute, namely the potential for infinite modification of body structure for different purposes. To confer advantages in special habitats these include variety in the form and size of mandibles, different forms of legs and ovipositors, and adaptations of the wings as membranous organs for flight or leathery wing cases for protection. Lastly, their small size creates advantages in the struggle for existence since the nutritional needs of individuals are relatively small.

STRUCTURE OF INSECTS

The body segments of adult insects are usually arranged in three principal regions, the head, thorax, and abdomen. The head (Fig. 4.1) bears a pair of *antennae* or 'feelers' while the three pairs of legs are all carried on the thorax. It is characteristic of insects that, during their development, they pass through distinct changes of form, i.e. metamorphosis, so that immature stages are often strikingly different from the adults. In the butterflies and moths, for instance, the grub, or caterpillar, is quite unlike either the chrysalis or the adult. In these developmental forms the distinguishing features typifying the adults are often obscured. The caterpillar has no obvious division into thorax and abdomen, though the former is recognizable by the presence of three pairs of jointed legs: unlike the adult, the abdomen carries thick fleshy legs. Fly maggots have no legs and, to the casual glance, seem headless. The chrysalis of a butter-

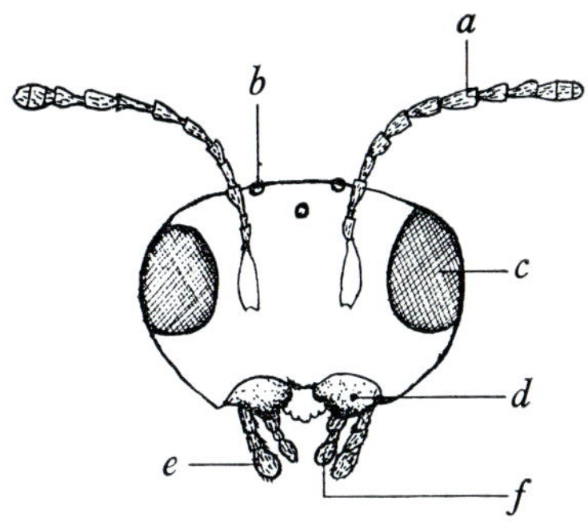

FIG. 4.1. Head of an insect with biting mouthparts: (*a*) antennae; (*b*) ocelli; (*c*) compound eyes; (*d*) mandibles; (*e*) maxillary palps; and (*f*) labial palps.

fly, though immobile, has the outline of the adult legs, wings and antennae visible through the outer skin. In some flies, however, this stage is passed within the shelter of the toughened last larval skin or 'puparium' which has no sign of appendages on its brown, barrel-shaped surface.

External structure

The head (Fig. 4.1)

An insect's head bears the eyes, antennae and mouthparts, and it contains the brain which is the sensory centre of the organism. The antennae or feelers, arising between or in front of the eyes, are organs of touch and smell. They commonly consist of a string of many small segments but they may take other forms. In male moths, for instance, they are feathery or *plumose* and in cockchafers they have a leaved or *lamellate* terminal club. The eyes are of two kinds, the *ocelli* or simple eyes, and compound eyes. The minute ocelli, usually three in number, are situated above the antennae and are only capable of perceiving changes in light intensity and do not form images. The large compound eyes, on the other hand, are situated on either side of the head and consist of many identical lenses arranged like a honeycomb. Each lens, usually hexagonal in form, perceives only a small portion of the field of view but, in combination, they provide the insect with a most effective means of detecting movement. Any attempt to catch a resting insect in the hand will illustrate how sensitive it is to even the slightest movement or shadow.

Though the form of the mouthparts varies according to the type of food and the feeding habits, there are two basic types, one adapted for biting and the other for piercing and sucking. The biting type, adapted for feeding on solid food, is found in the crickets, earwigs and beetles. Here the mouth parts consist of an upper lip or *labrum*, which acts as a roof to the mouth cavity, and a pair of *mandibles* or jaws, for crushing food. Behind the mandibles a pair of *maxillae*, provided with jointed appendages or *palpi*, serve to hold and chew the food. The lower lip or *labium*, bearing as many as four lobes and an additional pair of palpi, is principally sensory in function.

Where insects are adapted for sucking a liquid diet there may be considerable modification of these basic appendages. In butterflies and moths, for instance, the maxillae become a long flexible tube or *proboscis* which is coiled under the head when not in use. The proboscis of the housefly is a complicated organ adapted for both licking and sucking by a modification of the maxillae and by increased development of the membranous areas of the labium into a fleshy organ. The most important form of the sucking mouthparts as far as the glasshouse grower is concerned is that found in the order *Hemiptera* which includes the aphids, whiteflies and scale insects. Here the mandibles and maxillae have become long flexible stylets which are fitted tightly together and rest in a grooved labium almost as long as themselves. When not in use this sheath, or *rostrum*, is bent back under the head, but during feeding is held vertically with the labium bent to permit the stylets to penetrate the plant tissues. The rostrum is very short in the scale insects, and the long stylets, which exceed the length of the scale, are withdrawn into, and coiled within, an internal pocket. Each maxillary stylet is grooved into two channels along its inner face. Along one of the two fine tubes so formed pass the salivary juices which serve to break down cell walls, etc., and the partially digested plant juices are sucked up the other.

The thorax

The middle region of the body is composed of three segments which are usually difficult to identify. Each segment bears a pair of legs which are variously modified, e.g. for digging (mole cricket) and for running (cockroach). Each leg is subdivided into five distinct parts (Fig. 4.2). A basal portion or *coxa* unites the leg to the thoracic segment and is hinged to the *femur*, or thigh, by a small joint called the *trochanter*. The femur carries the shank, or *tibia*, to which is attached the jointed foot or *tarsus*.

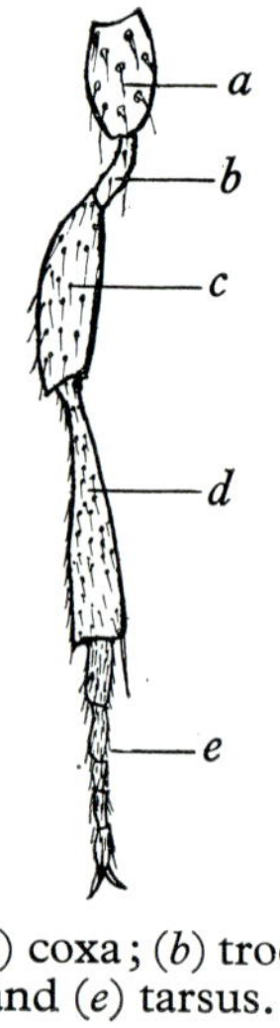

FIG. 4.2. Leg of an insect: (*a*) coxa; (*b*) trochanter; (*c*) femur; (*d*) tibia; and (*e*) tarsus.

The second and third thoracic segments bear the wings, usually four in number but occasionally, as in the true flies, only two. In some primitive insects, like springtails, wings are always absent. The wing membrane is usually strengthened by a framework of veins with a pattern which provides important features for classification. In *Lepidoptera* this pattern, though present, is obscured by a covering of coloured scales. In the beetles the front pair of wings is modified into wing cases or *elytra* which protect the membranous hind pair when these are at rest: the cockroaches have toughened, rather parchment-like, protective forewings called *tegmina*.

In most insects possessing two pairs of wings these are coupled together during flight so as to act in unison. Various structures have been evolved to achieve this, varying from that in bees, where a series of hooks on the hindwing engages in a fold in the forewing, to that in moths where a stiff bristle on the hindwings locks into a hooked catch on the forewing.

The abdomen

The hind-body is typically divided into eleven segments, though fewer are commonly visible. The anus is usually terminal and in many species there are a pair of *cerci* or sensory appendages on either side of it. These cerci assume many forms ranging from the well-known forceps of earwigs to the long filamentous tails of the mayflies so familiar to anglers.

The males of many insects carry a pair of *claspers* on the ninth abdominal segment to hold the female during mating. These structures are especially striking in the *Lycoriidae* or fungus gnats (see Fig. 9.12).

The female abdomen may also carry a prominent appendage, the *ovipositor* or egg-laying organ. This may, as in many parasites, be long and needle-like, or broad and scimitar-shaped as in the long-horned grasshoppers.

Internal structure

Digestive system (Fig. 4.3)

The alimentary canal can be considered as a simple tube with three distinct regions, two of which, the fore- and hindgut, are invaginations of the integument and are therefore lined with cuticle. At the termination of the foregut of insects that feed on solid food the muscular *proventriculus* serves as a sieve specially adapted to crush food particles by an arrangement of teeth and circular muscles. The *rectum* of the hindgut often carries *papillae* which assist the maintenance of water balance by absorbing water from the faeces. At the junction of the hindgut with the midgut are a group of long outgrowths, the *Malpighian tubes*. These vary in number from 0 to 120 and are the main excretory organs absorbing waste materials from the blood which surrounds them in the haemocoel. The midgut is lined with large epithelial cells which not only secrete enzymes to digest starch, sugars, fats and proteins, but also reabsorb the food material leaving the unwanted faecal matter to pass on to the hind-intestine. In some insects the surface area of the midgut is increased by the outgrowth of varying numbers of *caeca*.

Respiratory system

Air enters the body of most insects through small openings or *spiracles* which often show as dark spots on the side of the body; a muscular apparatus usually controls their opening and closing. From the spiracles a system of small tubes or *tracheae* ramifies throughout the body, becoming progressively finer until the air is conducted to individual cells. To keep the tube distended and so allow free movement of gases the tracheae are lined with cuticle laid down in characteristic spirals.

Active insects such as bees contain air sacs to increase the rate of ventilation in the tracheal system by providing a larger volume of air flow during each respiration cycle.

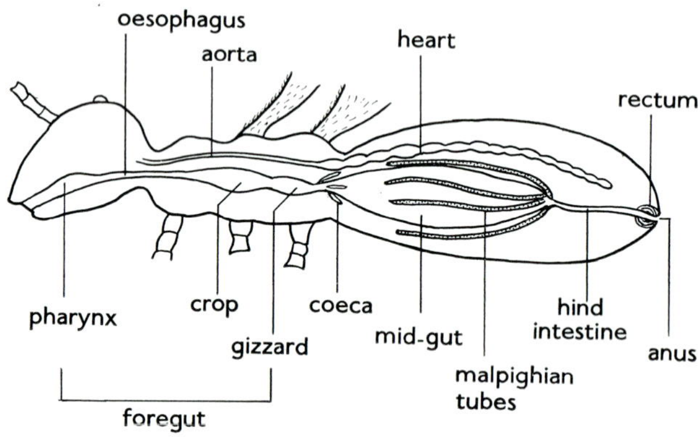

FIG. 4.3. Internal anatomy of an insect (diagrammatic).

Circulatory system (Fig. 4.3)

The internal organs of an insect lie free within the body cavity, or *haemocoel*, which is itself full of blood or *haemolymph*. The blood is circulated by being drawn into the dorsal heart which propels it forward along the aorta whence it returns to the haemocoel. All absorbed food materials are carried about the body by this circulatory process. Several different kinds of cells occur in insect blood but possibly the most important are the *phagocytes* which can digest tissue debris and bacteria. They play a most important role in the development of resistance to parasites and injury and may, on occasion, surround and completely destroy parasite eggs as well as heal wounds.

Reproductive system

The reproductive organs of insects comprise the *testes* of the male and *ovaries* of the female together with accessory glands associated with the ducts leading to the *penis* or *vagina*. The ovaries comprise a number of separate tubes or *ovarioles*. There are usually between 4 and 8 of these tubes but occasionally, as in queen termites, there may be several hundred. The eggs develop within these tubes in a successive series, the youngest egg cells being formed in the *germarium* at the apex of the tube. As they develop, eggs pass down the ovariole, absorbing their associated *nurse cells* en route, until mature and ready for oviposition. The female reproductive organs are often surrounded by irregular masses of fat-body which become progressively depleted as these food reserves are called upon to provide nutrition for the developing eggs.

There are several variants in the method of reproduction which pose special pest problems in protected crops. In the aphids and scale insects the females are capable of reproducing without fertilization so that males are absent or very rare. This short-circuiting of the normal sexual cycle eliminates the need for elaborate mating behaviour so commonly associated with normal sexual reproduction. It further encourages the development of genetically isolated communities which, in turn, tends to increase the complexity of local patterns of resistance to insecticides.

Simplification of the reproductive process is taken still further by the mushroom cecids whose larvae are able to reproduce asexually as long as the food source is adequate, by a process known as *paedogenesis*.

In aphids the egg stage is passed within the parent's body, the females giving birth to larvae, a process known as *vivipary*. In one group of mites (*Pediculopsis* sp.) this process is carried even further; the whole life cycle occurs within the female, and adult males and females emerge from the mother.

METAMORPHOSIS

Insects are unlike most animals in that their growth is a discontinuous process involving short periods of growth separated by moults, or *ecdyses*. During a

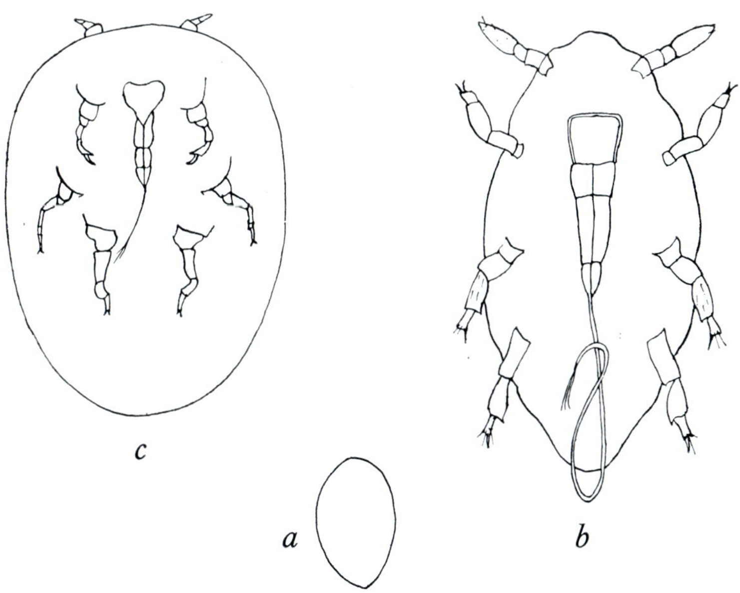

FIG. 4.4. Direct metamorphosis (scale insect—Coccidae): (*a*) egg; (*b*) nymph (venter); and (*c*) adult (venter).

moult the old skin (or *exuvium*) is shed to reveal a soft, new cuticle which is, initially, elastic to permit increase in size and changes of shape. The number of moults is variable. Most fly larvae moult only twice during development but the cockroach sheds its skin seven times. The intervals between successive moults are known as *stadia* while the form assumed in any particular stadium is called an *instar*.

There are two basic types of metamorphosis recognized in insects, one simple or *direct*, and the other complex, or *indirect*.

Simple metamorphosis (Fig. 4.4)

Crickets, cockroaches, earwigs, thrips and plant lice hatch from the egg in a comparatively advanced stage of development. Newly hatched forms, referred to as *nymphs*, lack wings and are of course smaller than the adult which they outwardly resemble. The size and the degree of development of the wings which, in the nymphs, are visible as *wing buds* on the thorax, increase with each moult.

Complex metamorphosis (Fig. 4.5)

Among the moths, beetles and flies, the developmental stages are strikingly different from the adult. They lack wing buds and the compound eyes which are consistently present in nymphs. In indirect development a *larva* hatches from the egg and may be one of three generalized types: firstly, the caterpillar type, associated with moths and sawflies, which has abdominal as well as thoracic legs; secondly, the type which has well-developed legs only on the thorax, e.g. the active predatory larvae of Carabid ground beetles, the horny wireworms and the fleshy but relatively inactive, cockchafer larvae; and thirdly, a group of legless larvae associated with the weevils, flies and many parasitic insects.

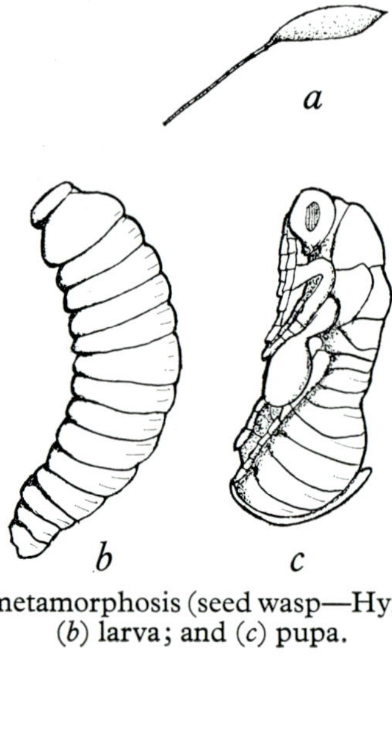

FIG. 4.5. Indirect metamorphosis (seed wasp—Hymenoptera): (*a*) egg; (*b*) larva; and (*c*) pupa.

After passing through several larval instars growth is completed and the insect enters an inert resting stage, the *pupa*, during which the complex processes of metamorphosis into the mature adult are completed.

There are three basic types of pupae: the *exarate* or free pupae—a delicate form found in the bees, wasps, and beetles, in which the legs, antennae and wings are free; the *obtect* pupa—in which the appendages, while fastened to the body, are still visible (the chrysalis of butterflies being the best example); the third type is *coarctate*, in which the last larval skin is retained but hardened into a barrel-shaped protective case or puparium without any evidence of legs or wings.

SYSTEMATICS OF INSECTA

At the present time 29 orders of the class Insecta are recognized. For simplicity we shall only discuss those which are important in the environments afforded by protected cultivation. The class is divided into two subclasses, the *Apterygota* and *Pterygota*. The former group comprises those primitive insects which have never evolved wings, and which develop with almost no metamorphosis: they possess abdominal appendages in addition to terminal cerci. The Pterygota are nearly all winged in the adult stage and develop through a series of immature forms unlike the adults.

Apterygota

COLLEMBOLA

Springtails have relatively large heads with 4- or 6-jointed antennae. They lack compound eyes though most species have groups of ocelli at the sides of the head. The abdomen consists of only 6 segments which are covered with hairs or scales. Springtails are generally dull grey-brown in colour but some species may be white, red, yellow or even metallic green. In general shape most are elongate but the family *Sminthuridae*, which includes important agricultural pests, are almost spherical in form. Most of these small wingless insects are capable of jumping by means of a special organ called the *furcula*. This appendage, on the fourth abdominal segment, is normally held adpressed to the ventral surface of the body by a hook, or *hamula*, on the third segment but, when it is released, the violent downward movement thrusts the insect forward. The jumping forms are especially common in damp decaying vegetation.

Pterygota

The class Pterygota can itself be subdivided into the *Exopterygota* (where the metamorphosis is simple with the young stages, or nymphs, closely resembling the adults), and the *Endopterygota* (where the metamorphosis is complex and therefore includes larvae which are quite unlike the mature insects).

Exopterygota

Within the Exopterygota there are five orders of pest importance, the *Orthoptera*, *Dermaptera*, *Dictyoptera*, *Thysanoptera* and *Hemiptera*.

ORTHOPTERA: Crickets, Grasshoppers and Locusts

An order comprising medium to large-sized insects with well-developed mandibulate, biting mouthparts. The femora of the hindlegs are usually enlarged to accommodate the large muscles used for jumping or running. The forewings, called *tegmina*, are modified and hardened to protect the membranous hindwings when the insect is at rest. The hindwings are folded like a fan below the tegmina. The abdomen usually bears a pair of short, jointed cerci.

The males of many species possess sound-producing stridulatory organs, which are used to attract mates. These organs are usually on the wings and sound is produced either by rubbing specially adapted portions of the tegmina together or by rubbing the tegmina against a row of teeth on the femora.

In the grasshoppers and locusts the eggs are laid, in groups of up to 100, within an egg-pod, or *oötheca*.

DERMAPTERA: Earwigs

These elongate, flattened insects have biting mouthparts typical of omnivorous feeders. The forewings are modified into short leathery tegmina quite devoid of veins. The abdomen is terminated by a characteristic pair of curved, horny forceps. There are five British species although a few others have become established in hot-houses.

Earwigs show an unusual degree of parental care for the young. The female overwinters with her eggs and subsequently stays with the young nymphs until they are old enough to look after themselves.

THYSANOPTERA: Thrips

Thrips are very small, slender-bodied insects with 6- to 9-jointed antennae situated close together on the head. There are no abdominal cerci and the four narrow wings are each fringed with long hairs. They are commonly yellow-brown or black in colour.

The mouthparts, which have both piercing and sucking functions, consist of stylets operating within a short cone, or *rostrum*, projecting vertically from the head. The tarsi terminate with a protrusible vesicle which, when extended, allows the insect to grip any kind of surface.

Parthenogenesis (reproduction without fertilization) is frequent among the 150 or so British species. Most species are plant feeders but a few are predacious on aphids and mites.

HEMIPTERA: Plant Bugs

Without exception the plant bugs possess mouthparts for piercing and sucking within a jointed rostrum. Compound eyes are well developed and the simple antennae have only a few segments. Two pairs of wings are almost always present but the front pair may be partially sclerotized with only the distal portion membranous. Such modified wings, called *hemelytra*, are found in the suborder Heteroptera.

Heteroptera

The Capsidae, a family of small, fragile, long-legged bugs, is the most important group of pests in this suborder.

Homoptera

In the suborder *Homoptera*, where the forewings have a uniform consistency and may be entirely membranous, occur the important families: *Cercopidae* or froghoppers; *Jassidae* or leafhoppers; *Psyllidae* or jumping plant lice; *Aleyrodidae* or whiteflies; *Aphididae* or greenflies; and *Coccidae* or scale insects.

Endopterygota

The Endopterygota includes four orders of familiar insects, many of which are important in protected culture: *Lepidoptera*, *Diptera*, *Hymenoptera* and *Coleoptera*.

LEPIDOPTERA: Butterflies and Moths

These insects have relatively large wings in which the venation is obscured by the covering of scales and hairs. These broad scales bear minute longitudinal striae which are responsible for many of the iridescent colours found on the wings of butterflies. The mouthparts take the form of a long coiled proboscis for feeding on plant nectars. When not in use this proboscis is coiled beneath the head. Antennae are long and many-jointed; at the apex they are clubbed in the butterflies but in male moths they are feathered or *pectinate*.

Metamorphosis is complete, the caterpillars having well-developed biting mouthparts, three pairs of thoracic legs and five pairs of abdominal pseudopods. The latter have minute crochets (hooked hairs) on the soles. The pseudopods are usually found on abdominal segments 3 to 6 with claspers on the apical segment. In the 'looper' caterpillars of the family *Geometridae* only the claspers and those on segment 6 are present, while in some *Noctuidae* the pseudopods of segments 3 and 4 are absent.

The adults, which largely feed on nectar, or slime fluxes exuding from trees, do no damage, but the caterpillars, particularly those of moths, are destructive to plants, stored and manufactured products, and even timber.

DIPTERA: Flies

These are readily recognized as they have only one pair of membranous wings, the hind pair being replaced by minute, knobbed balancers called *halteres*. The mouthparts are highly modified, usually for licking and sucking, by the expansion of the labium into a pair of fleshy lobes. There are no abdominal appendages. The relatively large, mobile head bears large compound eyes. The antennae vary from the simple, many-jointed, flagellate type found in fungus gnats to highly modified organs with only two or three basal joints and a terminal bristle as in the Agromyzid leafminers.

The larvae have no recognizable legs, movement being achieved by telescopic contractions and extensions of the body during which small ventral swellings give the organism purchase on the substrate. In some cases, e.g. the fungus gnat, there is a well-developed head with biting mouthparts, while in others, e.g. house flies, it is completely absent. The pupae range from the free type found in the crane flies to the *obtect*, barrel-shaped, puparium of the bluebottle.

About 5000 British species are known and include many important pests of plants and animals. A few, such as the larvae of hover flies and certain *Anthomyiidae*, are predators and parasites.

HYMENOPTERA: Bees, Wasps and Ants

The members of this enormous order comprising some 100,000 species may be recognized by the possession of two pairs of membranous wings, the fore and hind pairs being locked together in flight by a series of minute hooks along the leading edge of the hindwing.

The Hymenoptera vary in size from minute to quite large insects. Excluding the sawflies and wood wasps they all have a narrow 'waist' (*petiole*) between the thorax and abdomen. The mouthparts are generally adapted for biting. The female abdomen terminates with a needle-like ovipositor which, in some parasitic wasps, is much longer than the body.

The larvae, which have well-developed heads, are usually legless but the sawflies have caterpillar-like larvae with as many as seven pairs of prolegs. The pupae are all *exarate*, i.e. the limbs are visible, though the pupae are frequently hidden within a cocoon.

Many Hymenoptera such as the bees, wasps and ants, have interesting social habits involving several castes of individuals adapted for specific duties. But for the most part they are parasites and, as such, will undoubtedly play an increasing role in commercial pest control.

COLEOPTERA: Beetles

Beetles are a familiar order of insects easily recognized by the highly modified forewings or *elytra*, which serve as horny covers to the membranous hindwings normally extending over most of the thorax and abdomen. In some species, i.e. certain weevils and ground beetles, the elytra are permanently joined and the hindwings absent. The mouthparts are typically mandibulate, and in the stag beetles, denizens of rotting wood, these are excessively developed into massive 'horns'. The weevils have their mouthparts at the apex of a 'beak' or *rostrum*, which is a striking prolongation of the head. In beetles the prominent antennae are variable in form, consisting of several to many segments, they are frequently clubbed. Most beetles have legs adapted for rapid movement but, in some species, they may be modified for special purposes such as digging (the dung beetles) or leaping (the flea beetles).

The larvae are variable in form but all have a well-developed head and biting mouthparts. While some (weevil grubs) are legless, most have three pairs of thoracic legs. The active predatory larvae of the ground beetles (Carabidae) and rove beetles (Staphylinidae) possess legs adapted for running. In the fleshy cockchafer grubs, the legs, though strongly developed, are only used to assist limited vertical movements within soil.

The pale, soft pupae are exarate, all the adult appendages lying free outside the body.

The Coleoptera is the largest insect order and contains species adapted to a wide range of habitats and function. Beetles are chiefly found on, or near, the ground, many species living in soil, rotting vegetation, dung, carrion and dead or dying trees. The order includes many pests of plants and stored products.

REFERENCES

IMMS, A. D. (1947). *Insect Natural History*. Collins, London. 317 pp.
— (1957). *A General Textbook of Entomology*. 9th Ed. Methuen, London. 886 pp.
MORETON, B. D. (1958). *A Guide to British Insects*. Macmillan, London. 188 pp.
WIGGLESWORTH, V. B. (1965). *Principles of Insect Physiology*. 6th Ed. Methuen, London. 554 pp.

GLASSHOUSE PESTS

Bionomics and Control

Chapter 5

Insect Pests

Order COLLEMBOLA: Springtails
(Figs. 5.1 and 5.2)

Springtails are the most common insects found in soils. They range in size from 0·3 mm to more than 10 mm. All stages are wingless, and in most species, the fourth of the six abdominal segments carries a characteristic caudal springing organ which enables them to leap into the air. This leaping habit has led to many common names referring to them as 'fleas'. The spring, or *furcula*, is normally hidden below the body locked in a hook, or *hamula*, carried on the preceding body segment. The head has short, four-jointed antennae, a group of up to eight ocelli, or simple eyes, and biting mouthparts.

Some species are parthenogenetic: probably due to their greater longevity, females are normally preponderant even in bisexual populations. The males produce stalked spermatophores (a pellicle containing sperm) which, in chance contact with a female, burst mechanically to release sperms on her body surface which travel to the genital aperture by chemotactic response. Eggs may be laid singly, or in small clusters, according to species. During the early stages of embryonic development they increase in size and, within glasshouses, complete the incubation period in 5–10 days. Moulting continues even after maturity, which is normally attained at about the seventh instar, as many as 50 moults having been recorded.

The taxonomic difficulties involved in separating the many potential pest species make it impossible to indicate useful differences here, and reference should be made to the works of Gisin (1960) and Scott (1961).

While much remains to be discovered about Collembolan feeding habits these can be divided into several types, each apparently typical of different groups of genera. One large group feeds on fungal hyphae, bacteria and dead or decaying plant tissues, a second group on pollen grains, algae and spores, a third on pollen grains and living plant material, a fourth on live soil animals and a fifth on dead animals.

They are rarely pests though many species, and thousands of individuals, may be found in all glasshouse soils which are infrequently sterilized. However, some cases of primary damage to glasshouse crops, especially lettuce, have been recorded.

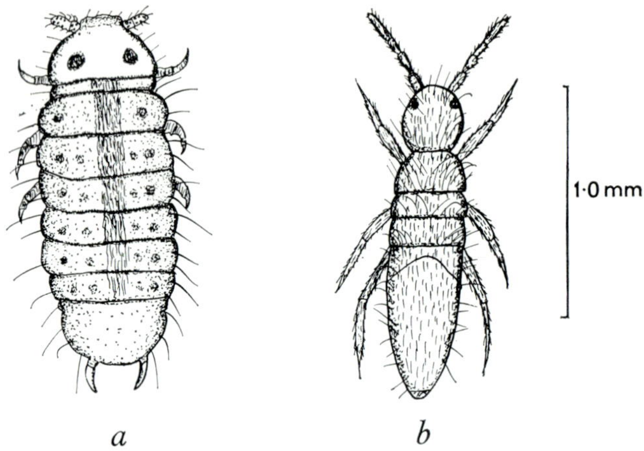

FIG. 5.1. Collembola (springtails): (*a*) *Achorutes armatus*; (*b*) *Lepidocyrtus* sp.

When damage does occur seedlings and young plants are principally affected. Injury to cyclamen, cineraria and *Zantedeschia* sp. has been reported. Brown (1954) records feeding of *Onychiurus stachianus* Bagnall, *O. nemoratus* Gisin

FIG. 5.2. Collembolan adult (×50). (Courtesy of Rothamsted Experimental Station.)

and *Anurida granaria* Nicolet on tomato seedlings. Where seedlings in pots are placed directly on the soil, Collembola may enter the pots through the drainage holes. Following the development of a pin-sized hole at soil level, caused by scraping of the epidermis and other superficial tissues, affected seedlings topple over. Damage to the stem immediately above and below

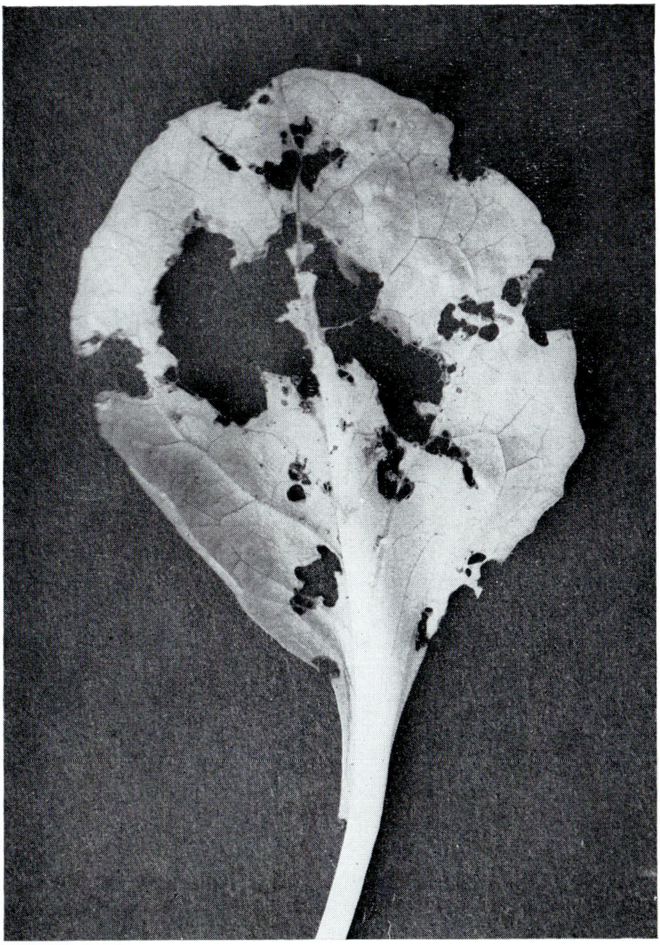

FIG. 5.3. Springtail damage to lettuce. (Courtesy of Glasshouse Crops Research Institute.)

the soil surface prevents the surviving plants making growth. The root hairs are eaten off and fresh roots eaten as fast as they appear. Similar damage may occur on direct-planted lettuce seedlings.

Leaves of lettuce or cucumber, in contact with the soil surface, may be minutely pitted or completely skeletonized by species such as *O. armatus* Gisin and *Bourtletiella signata* (Nic.) Agren (Fig. 5.3). Similar, smaller holes have been gnawed in young chrysanthemum leaves. *Heteromenus nitidus* Templeton and *Orchesella* spp., breeding in the moss used to culture orchids, may cause damage to germinating seedlings or developing pseudobulbs. While most glasshouse Collembola are saprophytes a few are active predators including some *Onychiurus* sp. which feed on eelworm cysts.

Control

When springtails are numerous but obvious signs of damage are not apparent, it is difficult to decide whether control measures are justified. Also, it is most important to differentiate between symphilids and non-jumping springtails as the latter tolerate parathion.

If many springtails are present in borders about to be planted with lettuces or used for standing out boxes of seedlings, precautionary control is necessary, though it is doubtful whether treatment of soils to be planted with tomatoes is justified unless previous experience suggests that damage is likely.

Springtails are easily controlled by applying a gamma-BHC dust shortly before planting and lightly raking it into the surface soil. The gamma-BHC content of dusts varies from 0·2–0·65% and they should be applied at rates of 1–$\frac{1}{3}$ oz/yd^2 depending on the content of active ingredient. Equally satisfactory results are obtained by spraying the surface of the soil with 0·005% gamma-BHC or 0·01% malathion, using about 2 gal/100 ft^2. The best results are obtained by either method when the surface soil is moist. Applications of gamma-BHC dusts or sprays, without raking in, give good control of springtails attacking newly planted lettuces. Springtails in cucumber beds may be controlled by watering the beds with 0·005% gamma-BHC at $\frac{1}{2}$–1 gal/yd^2. The foliage should not be wetted.

If boxes or pots containing seedlings are infested, they should be lightly watered with 0·005% gamma-BHC, taking care to avoid wetting the foliage of plants susceptible to injury by this insecticide. Dusts are less phytotoxic than sprays, and light dustings, repeated if necessary after 7–14 days, can be given to most seedlings other than tomato and cucumber. As an alternative, particularly when the application of dusts or liquids is impractical, overnight fumigation with lindane smokes will effect control if repeated several times at intervals of about 10 days. These smokes should not be used on tomato or cucumber seedlings.

Orchids attacked by springtails, or harbouring objectionably large populations in the moss culture medium, can be sprayed with 0·01% gamma-BHC.

Wet conditions favour springtails and where the soil in mixed-plant glasshouses remains wet due to poor drainage this should be remedied.

REFERENCES

BROWN, E. B. (1954). Springtail damage to tomatoes. *Pl. Path.*, **3**, 87–8.

CHRISTIANSEN, K. (1964). Bionomics of Collembola. *A. Rev. Ent.*, **9**, 147–78.

CRÜGER, G. (1964). Damage to young vegetable plants by the globular springtail *Bourletiella signata* (Nic). *NachrBl. dtsch. PflSchutzdienst, Braunschw.*, **16**, 106.

GISIN, H. (1960). *Collembolenfauna Europas*. Museum D'Histoire Nat. Geneva. 312 pp.

HALE, W. G. (1965). Observations on the breeding biology of Collembola I & II. *Pediobiologia*, **5**, 146–52; 161–71.
SCOTT, H. G. (1961). Collembola: Pictorial keys to Nearctic Genera. *Ann. ent. Soc. Am.* **34**, 104–113.

Order ORTHOPTERA: Crickets

Family GRYLLIDAE

House crickets may enter glasshouses along the pipe ducts leading from the boilerhouses, and occasionally damage cucumber and other seedlings. They are omnivorous and may eat the stems of many plants at the soil surface; aerial roots of orchids and the blooms of chrysanthemum, cineraria and orchids are frequently damaged.

The common house cricket *Gryllus domesticus* (L.) is about $\frac{3}{4}$ in. long with two pairs of wings, the front pair, or tegmina, being modified as covers for the membranous hind pair. The body is oblong and the hindlegs adapted for leaping. They characteristically chirp, or stridulate, by rubbing a series of small teeth on one tegmen against a smooth area on the other. Another species *Gryllodes sigillatus* (Walker) has become established in some hot-houses.

Family GRYLLOTALPIDAE: Mole Crickets

The single British species, *Gryllotalpa gryllotalpa* L., is $1\frac{1}{2}$–2 in. long and readily distinguished by the conspicuous, elongate thorax and remarkably expanded and toothed forelegs modified for digging (Fig. 5.4). It is almost entirely a subterranean insect and is said to eat through the stems of any plant encountered as it tunnels, in more or less straight tracks, through the soil.

REFERENCE

HINCKS, W. D. (1949). Dermaptera and Orthoptera. Handbooks for the Identification of British Insects. *R. ent. Soc., Lond.*, **1**, Pt. 5, 1–20.

FIG. 5.4. Adult mole cricket (*Gryllotalpa gryllotalpa* L.) ($\times 1\frac{1}{2}$). (A Shell photograph.)

Family GRYLLACRIDIDAE: Camel Crickets

Tachycines asynamorus Adelung, an apterous, grasshopper-like insect, is widely established as a breeding species in British glasshouses. Slightly over ½ in. long, the body has an upturned ovipositor ½ in. long. The remarkably long antennae are almost three times the length of the body. Camel crickets are predominantly yellowish-grey to green in colour with dark, broad, transverse stripes on both body and legs. *T. asynamorus* is extremely active and, while largely carnivorous, may feed on the young growth of many plants. Concealed in crevices and beneath leaves during the day it becomes active only at night. Over a period of about a month the female lays 200–300 eggs about 1 cm below the soil surface. The larvae hatch in 3–4 months and mature about 7 months later.

REFERENCE

BRETTYER, C. R. (1950). Die Gewachshausheuschdreke (*Tachycines asynamorus* Adelung). *Abh. braunschw. wiss. Ges.*, **2**, 13–39.

Control of orthopteroid pests

Rubbish in which house and camel crickets shelter should be removed from the glasshouse, particularly from the vicinity of the heating pipes, and either a DDT, gamma-BHC or malathion dust liberally applied to the pipe ducts and similar warm, dry places frequented by crickets. It is rarely necessary to give overall treatments for house crickets, but where localized insecticide applications do not eradicate the pest the dusts should be applied to the surface of the soil, paths, etc., at ½ oz/yd^2, and if the crickets are eating blooms or aerial roots, to the plants also. Care should of course be taken to ensure that the plants are tolerant (see p. 382). Smoking with DDT or gamma-BHC, and aerosols of DDT (3 g DDT/1000 ft^3) or gamma-BHC (0·7 g/1000 ft^3) have given good results in the evening.

An alternative method of control is by chlordecone pellets or a bait prepared by mixing 28 lb of bran with 1 lb of Paris green (copper aceto-arsenite) together with sufficient fruit juice or mashed fruit to give a friable product. The bait should be sprinkled over areas where crickets congregate.

Mole crickets can be destroyed by watering an area, in line with destroyed plants, with 0·05% DDT or 0·01% gamma-BHC, using about 2 gal/yd^2.

Order DERMAPTERA: Earwigs

(Fig. 5.5)

Several species have been recorded from glasshouses. They are elongate creatures with short, leathery tegmina below which, after folding fan-wise, the hindwings are tucked away and then turned back on themselves along transverse folds. The unjointed cerci or 'pincers' are modified into horny forceps, almost straight in the females but usually sharply curved in males. They are mainly used for defensive 'display' but also play a part in copulation.

Earwigs are largely nocturnal in habit, hiding during the day in soil, herbage, flowers and hollow canes, or below stones. The adults can severely damage the flowers and foliage of chrysanthemum, cineraria, dahlia, and other ornamentals by gnawing the petals. Spoiled blooms have a ragged appearance. Leaves may also be skeletonized. Groups of 20–90 pale, oval, smooth eggs are laid in depressions in the soil. The female rests over them for the 4–6 weeks of incubation. The 'family' remain together until the pale coloured, apterous nymphs which resemble the adults are able to fend for themselves. Nymphs are, largely, subterranean.

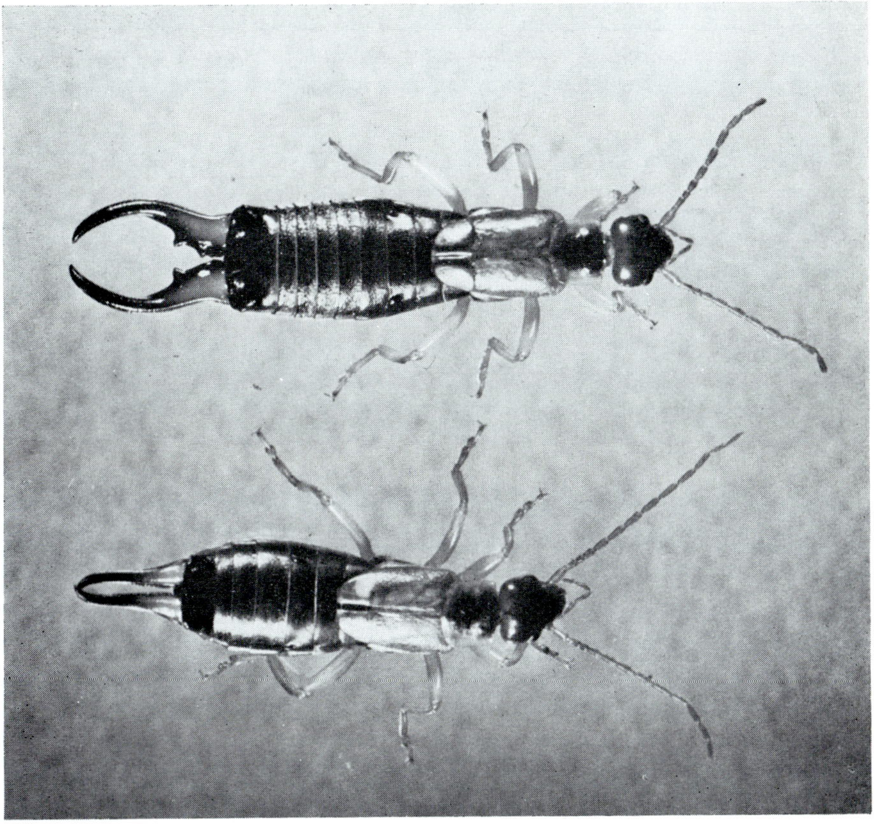

FIG. 5.5. Earwigs (*top*, male; *bottom*, female) ($\times 6$). (A Shell photograph.)

Control

Earwig damage to chrysanthemum flowers is more common on plants grown out of doors until September and then placed under glass, than on year-round or direct-planted chrysanthemums. Before housing outdoor-grown chrysanthemums it is desirable to spray them with an insecticide which will kill both earwigs and other insects. Several proprietary mixtures containing petroleum, DDT and a fungicide, or mixtures of DDT with gamma-BHC or malathion are suitable, if applied as H.V. sprays. It is, however, desirable to achieve direct contact with these insecticides and this normally means spraying after dark.

In glasshouses control measures should be taken at the first sign of damage to flowers, the choice of pesticide and methods of application depending on circumstances. To avoid detriment to the appearance and value of flowers by visible residual deposits, emulsifiable-concentrate formulations of insecticides should be used. High volume sprays, as advised for use before housing chrysanthemums, containing 0·01% gamma-BHC or 0·08% TDE give the best control but may be undesirable when flower diseases are present or when they might provide conditions conducive to fungal attack. Under such conditions atomized gamma-BHC (1 g/1000 ft^3) or smokes should be applied in the evening. If simultaneous control of caterpillars is required DDT should be combined with the BHC.

REFERENCE

HINCKS, W. D. (1949). Dermaptera and Orthoptera. Handbooks for the Identification of British Insects. *R. ent. Soc., Lond.*, **1**, Pt. 5, 1–20.

Order DICTYOPTERA: Cockroaches
(Fig. 5.6)

Family BLATTIDAE

Several species have become established in glasshouses and occasionally cause economic damage. As a group they may be recognized by the long, many-

FIG. 5.6. Oriental cockroach (*Blatta orientalis* L.), nymph (×5). (A Shell photograph.)

jointed antennae, biting mouthparts, flattened body with the head covered by the shield-like pronotum, and long legs adapted for rapid movement. The commonest species, *Blatta orientalis* L., has the tegmina shorter than the abdomen, distinguishing it from *Periplaneta americana* L. and *P. australasiae* Fabr. in which the tegmina are longer than the abdomen. A smaller tropical species *Pycnoscelus surinamensis* (L.), which is parthenogenetic and viviparous, is sometimes found in temperature greenhouses. A key to thirteen species is given by Hincks (1949) (see p. 72). The eggs are enclosed in a sclerotized oötheca containing two rows of 'pockets' each containing an egg. In some species the oötheca is carried protruding from the body for only a short time, in others for the whole incubation period. On hatching the oötheca splits into two halves to liberate the nymphs. Maturity is reached within 9 months.

Pycnoscelus surinamensis damages the leaves of *Sansevieria* sp. just above ground level. All cockroaches may cause losses to newly sown seed and to flowers of both chrysanthemums and orchids but foliage and stems of many plants, as well as aerial roots of orchids, may be attacked.

Control

Although strains of cockroaches resistant to BHC and DDT exist elsewhere there is no evidence that they occur in glasshouses, and the measures advised on p. 70 for control of house crickets should prove satisfactory. If roaches are resistant to BHC or DDT, Paris green and bran baits or malathion dust should be used. Cockroaches are susceptible to propoxur and a 1% dust, or bait specially formulated by the manufacturer, can be used effectively as a ground treatment.

REFERENCES

GUTHRIE, D. M. & TINDALL, A. R. (1968). *The biology of the cockroach*. E. Arnold, London. 408 pp.

HAAS, G. (1965). Thirty months' experience with Baygon at the Zoological Gardens, Wuppertal. *Pflanzenschutz Nachrichten Bayer, Leverkusen*, **18**, 82–92.

Order THYSANOPTERA: Thrips
(Fig. 5.7)

'Thrips' are small, slender insects, 1–2 mm in length, usually associated with the flowers or foliage of plants. Many, of tropical or subtropical origin, have been introduced into British glasshouses where they are pests of ornamental plants. These species cannot survive outdoors in Britain. Other temperate species have become established in glasshouses, or enter them in large numbers during the summer from outside plants. They may cause serious injury to many crops including cucumbers, tomatoes, carnations, roses and other flowering plants.

The Thysanoptera are divided into two suborders: the *Terebrantia*, in which the female possesses a saw-like ovipositor with which the eggs are laid *within* the plant tissues, and the *Tubulifera*, in which the ovipositor is absent and the eggs are laid *on the surface* of the plant tissue.

With the exception of the lily thrips (Tubulifera), which has two pupal instars, all recorded British glasshouse species are Terebrantians, their life cycle comprising the egg, two larval, one prepupal, and one pupal instar as well as the adult. The latter, which vary from yellow to brown or blackish-brown (*Heliothrips*, *Liothrips*), have two pairs of narrow wings fringed, along their margins, with long, fine hairs. When at rest, the wings are laid parallel along the back. The forewings may be hyaline, shaded grey or brown or even hyaline with shaded cross bands (*Hercinothrips*). In a few species the wings may be either fully developed (macropterous), much reduced in size (brachyp-

FIG. 5.7. Lily thrips (*Liothrips vaneecki* Priesner), adult (×50). (A Shell photograph.)

terous) or absent altogether (apterous). The antennae are 6- to 8-segmented. Females are usually more numerous than males and are distinguished by their larger size and pointed apex of the abdomen, but in some species (*Heliothrips* and *Thrips tabaci*), males are rare or unknown; reproduction is then parthenogenetic.

The duration of the life cycle is directly influenced by temperature and humidity. MacGill (1927) studied the life cycle of *T. tabaci* under glasshouse conditions at a mean temperature of 66 °F (19 °C) and a mean relative humidity of 84·5%. The life cycle is typical of most Terebrantian species. Each female lays about 60 whitish, reniform, eggs (0·2 × 0·1 mm) singly, through a slit cut by the ovipositor, within leaf or flower, tissues. One end of the egg protrudes slightly from this slit to facilitate the escape of the newly hatched larva. Eggs hatch after about 8 days, into translucent white, or pale yellow, first stage larvae about 0·4 mm long; the head is abnormally large and the eyes bright red. At first gregarious, the larvae later become scattered over the leaf surface. The next larval stage varies from pale to deep yellow. After 10–14 days the fully fed larvae conceal themselves in the soil to undergo a short (1–2 days) prepupal stage before changing to the pupa. A few (*T. fuscipennis*) undergo these changes on the food plant. During this time the insects do not feed but remain capable of limited movement. The prepupa may be recognized by the shorter wing pads and the antennae which are not, as in the pupa, turned back over the head. The pupa matures to the adult in 4–7 days. Under glasshouse conditions the total life cycle is 23–31 days so there may be at least seven generations during the summer.

Thrips are gregarious and large numbers of adults of the same, or even different, species may often be found together on a leaf or within a single flower. Alternatively, females of one species appear to avoid those of another; it is rare to find females of *T. fuscipennis* associating with females of *T. tabaci*, though males of the former congregate readily with the parthenogenetic females of the latter. Thrips rarely fly in glasshouses, though some are able to jump when disturbed.

Adults and larvae feed by imbibing sap after piercing plant tissues with their mouthparts. Tissue surrounding the feeding punctures dries out giving the leaves, or petals, a silver-flecked appearance. Heavy infestations cause necrosis of the foliage and render flowers unmarketable; in severe cases, malformation of the blooms may occur. Further damage to hot-house ornamental plants by *Heliothrips* and allied species is caused by their habit of depositing large amounts of faecal matter on the foliage which then becomes invaded by 'sooty' moulds. In addition to the direct injury caused by feeding, thrips have attracted attention as being possible vectors of plant viruses; *T. tabaci*, for instance, transmits spotted-wilt virus of tomato.

Indigenous outdoor species overwinter, either as mature larvae or adult females, in sheltered sites in the soil, or under the bark of trees and shrubs. The duration of hibernation depends on the temperature and may last 4–5 months. Males seldom hibernate and they mostly succumb in colder weather. In glasshouses, females of *T. tabaci* overwinter in soil whereas those of *T. fuscipennis* hibernate in the glasshouse structure or beneath the bark of rose bushes. Where heat is maintained and suitable plants are available, some species may continue to feed sparingly and lay eggs throughout the winter. Since Morison (1957) provides a key to all the many species which may breed in, or accidentally enter, glasshouses only brief descriptive notes are given.

Suborder Terebrantia

SPECIES OF TROPICAL, OR SUBTROPICAL, ORIGIN EXISTING IN BRITAIN ONLY IN HEATED GLASSHOUSES INTO WHICH THEY HAVE BEEN INTRODUCED BY COMMERCE

Heliothrips haemorrhoidalis Bouché (greenhouse thrips) is about 1·8 mm long. The body surface is strongly reticulate and blackish-brown with the tip of the abdomen lighter. The pale yellow antennae are 8-segmented with the terminal segment long and needle-like. The legs are pale yellow. The forewings are hyaline and narrow with a broad base. Common and widely distributed under glass in Europe and America as well as in British glasshouses, *H. haemorrhoidalis* is probably the most widespread and economically important of the introduced species. Host plants include arums, azaleas, citrus, ferns, fuchsias, grape-vines and orchids.

Hercinothrips femoralis Reuter (banded greenhouse thrips or beet thrips) is 1·5 mm long, dark brown, with the head, fore-part of thorax, and tip of the abdomen paler; the entire surface is reticulate. The antennae are 8-segmented, yellow, but darkening to brown towards their distal end, where the terminal segment is long and needle-like. The forelegs are brownish yellow, the mid- and hindlegs have dark-brown femora but brownish-yellow tibiae and tarsi. The narrow forewings are pale with a shaded area at their base and two dark cross bands of which the proximal is about twice the length of the distal. At one time *H. femoralis* was common and widely distributed under glass in Britain, Europe and America. Host plants include arums, begonias, *Crinum* sp., orchids and *Tradescantia* sp.

Heliothrips errans Williams is 1·5 mm long with a dark-brown body. The surface is strongly reticulate. The 8-segmented antennae are mainly yellow with segments II and VI brownish, and the terminal segment long and needle-like. The forelegs are yellow, lightly shaded with brown, but the mid- and hindlegs are brown with both the apex and bases of the femora and tibiae yellowish. The forewings are shaded light brown with a narrow clear band near their base and another just before their apex. Recorded only from orchids in the southern half of England, no damage of economic importance has been reported.

Parthenothrips dracaenae Heeger (palm thrips). The body, about 1·3 mm long, is light brown. The head and thorax are strongly reticulate but the abdomen less so. The 7-segmented antennae are yellow, with segments VI–VII brown, and the terminal segment long and needle-like. The legs are mainly brown but the tibiae and tarsi are yellow. The forewings, which are distinctive and easily recognizable, are broad throughout their length and mainly hyaline with the surface membrane reticulate. Portions of the long veins are heavily shaded with brown at intervals. The foremargin has no fringe of fine hairs. Widely distributed in glasshouses throughout the world and in Britain, *P. dracaenae* has been recorded from widely scattered localities. Host plants include species of *Citrus, Crinum, Croton, Cycas, Dracaena, Kentia, Stephanotis* and *Tradescantia.*

Scirtothrips longipennis Bagnall. The body, 0·9 mm long, is pale yellow. The surface of the head and thorax is striate, the abdomen bears rows of conspicuous microtrichia laterally. The antennae, 8-segmented, are mainly yellowish-grey and have a short terminal segment. The legs are pale yellow and the forewings narrow, greyish-brown.

S. longipennis has been recorded under glass in many European countries, including Britain, and in America. Host plants include begonia, *Caladium* sp., cyclamen and *Ipomaea* sp.

Leucothrips nigripennis Reuter (fern thrips). Structurally similar to *S. longipennis*, this species has 7-segmented antennae. Of limited distribution in glasshouses of Europe and America, it occurs on *Adiantum* sp., *Davallia* sp and *Pteris* sp. It is not a pest in British glasshouses, where there are only isolated records of its occurrence.

Anaphothrips orchidaceus Bagnall (yellow orchid thrips) is about 1·3 mm long with a yellow body, the surface of which is neither reticulate nor otherwise noticably sculptured. The 8-segmented, brown antennae have both the base and apex of segment III and the bases of segments IV and V paler. The legs are yellow. The forewings are pale-brown with a clear area near their base and an indefinite, darker cross band near the base of the second vein. Widely distributed on orchids, it causes brown patches on the green tuberous parts of the plants.

Chaetanophothrips orchidii Moulton (orchid thrips or banana rust thrips). The body (length 1·2 mm) of this species is yellow and devoid of sculpturing. The antennae are 8-segmented, yellow but with the apices of segments IV and V, the apical half of VI and the whole of VII and VIII brown. The forewings are brownish, with one pale cross band just beyond the base and another at the apex. Though widely distributed in glasshouses in Europe, America and Japan, there are only isolated records of its occurrence in Britain. Host plants include begonia, *Epiphyllum* sp., orchids and *Tradescantia* sp.

INDIGENOUS SPECIES

Taeniothrips atratus Haliday. *T. atratus* is 1·6 mm long, with the body and legs dark brown. The antennae are 8-segmented and uniformly dark brown. The forewings are narrow, brown, but paler at their base. Abundant in most parts of Britain on a very wide range of outdoor flowering plants, this species is a frequent inhabitant of glasshouses, and enters them in large numbers from June to September. Without necessarily breeding it may cause severe damage to carnation flowers under glass.

Taeniothrips simplex Morison (gladiolus thrips). The body is about 1·5 mm long, dark brown; the 8-segmented antennae are mainly dark brown, though segment III is paler and IV and V each have a pale ring near the base. The legs are dark brown with the apical portions of the tibiae and all the tarsi paler, yellowish-brown. The forewings are brown with a pale area near the base. *T. simplex*, which originates from warmer climates, is not at present a glasshouse pest but is included here on account of its importance on gladiolus and because it may be introduced into glasshouses on corms. With the increasing cultivation of freesias, another host, it could become a pest of this crop. It breeds in the open during the summer months but is unlikely to survive the winter outdoors in Britain. On gladiolus, *T. simplex* causes 'silvering' of the

FIG. 5.8. Damage caused by gladiolus thrips. (Courtesy of Glasshouse Crops Research Institute.)

leaves and flower petals (Fig. 5.8); occasionally flowers may become deformed or fail to expand their petals. Outdoors the insects descend into the soil in autumn and, provided the temperature does not fall below 50 °F (10 °C), may continue to feed and breed upon corms.

Thrips fuscipennis Haliday (rose thrips). *T. fuscipennis* is about 1·1 mm in length. The body is light greyish-brown to dark brown: the antennae are yellowish-brown and only 7-segmented. The legs are brown and the forewings greyish-brown though paler at the base. Widely distributed in Europe and Britain on a wide range of flowering plants in the open, including *Convolvulus* sp., larkspur, lupin and rose, this species is a serious pest of roses, cucumbers and peaches in glasshouses. On roses, the eggs are laid in the outer petals of the flowers as soon as the calyx sepals begin to separate. At this time, the parent females feed at the apex of the flower bud causing malformation of the petals of the opening bloom. Eggs may also be laid in the foliage of weak shoots. The cucumber flower is often chosen for oviposition and, on hatching, the larvae move to the foliage which may suffer much injury. In glasshouses, the duration of the life cycle is about the same as that of *T. tabaci*. Oviposition usually begins at the end of February, the first generation being completed at the beginning of April. It is suspected that, irrespective of temperature, many females hibernate during winter.

FIG. 5.9. Damage to cucumber by *Thrips tabaci*. (Courtesy of Glasshouse Crops Research Institute.)

Thrips nigropilosus Uzel (chrysanthemum thrips). The body is yellow and about 1·2 mm long. The antennae are 7-segmented, brown, with segment I yellow. The pale-yellow forewing may be more or less reduced in size. Although a serious pest of chrysanthemums and other plants under glass in Europe and parts of U.S.A., there is only a single British record. Speyer (1935) recorded brachypterous females damaging chrysanthemum cuttings in Kent. Extensive brownish-red markings appeared at the base of the petioles and the foliage was marked with brownish warty patches. *T. nigropilosus* is widely distributed outdoors in Britain, breeding mainly on *Plantago lanceolata* and *P. maritima*.

Thrips tabaci Lindeman (onion thrips) is about 1 mm long. The body is greyish-yellow to brown. The 7-segmented antennae are mainly yellowish-brown; the legs are yellow, shaded with brown. The forewings are uniformly pale brownish-yellow. *T. tabaci*, commonly named for its attacks upon onion foliage, is cosmopolitan and abundant on flowering plants. It is the commonest of British glasshouse thrips breeding upon the foliage of cucumber (Fig. 5.9), chrysanthemum and cineraria and within flowers such as arums, carnations, cyclamen and gerbera. Reduced pesticide usage on cucumbers may permit these thrips to become established in the foliage. In severe attacks the damage symptoms extend from slight mottling, to the formation of small 'windows' in the epidermis caused by the collapse of the superficial cells. As well as causing damage to the foliage and fruits of tomato, the species transmits spotted-wilt virus. Adults often collect in chrysanthemum flowers and cause damage which is most apparent on bronze varieties. During winter, eggs may be laid in the petals of rose blooms grown under glass, but the insect is unimportant on this crop. Feeding on the petals frequently causes silvery flecks on the flowers of carnations. Reproduction is almost entirely parthenogenetic, males being rare. Pupation is normally in the soil and, under favourable glasshouse conditions, the insect continues to breed slowly throughout the winter.

Suborder Tubulifera

Liothrips vaneeckei Priesner (lily thrips) is 2–2·5 mm long with 8-segmented antennae (Fig. 5.7). Segments I–II and VII–VIII are dark brown, III yellow, IV–VI yellow at the base but brown apically. The long, slender legs are variegated orange-yellow and brown. The strap-shaped forewings are pale brown, widest near the base and narrowest at the apex. Widely distributed outdoors in Europe, Asia and America, this species seems to be confined mostly to Southern England, to which it was probably introduced on imported bulbs. It is harmful to lily bulbs under glass and in the open, and has been found on orchids. The insect is gregarious and breeds between the scales of the bulbs. In heavy attacks, the scales become flabby and the outer ones thin and papery; even so, the aerial parts of the plants are not attacked. It overwinters from November to March as an adult, or second instar larva, and according to conditions, there are 4–7 generations a year.

Control of thrips

Most organochlorine and organophosphorus insecticides are toxic to thrips, as are nicotine and pyrethrum. DDT and BHC have the merit of low cost and can be conveniently applied as aerosols, smokes and dusts. Consequently they are recommended as the standard control for thrips on all plants other than those which are intolerant to these insecticides, and, for DDT dusts, where the required period cannot be given between application and harvesting of edible crops. These insecticides are sufficiently persistent to kill many thrips which are protected from immediate contact by concealment beneath the flower petals or leaf sheaths. Applications, at 10- to 14-day intervals, of either 5% DDT or 0·6% gamma-BHC dusts at 15–20 lb/acre, or aerosols and smokes of DDT (3·5 g/1000 ft^3) or gamma-BHC at (0·7 g/1000 ft^3) will control the pest. Sprays containing 0·1% DDT or 0·01% gamma-BHC are equally effective when applied at these intervals but may cause injury to flowers, especially when environmental conditions are conducive to the infection of flowers by fungi such as *Botrytis cinerea*. The foliage of some varieties of carnation, particularly Ashington Pink and related varieties, is susceptible to injury by aerosols or emulsifiable-concentrate sprays of gamma-BHC. A strict look-out on flower crops and ornamental plants should be maintained, particularly in the late summer and autumn, otherwise much damage may be caused by thrips entering glasshouses. Unsaleable flowers should be destroyed to prevent them harbouring the pest.

Malathion aerosols (2·5 g/1000 ft^3) and sprays (0·1%), and diazinon aerosols (1·8 g/1000 ft^3) and sprays (0·02%) also give good control of thrips, and are suitable for use on established cucurbits and many other plants susceptible to injury by DDT and BHC. Except where thrips are present in open flowers, or continue to enter glasshouses from outside, complete control is normally obtained by two applications with a 10- to 14-day interval. Combined malathion and DDT atomizing concentrates are available and may be advantageous where other pests such as aphids are also present. Should attacks of thrips occur on seedlings or very tender young plants, including cucumbers, parathion smokes (0·5 g/1000 ft^3) or sprays (0·01%) provide a satisfactory control in most cases. Aerosols of dichlorvos (1·0 g/1000 ft^3) or naled (1·5 g/1000 ft^3) vaporized from the heating pipes, or by other means, will also control thrips but two or more applications at 10- to 14-day intervals are usually required if these non-persistent insecticides are used.

The regular use of diazinon to control red spider mite normally controls thrips. Petroleum emulsion sprays are moderately effective, but the addition of nicotine (0·05%) provides excellent control of thrips on foliage. Very frequent application of smokes or sprays of nicotine, or pyrethrum dusts or sprays are required to control thrips on flowers such as roses, chrysanthemums and carnations. The susceptibility of *Phytoseiulus riegeli* (p. 16) to organophosphorus compounds will necessitate the use of alternative, non-acaricidal insecticides for the control of thrips on cucumbers where this predator is used to control red spider mite. Soil drenches of BHC or diazinon appear to be effective.

As gladiolus thrips overwinter on corms any badly infested material should be rejected. To eradicate the pest, corms should either be dusted with a 5% DDT or 0·6% BHC dust before storing or, alternatively, they may be fumigated by sprinkling with powdered naphthalene. This treatment must not be done in airtight containers. Soaking the corms, just before planting, for 3 hr in a solution of 'Lysol' (5 fl. oz/4 gal) has been suggested as a treatment for infested corms which were not treated at storage. Immersion of lily bulbs for 1 hr in water maintained at 111 °F (44 °C) should control both lily and gladioli thrips. As thrips feeding between the scales are protected from contact, dusting with DDT may not be effective. Gamma-BHC may be more effective because it is more volatile than DDT.

REFERENCES

AITKENHEAD, P. (1951). The Gladiolus Thrips—a pest new to Britain. *Agriculture*, **57**, 517–23.

HODSON, W. E. H. (1935). The Lily Thrips (*Liothrips vaneeckei* Priesner). *Bull. ent. Res.*, **26**, 469–74.

MACGILL, ELSIE I. (1927). The Biology of *Thysanoptera* with reference to the cotton plant II.—The relation between temperature and life-cycle in a saturated atmosphere. *Ann. appl. Biol.*, **14**, 501–12.

MORISON, G. D. (1947). *Thysanoptera* of the London area. *Lond. Nat.* (Suppl.), **26**, 1–36.

— (1948). *Ibid.*, **27**, 37–75.

— (1949). *Ibid.*, **28**, 77–131.

— (1957). A review of British Glasshouse *Thysanoptera*. *Trans. R. ent. Soc. Lond.*, **109**, 467–534.

SPEYER, E. R. (1934). Some common species of the genus *Thrips* (*Thysanoptera*). *Ann. appl. Biol.*, **21**, 120–52.

— (1935). Thrips etc. *Rep. exp. Res. Stn, Cheshunt*, **1934**, 71–2.

— (1951). Gladiolus Thrips (*Taeniothrips simplex* Mor.) in England. *Proc. R. ent. Soc. Lond.*, (B), **20**, 53–62.

SPEYER, E. R. & PARR, W. J. (1941). The external structure of some Thysanopterous larvae. *Trans. R. ent. Soc. Lond.*, **91**, 559–635.

Order HEMIPTERA: Plant bugs

Suborder Homoptera

Family CICADELLIDAE (=JASSIDAE)

Zygina pallidifrons Edwards (the glasshouse leafhopper). The adult is a slender, pale-yellowish insect about 3 mm long with wings longer than the body (Fig. 5.10). Two dark areas form V-shaped bands across the back when the wings are folded. The head is pale yellow with dark, prominent eyes between which there are two dark spots. The thorax is dusky with two distinctly darker spots between the forewings; the legs are long and pale yellow. Mature hoppers are extremely active in sunny weather, leaping into flight when disturbed.

Eggs are inserted into secondary veins on the underside of the leaves by the saw-like ovipositor of the female. They are white, elongate, and slightly curved (0·5–0·7 mm long), and are usually laid singly (although occasionally they may occur in groups of up to eight). The egg, when first laid, is not easily seen but within a few days a minute swelling is discernible at each oviposition site.

FIG. 5.10. Leafhopper adult (*Typhlocyba rosae* L.) (×8). (A Shell photograph.)

The incubation period is less than 1 week in hot weather, but may extend to over 5 weeks in cooler conditions. Newly hatched nymphs are inactive, whitish and almost transparent. Five instars are completed before adulthood, the nymphs becoming increasingly active with age. The wing pads first become visible in the third instar. Young hoppers feed almost exclusively on the undersides of the leaves where the cast skins, or exuviae, remain for a considerable time as, during moulting, the head is anchored to the leaf by the inserted stylets. Cast skins are white and have led to the insect being called 'ghost fly' by some growers. The length of the nymphal period varies from 19–50 days. Breeding is continuous, both adults and developmental stages being present on the plants at any one time. *Z. pallidifrons* may attack important commercial plants, including forced mint, chrysanthemum, tomato and cucumber in addition to a wide range of ornamentals including *Calceolaria* sp., fuchsia, *Nicotiana*, *Pelargonium* sp., primulas, salvias and *Verbena* sp. Only the leaves are attacked; the leafhopper inserts its stylets into the lower surface to penetrate as far as the palisade parenchyma. As the cell contents are sucked out a minute blanched area appears on the upper leaf surface and, as feeding continues, these spots become confluent (Fig. 5.11). In severe attacks the leaves appear variegated or even completely blanched. Where seedlings are damaged growth may be severely checked.

The glasshouse leafhopper may be parasitized by a minute wasp *Anagrus*

FIG. 5.11. Injury to rose leaves by leafhoppers. (A Shell photograph.)

atomus (L.) (Hymenoptera: Mymaridae). The adult female (the males are rarely seen) is about 0·6 mm long, with long marginal setae on the distinctly angled forewings. The antennae have nine segments which increase in size towards the apex and terminate in a distinct club. *A. atomus* is very active, running rapidly over the leaves and making short flights from leaf to leaf. The female reproduces parthenogenetically and oviposits within the eggs of her host. Although several eggs may, on occasion, be laid within a single host egg only one parasite can mature therein. Except for the two curved, strongly chitinized mandibles and a characteristic appendage on the posterior body segment the young parasite larva is transparent. About a week after hatching, while in the second instar, the posterior appendage disappears and the body reddens and indicates that the leafhopper eggs are parasitized. The parasite's life cycle is complete in about 16 days; it breeds continuously at suitable temperatures and could probably be used for biological control if artificially introduced.

Two other leafhoppers *Eupteryx melissae* Curtis and *E. aurata* L. also damage chrysanthemums. They are normally pests only where chrysanthemums become infested outdoors, followed by rapid increase of hoppers when the plants are brought under glass for autumn flowering. *E. melissae* has three black spots on the crown of the head and six on the thorax of which the inner four tend to coalesce into a dark area. The wings are pale green mottled with brown. *E. aurata* has only two, occasionally confluent, spots on the head while, on the thorax, are a pair of reniform black spots near the front margin. The wings are greenish-yellow with a wide, irregular, brownish stripe down the middle.

Control

Hygienic measures and pesticides to control other, more important, pests should eliminate leafhoppers from most glasshouses. They are frequent on plants which are taken into glasshouses continuously occupied by a miscellany of ornamental plants.

Control is readily achieved with several insecticides, the choice being governed by considerations of safety to the plants, the presence or absence of other pests, cost and convenience.

Even though two or three applications may be necessary to eradicate the pest, smokes of DDT, gamma-BHC (or mixtures of these), or nicotine, are most convenient for both commercial nurseries and amateurs. Where equipment for application is available, atomized solutions (aerosols) of DDT, gamma-BHC or malathion at rates equivalent to 3 g, 1 g and 2·5 g of active ingredient respectively per 1000 ft^3 may be preferred. The interval between applications should be about 14 days at above 60 °F (16 °C) increasing to 28 days at 45 °F (7 °C).

During warm weather, when the incubation of the eggs is short, complete control can usually be obtained, provided the under-surfaces of the leaves are thoroughly wetted, from a single H.V. spray of DDT at 0·1%, gamma BHC at 0·01% or malathion at 0·1%. Leafhoppers are also susceptible to nicotine, parathion and pyrethrum though more than one spray of the less persistent chemicals nicotine, malathion or pyrethrum may be required.

Where other pests do not influence the choice of the insecticide, DDT is preferred for use on ornamental crops; on edible crops, malathion is more suitable with gamma-BHC as an alternative. If there is neglect to control leafhoppers before sowing, attacks on seedlings may be severe. The choice of a suitable insecticide is then restricted by the susceptibility of the seedlings to damage by pesticides. Atomized concentrates should not be used unless previous experience has shown them to be safe. DDT (not on cucurbits) or parathion smokes or sprays are least likely to cause injury.

The choice between DDT, gamma-BHC or malathion for controlling *E. melissae* or *E. aurator*, should these occur on outdoor chrysanthemums about to be taken under glass, depends on whether aphids, leafminers and caterpillars are also present. If it is necessary to control caterpillars, DDT should be used, otherwise gamma-BHC or malathion.

REFERENCES

FOX WILSON, G. (1938). The Glasshouse Leaf Hopper (*Erythroneura pallidifrons* Edw.). *Jl. R. hort. Soc.*, **43**, 481–4.

MACGILL, E. I. (1934). On the biology of *Anagrus atomus* Hall., an egg parasite of the leaf-hopper *Erythroneura pallidifrons* Edwards. *Parasitology*, **26**, 57–63.

Family CERCOPIDAE

Philaenus spumarius L. The cuckoo-spit insect or common froghopper is commonly introduced into glasshouses when chrysanthemums are brought in for flowering. *P. spumarius* (6 mm long) is larger than a leafhopper, usually brown, but variable in colour. At the apex of the hind tibiae, as in other *Cercopidae*, there is a circle of teeth which contrast with the isolated spines in *Cicadellidae* (p. 82). *P. spumarius* lays batches of 1–30 eggs on the stems. These overwinter before hatching into nymphs which surround themselves with froth-like 'spittle' (Fig. 5.12), produced by secretions from the alimentary

FIG. 5.12. 'Spit' produced by common froghopper (*Philaenus spumarius* L.). (A Shell photograph.)

canal. There are five nymphal instars which, depending on temperature, are completed in 4–13 weeks. Froghopper damage is similar to that caused by leafhoppers but the unsightly masses of froth are objectionable. Occasionally newly emerged adults may damage young shoots of chrysanthemums.

Control

Only where exceptionally heavy infestations occur on chrysanthemums is it necessary to control froghoppers; the masses of froth are sufficiently conspicuous for workers to destroy the nymphs when trimming and disbudding. They can also be dislodged from plants by a jet from a hose.

Severe infestations should be sprayed with DDT, gamma-BHC or malathion as for leafhopper. Smokes and aerosols are unsatisfactory as the froth protects the insect; forcible spraying is required to remove this covering.

REFERENCES

HALKKAO, O., RAATIKAINEN, M., VASARANA, A. & HEINON, L. (1967). Ecology and ecological genetics of *Philaenus spumarius* (L.). *Ann. Zool. Fenn.*, **4**, 1–18.

Family ALEYRODIDAE

Trialeurodes vaporariorum Westwood, the glasshouse whitefly, is an important pest of ornamentals (arum, begonia, *Calceolaria* sp., *Coleus* sp., freesia, fuchsia, *Iresine* sp., pelargonium, primula, salvia, verbena) and on the main vegetable crops (cucumber and tomato). The adults, about 1 mm long, are gregarious, collecting in vast numbers on the upper foliage of plants. The body and wings are covered with minute spirals of wax giving the wings a pure white coloration and the yellowish body a 'mealy' appearance. Reproduction is largely parthenogenetic, although races exist in which unfertilized females lay only male-producing eggs so that mating is necessary to produce females. Males are identical with females but smaller. The adult female lives for 30–40 days and is attracted to the apical leaflets of its host plant by flying upwards towards green surfaces, especially those with a more yellow hue. So young foliage is chosen for feeding. Once settled, the rostrum is inserted into the leaf through a stoma or between epidermal cells. The path of the stylets is intercellular and, although some parenchyma cells are pierced, sap is drawn from the phloem of the vascular bundles. The reproductive rate is affected by the nutrient status of the host; the better the host plant is fed, the faster the whiteflies reproduce.

The eggs, 0·2–0·25 mm long, are yellowish when first laid, but after 2 days become black. They are attached to the leaf surface by a short stalk and normally occur on the undersides of the leaves though a few may be laid on the upper surface, or even on the main stem. When ovipositing the female usually leaves her rostrum in the tissues and pivots her abdomen around in a circle (Fig. 5.13). Hence, on smooth-leaved plants, where such movements are uninterrupted, a complete circle of 20–40 eggs may be laid. On pubescent hosts the circles become arcs but on crops such as tomato the eggs are scattered irregularly over the leaf surface. Under favourable conditions females lay between 150 and 500 eggs at an average rate of more than 25 per day.

The incubation period of the eggs varies from 30 days at 46 °F (8 °C) to 4 days at 86 °F (30 °C), under average glasshouse temperatures they hatch within 10 days. The newly hatched larva is flat, 0·3 mm long, and light green in colour with bright-red eyes. It has functional legs and moves about for 2–3 days before settling to feed alongside a leaf vein. The larva moults after 3–4 days into the second instar nymph which flattens out on the leaf and becomes even more transparent than the larva; legs and antennae become vestigial. These immobile stages are called 'scales' (Fig. 5.14). Within 3 days

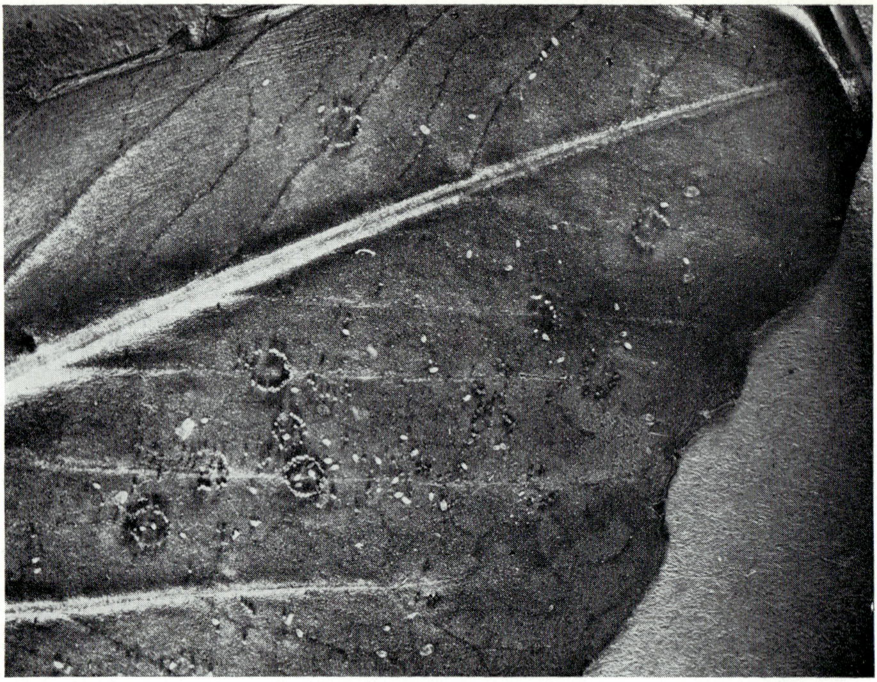

FIG. 5.13. Eggs of whitefly (*Trialeurodes vaporariorum* Westw.) (×5). (A Shell photograph.)

the nymphs moult into a third instar (0·5 mm), slightly larger than the foregoing stage. Within a further 3–4 days the nymph moults again to the pupa. At first this is whitish-green, and closely set to the leaf surface, but, as it matures, it is lifted from the leaf surface by submarginal wax rods which eventually form a striated palisade. Unlike the other stages the pupa has 5–8 pairs of prominent dorsal wax hairs and shorter marginal processes. Until recently pupae which lacked these processes were believed to be a distinct species *T. sonchi* Kotinsky, but it is now known that this is merely a difference caused by the host plant, and that many erroneous synonymies have arisen in the *Aleyrodidae* in this way (Hussey & Gurney, 1957). At 70–80 °F (21–27 °C) the adult normally emerges in about 6 days through a T-shaped opening in the dorsal surface of the pupa. The complete life cycle therefore occupies 3 weeks at 70 °F (21 °C) or 4 weeks at 60 °F (15 °C).

The glasshouse whitefly seriously disfigures the host plant when black moulds (*Cladosporium* spp.) grow on the honeydew excreted by the adults and nymphs. All developmental stages of *Aleyrodidae* are provided with a special apparatus, the 'vasiform orifice' which ejects honeydew away from the anus. The orifice is a depression of the abdominal segment with an unchitinized and very elastic floor to which powerful muscles confer considerable mobility. The elasticity of this membrane ensures the forceful ejection of liquid and semi-

FIG. 5.14. Tomato leaf infested with adults and scales of *Trialeurodes vaporariorum* together with black scales which have been parasitized by *Encarsia formosa* (× 12). (J. Hesling.)

solid faecal matter in different directions away from the body. Adults produce about 10 droplets per hr, pupae 25 per hr and larvae 8 per hr. Assuming that a female lays 120 eggs the amount of honeydew produced by her progeny during their development would cover 20 tomato leaflets. If the relative humidity rises above 90% for 70 hr (cumulative)—more than 7 consecutive nights—black moulds may develop where sufficient honeydew has accumulated. The mould forms a black felt over the leaves, infested plants become unsightly and fruits have to be cleaned before picking, so adding to the production costs. As affected foliage becomes chlorotic, or dies, infested plants lose colour and generally deteriorate. Reduction in tomato yield has been shown to occur when the whitefly population reaches a density of about 70 scales per $\frac{3}{4}$ in. disc on the apical leaves. With such numbers only about 30% of the fruits normally show any fungal growth. The maximum population which can be tolerated without any disfigurement of fruit is 20 scales per disc on these leaves.

Occasionally whiteflies may be attacked by entomogenous fungi. The most important of these is *Cephalosporium aphidicola* Petch, which covers its host with a fine weft of hyphae bearing capitate conidiophores. The first indication that an adult is infected by the fungus is the appearance of two large 'bundles' of wax filaments on the ventral side of the abdomen. Within 12 hr the fly becomes inert and 3 days later is covered with mycelium. Even under low magnification, the characteristic arrangement of conidial heads, at right angles to the main hyphal strands, is clearly visible. Only the eggs and mobile larvae are unaffected by the fungus. The conidia germinate readily only in a saturated atmosphere. If a large proportion of the spores are to germinate these conditions must be maintained for at least 20 hr. There is, therefore, little practical possibility for the utilization of this fungus for biological control, since the optimum conditions would also encourage many phytophagous fungi.

A small chalcid wasp, *Encarsia formosa* Gahan, is known to parasitize the scales of *T. vaporariorum*. It is only 0·6 mm long with a dark-brown head, black thorax with yellow sides, and a uniformly yellow abdomen with an ovipositor extending conspicuously beyond the tip. The legs are brownish-yellow. The male, which occurs only rarely, develops where an *Encarsia* egg is laid hyperparasitically in an *Encarsia* larva and is readily distinguished by the dark-brown abdomen. *E. formosa* is an active insect and feeds on the honeydew produced by the host. Indeed, experiments have shown that the adults, which normally live 28 days, survive only 4–5 days in the

FIG. 5.15. Female *Encarsia* probing a whitefly pupa. (J. Hesling.)

absence of honeydew. The ovipositing female walks rapidly over the leaf surface seeking host scales with her antennae (Fig. 5.15). Once located, the female mounts the scale and thrusts her ovipositor inside. The tip is moved about within the host tissues in search of other *Encarsia* eggs which may have been laid previously. It is unusual to find a parasitized scale containing more than one *Encarsia* egg though as many as nine have been found where the parasites were very numerous. A yellowish exudation, which hardens within a few hours to a deep-brown or black pellet, emerges from each oviposition puncture. All inactive, nymphal instars of the whitefly are subject to attack and are either killed and turn brown, or are parasitized and turn black. Whether the host is successfully parasitized, or killed by frequent ovipositor probes, is influenced by the density of the parasite adults and the stage of host attacked; hosts tend to be killed at high parasite densities, but to be parasitized at lower; while younger hosts tend to be killed older ones are parasitized. Each female parasite lays between 50–100 eggs (0·08 mm long) in the sedentary scale stages of *Trialeurodes* but never in the pupa. The incubation period is 4 days. The white, limbless parasite larva rapidly consumes its host and, after about 8 days, the host scale turns black. After 10 days the adult emerges through a circular opening cut on the dorsal surface of the host scale.

The interaction between *Encarsia* and the glasshouse whitefly is dominated by temperature. At 65 °F (18 °C) the fecundity of the host is ten times that of the parasite while the rate of development is equal. But, as the temperature rises to 80 °F (27 °C), the fecundity of the parasite increases to parity with its host and, in addition, the rate of development of the parasite becomes twice as fast as the whitefly. The parasite therefore tends to control *Trialeurodes* more effectively on cucumbers than it does on tomatoes and other crops grown in houses where the temperature is high only during the warm summer months. Experimentally, it has been shown that, above 70 °F (21 °C), economic control of low, uniformly distributed, whitefly populations can be obtained if two parasites are introduced for every square foot of planted area.

The value of *Encarsia* was recognized in pre-war years and the Cheshunt Experimental Station in Hertfordshire developed mass rearing techniques which resulted in the supply of 1½ million parasites annually. The development, in the post-war era, of DDT and other effective insecticides greatly reduced the whitefly problem and, except in botanic gardens, the parasite largely disappeared. More recently the development of insecticide resistance in mites and aphids has renewed interest in control techniques which integrate both biological and chemical agencies.

Three other species of whiteflies, *Aleuropteridis filicicola* Newstead, *Aleurotulus nephrolepidis* Quaintance and *Filicaleyrodes williamsi* Trehan, occasionally occur on ferns under glass.

Control

Hydrogen cyanide and tetrachlorethane, at one time extensively used as fumigants against whitefly, have been superseded by newer, less toxic compounds. Provided the host plants are tolerant, malathion and DDT are the most suitable insecticides; the former being the more effective against scales when aerosol applications are made. Both appear to be equally effective when applied as H.V. sprays.

Aerosols of malathion and DDT at 2·5 g and 3·0 g/1000 ft^3 respectively can be applied with safety to most established plants, and provide a convenient method of control where suitable equipment is available. Alternatively, smokes of DDT (3·5 g/1000 ft^3), parathion (0·25 g/1000 ft^3) or sulfotep (1·0 g/1000 ft^3) can be used; those containing parathion or sulfotep are recommended for use on young cucumber plants and on seedlings of most plants. Canisters, for smoking with DDT, are convenient and readily obtainable by non-professional growers. Atomizing (aerosol) concentrates containing a mixture of DDT and gamma-BHC are also available and are convenient for controlling many other insects. Where the plants are also infested with a parathion-susceptible strain of red spider it may, in some districts, be economic to employ a contractor to apply parathion aerosols.

Complete control cannot be achieved by a single application of either aerosol or smoke as some scales are not killed, consequently two or three applications are necessary. When the mean temperature is about 70 °F (11 °C) the interval between applications should be about 7 days.

High volume sprays of DDT (0·1%) or malathion (0·1%) as emulsifiable concentrates are highly effective, and, if spraying thoroughly wets both upper- and under-leaf surfaces, not more than two applications are required. Malathion should be used on edible crops.

If thoroughly applied, petroleum emulsion sprays (1 gal/100 gal) are also effective against whitefly, but their use is economic only where circumstances dictate their use against red spider. Other insecticides, which may be advantageous when other pests are present, are dichlorvos and naled (Ortho-Dibrom).

REFERENCES

BURNETT, T. (1964). Host larval mortality in an experimental host-parasite population. *Can. J. Zool.*, **42**, 745–65.

— (1967). Aspects of the interaction between a chalcid parasite and its aleurodid host. *Can. J. Zool.*, **45**, 539–78.

HUSSEY, N. W. (1958). Notes on a fungus parasitic on Greenhouse Whitefly. *Pl. Path.*, **7**, 71–2.

HUSSEY, N. W. & GURNEY, B. (1957). *Trialeurodes sonchi* Kotinsky a synonym of *T. vaporariorum* Westwood. *Ent. mon. Mag.*, **93**, 254.

LLOYD, L. (1922). The control of Greenhouse Whitefly with notes on its biology. *Ann. appl. Biol.*, **9**, 1–34.

MOUND, L. A. (1966). A revision of the British Aleyrodidae (Hemiptera: Homoptera). *Bull. Br. Mus. nat. Hist.*, **17**, (a), 399–428.

PARR, W. J. (1968). Biological control of Greenhouse Whitefly by the parasite *Encarsia formosa* on Tomatoes. *Rep. Glasshouse Crops. Res. Inst.* **1967,** 137–41.

SPEYER, E. R. (1927). An important parasite of the Greenhouse Whitefly (*Trialeurodes vaporariorum* Westwood). *Bull. ent. Res.*, **17**, 301–8.

SUPERFAMILY *COCCOIDEA*

Scale insects and mealy bugs constitute an important group of pests affecting perennial glasshouse-grown ornamental plants. They are of most importance on vines, roses or collections of ornamentals rather than in the large flower-producing industry, which largely comprises carnations and chrysanthemums.

Many scales are potential pests and have been introduced on plants from the tropics. They are usually small and inconspicuous and by the time an infestation has become obvious, numbers are so great that the host plants become stunted and the foliage chlorotic. The primary injury is loss of sap. The large quantities of honeydew produced frequently induce a prolific growth of sooty moulds which cause secondary damage by reducing the rate of photosynthesis.

As soon as the female scale insect matures, eggs are formed; this process continues until the body of the mother is depleted and a large mass of eggs is concealed beneath her hard, empty, shell. In fact, the adult female is reduced to an egg-producing machine in which almost all internal organs are sacrificed. Each female produces more than 1000 oval eggs which are often covered with minute waxy secretions from her wax glands. Each egg hatches to a pale, translucent yellow, first nymphal or 'crawler' stage, which is quite flat with three pairs of legs and often bears prominent dark eyespots. The crawlers move about actively; in some cases the mother scale is so firmly anchored to the substrate that they are unable to emerge, and the brood chambers become full of dead crawlers. Those which do escape settle down on, or near, veins on the undersides of the leaves or on the stems; during this stage they are easily spread by other insects or workers tending the plants. They then moult to a second, quiescent stage which, as it matures, shows sexual dimorphism. While the female scales tend to become circular, the males increase in size and elongate markedly. At the third nymphal stage the female scale changes from a flattened creature to one with a pronounced convex surface surrounding a hollow brood chamber within which the eggs are laid. This chamber is lined by woolly, wax secretions and, during the egg-laying period, continues to increase in size until it occupies almost the whole of the insect's body.

Throughout the feeding period, honeydew is produced through a specially adapted apparatus in which the anal plates are so hinged that they can be both elevated and separated, thus letting a bubble of viscous honeydew ooze out. This sweet substance is much sought after by ants, which are often found crawling over the female scales, and may carry young scales from plant to plant. The scale insects suck sap from the deep-seated phloem cells of the host with extremely long stylets which, when withdrawn, are kept in a special cavity at the base of the short proboscis.

After the second moult the male pupae form within the nymphal skin and long caudal filaments—brilliant white, waxy appendages which exceed the length of the insect's body—project on either side of the penis. The mature male has no functional mouth or indeed any vestige of a stomach or intestine. It possesses a single pair of membranous wings with a stout vein a short distance from the front margin and a more slender one parallel to the posterior wing margin. The halteres terminate in hooks which lock in pockets on the posterior margin of the wings. Males live 3–4 days and fly only a few feet at a time since the wings, though large, are frail. They mate with only a single female and, after copulation, usually die, *in situ*, on the dorsum of the female. Mated females give rise to both males and females, while unmated mother scales produce only females. In most species males comprise less than 5% of the population.

Mealy bugs, though closely related to scale insects, differ in possessing a covering of flocculent, white, waxy threads instead of a hard scale. They are active throughout life and produce groups of eggs which are either enclosed in an egg sac composed of secretions produced by the adult female or may be carried about between long plates of wax at the posterior end of the female's body. The newly hatched nymphs remain within this sac for a short time before emerging, and after about 12 hr activity they settle for feeding. The adult males are very sluggish and, as in the scale insects, do not feed.

All scale insects and mealy bugs breed rapidly in the warm, humid environment of glasshouses and must be considered potential pests; as more than 60 exotic specimens have been recorded under glass in this country, it is only possible to discuss here a few of the most commonly occurring forms.

SCALE INSECTS

Dynaspidiotus britannicus Newstead (= *Aspidiotus hederae* Vallot) (The Oleander Scale) forms roughly circular, light-brown, slightly convex scales

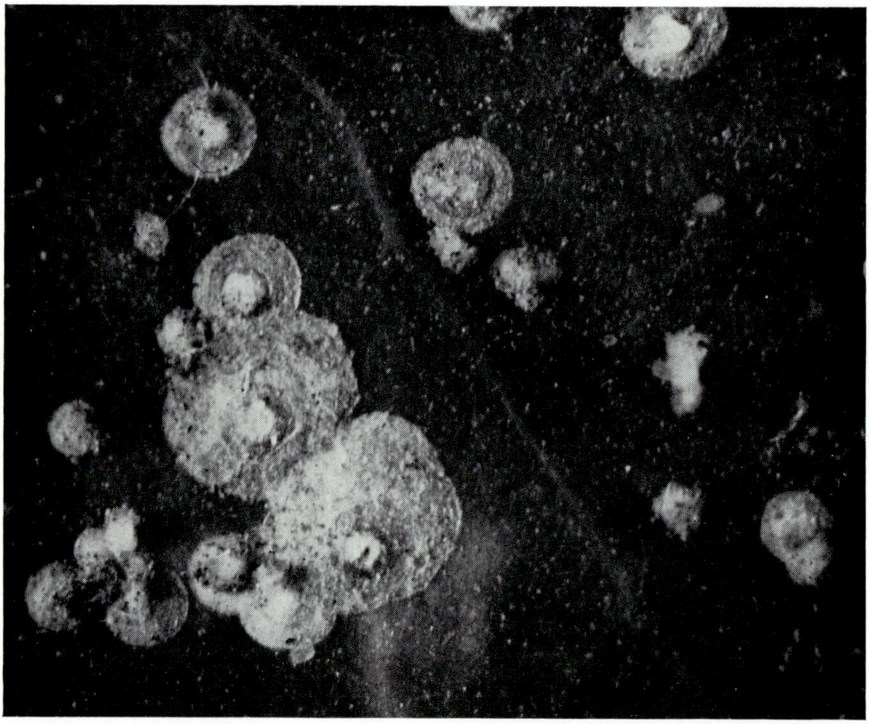

FIG. 5.16. *Dynaspidiotus britannicus* Newstead: Oleander scale (× 10). (Courtesy of R. Jefferson, University of California.)

whose diameter varies between 1·8 and 2·0 mm and beneath which shelters the yellow female (Fig. 5.16). *A. hederae* is a polyphagous species on plants of tropical and subtropical origin, and under British glasshouse conditions is known from *Phoenix*, *Kentia*, and other palms, *Asparagus sprengeri* and species of *Acacia*, *Aucuba*, *Dracaena*, *Erica*, *Fatsia*, *Ficus*, and *Laurus*.

Hemiberlesia rapax Comstock (=*camelliae* Signoret) (the camellia or greedy scale) is circular, 2·0–2·2 mm in diameter, but markedly convex, indeed almost conical in elevation. The central larval cast-skin is deep brown and so contrasts strongly with the mother scale, which is a yellowish-coffee colour. The ventral wall of the scale is very thick and pure white. *A. rapax* is a polyphagous species, and, in hot-houses, infests begonias, camellias, figs, cacti and orchids.

Aulacaspis rosae Bouché (the scurfy rose scale) can cause severe damage to roses under glass. It is normally found on the woody parts of the bushes, which may become enveloped in a scurfy covering of scales, though smaller colonies may also occur on the young growth. The female scale is flat, like mother-of-pearl, with the red-brown larval cast-skin usually eccentrically placed. It is quite a large species, 2·2–3·0 mm in diameter, although both size and shape may be modified in dense colonies. The male scales are elongate, only 1 mm long, and, as in many other Diaspididae, are pure white with three dorsal longitudinal keels. The mature female coccid, beneath the scale, is wine red and is characterized by the rounded and enlarged cephalothorax, which is much longer than the abdomen.

Diaspis boisduvalii Signoret (the orchid scale), lives on a wide range of host plants including palms (*Cocos*, *Kentia*, *Phoenix*, *Rhapis*), orchids (*Cattleya Cymbidium*, *Epidendrum*) and various species of cacti and Bromeliaceae, is one of the commonest greenhouse scales. It tends to feed more commonly on the leaves than the woody parts of the plants, though the stems of orchids may become completely encrusted. The partially transparent female scale is greyish-yellow, flat, and circular. However, it is possibly most readily recognized by the cottony mass of wool produced by clusters of males. The cast-skin of the larva is the same colour as the adult scale but eccentrically placed. Infested plants have a 'dusty' appearance due to the colour of the massed scales.

Pinnaspis aspidistrae Signoret (the aspidistra or fern scale) is a polyphagous species from the tropics of the New World which may be found under glass on *Aspidistra* sp., *Boehmeria* sp., ferns, *Nephrolepis* sp., *Ophiopogon* sp., palms (*Kentia*), and *Ruscus* sp. The female is a flattened, pyriform scale shaped rather like a mussel. It is shining reddish-brown. The male scales, which often outnumber the females, are pure white and roughly rectangular in shape (Fig. 5.17). Occasionally, individual *Aspidistra* plants may be completely covered with these scales which are generally located on the leaves between the main longitudinal veins.

Parlatoria proteus Curtis (the cattleya scale) is oval, flattened and generally straw-coloured with an eccentrically placed larval cast-skin. The largest width is between 1·2 and 1·6 mm. The elongate male scales are light grey and devoid of dorsal keels. Dense colonies are formed at the leaf bases of palms and orchids.

Parthenolecanium corni Bouché (the peach scale) is the commonest scale on peaches and nectarines under glass, has also been recorded on amaryllis, *Nephrolepis* sp. and *Nerium oleander*. It is a polyphagous species which varies

FIG. 5.17. *Pinnaspis aspidistrae* Signoret: fern scale. (Courtesy of R. Jefferson, University of California.)

in colour both with host and age. The female scale is very large (3–6 mm long and from 1·5–4 mm wide) and is normally yellow-green to pale brown with an irregular pattern of darker marks (Fig. 5.18). As the eggs develop the scale darkens to a polished red-brown. Males are almost unknown. Production of eggs begins at the end of May and continues until July. The oval, white eggs are laid beneath the scale, between 1500 and 2000 being produced by each female. They hatch within 3 weeks to reddish-brown active nymphs which settle on leaf veins on the undersides of the leaves.

FIG. 5.18. *Parthenolecanium corni* Bouché on vine. (Courtesy of Glasshouse Crops Research Institute.)

Chloropulvinaria floccifera Westwood (the cushion scale) is immediately distinguished by the egg sac. The female is dull brown, transversely wrinkled and measures about 3 mm in diameter whereas the egg sac frequently attains a size of 10–15 mm. The sac adheres to the abdomen of the female and protrudes as a conspicuous, linear, white mass covering the brown eggs. This species is particularly troublesome and conspicuous on the underside of the leaves of camellias and orchids.

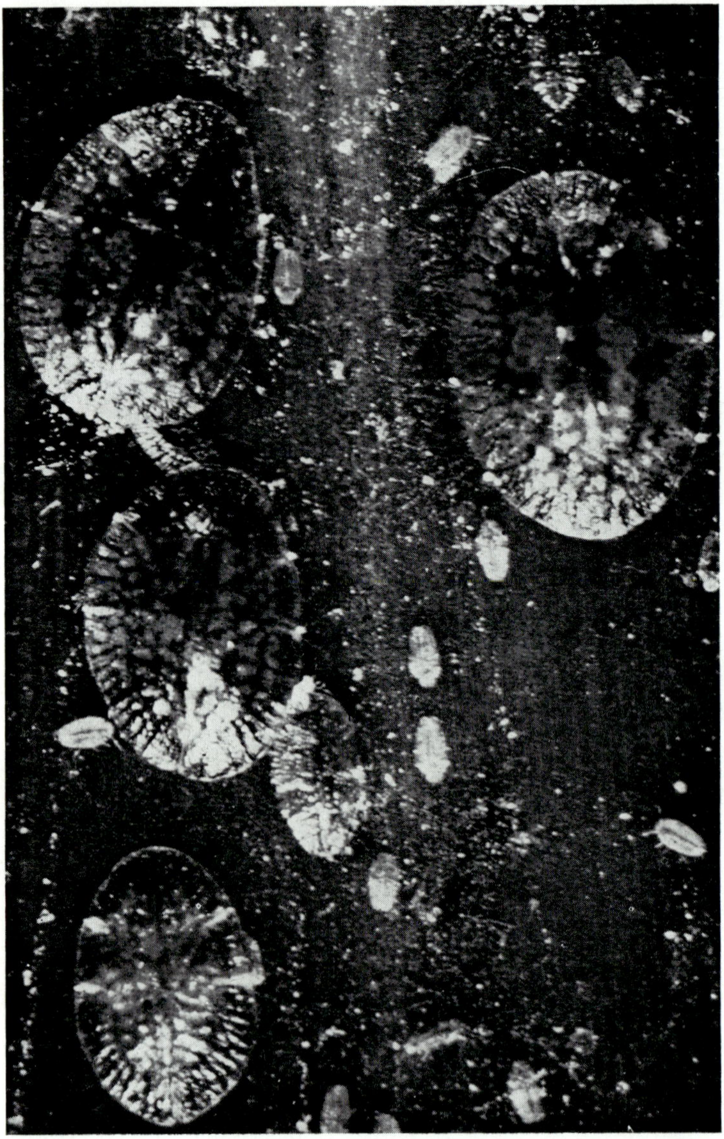

FIG. 5.19. *Coccus hesperidum* L.: soft scale (× 25). (Courtesy of R. Jefferson, University of California.)

Coccus hesperidum L. (the soft scale) is found throughout the warm and temperate zones and is a ubiquitous and persistent greenhouse pest on a vast range of plants, including amaryllis, camellia, citrus, figs, *Hibiscus* sp. and vine.

Soft scales feed exclusively on the phloem, their stylets piercing both sieve tubes and phloem parenchyma. Apparently, no cell-destroying enzymes are injected so that damage is minimal although abundant honeydew is produced and the foliage and fruit of infested plants may be contaminated by the unsightly growth of saprophytic fungi.

Coccus hesperidum appears to produce young viviparously but, in fact, the ovisacs peel off the young crawler larvae while they are still within the vagina of the mother. The empty, snow-white ovisacs remain as shrivelled pellets below the bodies of the females. The pink crawlers are active for 2 or 3 days and may move several feet before attaching themselves to a plant. Once attached, they lose their colour and feed. Moulting to the second instar occurs 8 days later. This instar produces copious quantities of wax both over its body and the surrounding plant tissues, giving the appearance of a scale. The larva moults to the adult female 17–21 days later. The size, shape and markings of the adult vary with the host (Fig. 5.19). For instance, on citrus, the female is flat and irregular in shape with radial markings, while, on *Laurus nobilis*, the scale is elongate-oval and yellowish-brown with no distinct markings. Whatever the colour pattern, the periphery of the scale is always lighter than the central zone and the anal furrow is well marked.

At 67 °F (19 °C) the period from birth until the first progeny are produced is 60 days, while the total life of a female is about 100 days. In hot-houses, with temperatures fluctuating between 65 °F (18 °C) and 77 °F (25 °C) six or seven generations are completed annually. Under these conditions the maximum number of progeny produced by each female has been estimated to range from 80 to 252.

Over 35 species of parasites of *C. hesperidum* have been recorded in the literature but the most important are *Encyrtus lecaniorum* (Meyr.) and *Microterys flavus* How. (Hymenoptera: Encyrtidae), *Coccophagus lycimnia* Walker and *C. scutelleris* Dalm. (Hymenoptera: Aphelinidae). In hot-houses, where the humidity is high, the pathogenic fungus *Cephalosporium lecanii* Zimm. is an important factor in control of soft scale.

REFERENCES

ANNECKE, D. P. (1966). Biological studies on the immature stages of the Soft Brown Scale *Coccus hesperidum* L. (Hemiptera: Coccoidea). *S. Afr. J. agric. Sci.*, **9**, 205–28.

SAAKYAN-BARANOVA, A. A. (1964). On the biology of the Soft Scale *Coccus hesperidum* L. (Homoptera: Coccoidea). *Ent. Obozr.*, **43**, 268–96.

Saissetia coffeae Walker (the hemispherical scale) is globular, smooth, regularly hemispherical and bright, shining brown. It has a pattern of ridges, in the form of the letter 'H', in relief on the dorsal surface. The size is variable and dependent on host plant, e.g. from 4·5 mm long on *Cycas* to only 2 mm on *Asparagus*. A large range of glasshouse ornamentals is attacked, including species of *Aspidium*, *Asplenium*, *Blechnum*, *Croton*, *Lygodium*, *Stephanotis*.

Figs and orchids may also be affected. This scale has three larval instars which may account for the low rate of parasitism found in glasshouses compared with that in soft scale.

MEALY BUGS

Mealy bugs are troublesome pests in floriculture and, while numerous kinds occasionally become established in glasshouses, the following are most common: *Planococcus* (= *Pseudococcus*) *citri* (Rosso), the common or greenhouse mealy bug; *Pseudococcus latipes* Green; *P. longispinus* Targ. (= *adonidum* Linn.), the long-tailed mealy bug; and *Nipaecoccus nipae* (Maskell), the palm or coconut mealy bug.

These mealy bugs may be conveniently distinguished by the form of the wax filaments: *P. citri* has short, stout, marginal filaments; *P. latipes* has long caudal filaments half the length of the body; *P. adonidum* has even longer filaments, the caudal pair often being longer than the body; *N. nipae* has a roundish, convex body covered with short pyramidal tufts of cream-coloured wax.

Planococcus citri is a widely distributed species which is common on amaryllis, *Anthurium* sp., asparagus, cacti (*Cereus peruvianus*), chrysanthe-

FIG. 5.20. *Planococcus citri* Rossi: greenhouse mealy bug (×12). (Courtesy of R. Jefferson, University of California.)

mums, *Coleus* sp., *Crassula* sp., *Eucharis* sp., ferns, gardenias, *Musa* sp., orchids, *Solanum* sp., *Stephanotis* sp., *Streptocarpus* sp. and vines. The mature female is elongate and, except round the periphery of the pink body where the marginal and caudal wax processes arise, covered with fine wax threads (Fig.

FIG. 5.21. *Pseudococcus longispinus* Targ.: long-tailed mealy bug (× 20). (Courtesy of R. Jefferson, University of California.)

5.20). The development is encouraged by high temperatures and humidities. The incubation period is 14 days at 62 °F (17 °C), decreasing to only 4 days at 90 °F (32 °C). The nymphs develop most rapidly at 82 °F (28 °C) maturing within 18 days. At 80 °F (27 °C) therefore, as many as 8 generations may be completed annually. When 4–4·5 mm long, the female secretes a compact egg sac in which between 100–150 pale-yellow, elliptical eggs (0·3 × 0·2 mm) are laid; each egg is separated from its neighbours by extremely fine wax filaments. Dense mealy bug populations cover the leaves and fruits with honeydew and often cause complete defoliation of the plants.

Pseudococcus latipes, probably the commonest British greenhouse mealy bug, attacks abutilon, amaryllis, azaleas, ferns, *Nevium* sp., pelargonium, and occasionally tomatoes, as well as vines.

Pseudococcus longispinus, which, apart from the long caudal filaments (Fig. 5.21), is smaller than either of the above, is particularly common on *Anthurium* sp., begonia, cacti (*Epiphyllum truncatum*, *Opuntia microdasys*, *Phyllocactus grandiflora*), *Croton* sp., *Cycas* sp., *Dracaena* sp., ficus, fuchsia, *Hoya* sp., *Jasminum* sp., *Kentia* sp., lilies, orchids, pelargoniums, *Phoenix* sp., *Phornium tenax* and *Zantedeschia* sp.

Nipaecoccus nipae. This tropical species occurs under glass on palms (especially *Cocos*, *Kentia*, *Kentiopsis* and *Sabal*). It is quite a small mealy bug, 3·5 mm long, globular in form, and salmon pink. The large, white wax cones create a characteristic stellate appearance on both dorsal and lateral surfaces. The mealy secretions associated with other mealy bugs are quite absent. Males are very common in this species, and the small, fleecy pupae (1·6 × 2·0 mm) may be recognized by their elongate form. The bugs tend to congregate at the bases of the leaves.

Pseudococcus fragilis Brain (=*gahani* Green) (the currant mealy bug) is an indigenous British species affecting a wide range of glasshouse plants, but is normally associated with *Ceanothus* sp., laburnum, red currant and sprouting potato tubers. The elongate-oval female (3·4 × 1·3 mm) is covered with a greyish, mealy type of wax secretion through which the dark cuticle presents a marbled appearance. There are a pair of stout wax processes posteriorly. An elongate (0·4 × 0·3 mm), marbled, light brown ovisac, composed of felted, brittle, wax threads is produced and contains about 100 eggs. When irritated, the females produce globules of claret-coloured liquid which rapidly harden. Only small quantities of honeydew are produced.

REFERENCE

WILLIAMS, L. (1924). On *Pseudococcus gahani* Green. *Ann. appl. Biol.*, **11**, 498–502.

Orthezia insignis Browne (the greenhouse orthezia or Kew bug) is dark olive-green (1·5 × 1·3 mm), and resembles an aphid rather than a mealy bug. It has 10 pairs of lateral white, waxy processes, and 12 pairs of similar processes arranged in two median, longitudinal, rows on the dorsal surface (Fig. 5.22). Young adults have processes projecting posteriorly from a sclerotized submarginal belt from which the white ovisac, constructed of brittle wax plates,

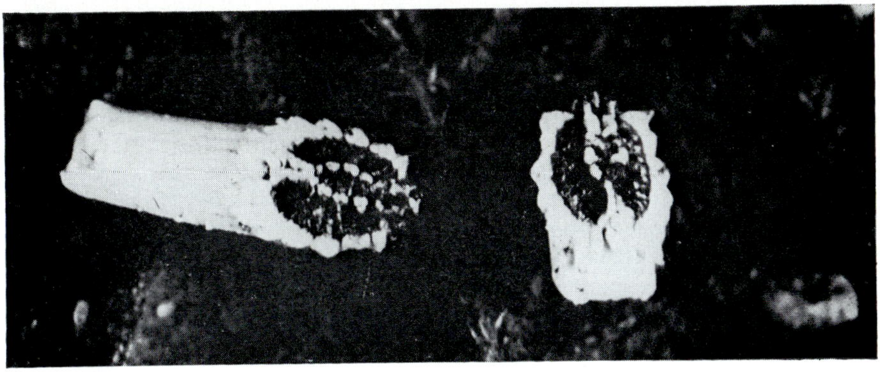

FIG. 5.22. *Orthezia insignis* Browne: greenhouse orthezia (×12). (Courtesy of R. Jefferson, University of California.)

is produced. Each female lays up to 80 eggs. The female moves freely, carrying this conspicuous white ovisac, a habit rarely found in other mealy bugs. The shortest incubation period is 23 days at 86 °F (30 °C), while at 68 °F (20 °C) hatching takes 30 days. The larva moults three times, and at 68 °F (20 °C) matures in about 55 days. Under glass in Britain, they complete four or five overlapping generations. Attacked plants are checked and subsequently turn yellow. In severe cases defoliation may occur. Plants affected include species of *Abutilon*, *Ageratum*, *Amaranthus*, chrysanthemum, *Coleus*, *Justicia*, *Lantana* and *Strobilanthes*.

REFERENCE

EZZAT, Y. M. (1956). Studies on the 'Kew-Bug' *Orthezia insignis* Browne. *Bull. Soc. ent. Égypte*, **40**, 415–31.

Other species of mealy bugs found in British glasshouses are listed below with their hosts plants:

Chorizococcus lounsburyi (Brain)	On roots and bulbs of Liliacea and *Amaryllidaceae*
Dysmicoccus wistariae (Green)	On *Wistaria*
Rhizoecus cacticans (Hambleton)	On roots of *Mammillaria*, *Phyllocactus* and various *Bromeliaceae*
R. dianthi (Green)	On roots of *Adiantum*, *Dianthus plumarius*, *D. barbatus*, and *Pelargonium*. The population growth of this and other root-infesting coccids is reduced in soils with abundant moisture
R. elongatus (Green)	On roots of *Aloe variegata*, *Mesembryanthemum* and *Phyllocactus*
R. falcifer Kunckel	
Spilococcus cactearum McKenzie	*Cactaceae*
S. filicicola Newstead	On the fronds of *Trichomanes spicatum*

Control

Sedentary insects, such as mealy bugs, are suitable for biological control techniques using parasites and predators. The work of Doutt (1952), who successfully controlled *Planococcus citri* on commercial crops of *Stephanotis*, is especially interesting. He liberated a parasite *Leptomastix dactylopii* Howard (Hymenoptera: Encyrtidae) and a ladybird *Cryptolaemus montrouzieri* Muls. One ladybird and 400 parasites per plant eliminated the pest from a single glasshouse, maintained at over 70 °F (21 °C), within 3 months.

Few critical trials have been made either to compare different insecticides for the control of the many species or to establish the safety of such chemicals, and their formulations, towards the host plants. These insects, which are difficult to control chemically, are a scourge of orchids and other perennial ornamental plants, particularly when these are grown as mixed collections in horticultural gardens, conservatories, and the like, for it is difficult to select an insecticide which is safe to use on a wide variety of plants.

Most of the coccids which attack glasshouse plants are tropical or subtropical species, not occurring outdoors in Britain, and so are almost invariably introduced on plants. It is obvious, therefore, that only plants on which no evidence of mealy bug or scale attack is discernible should be taken into a glasshouse containing potential hosts. Infested plants which are to be retained should be quarantined and treated until careful examination shows that no live insects remain, bearing in mind that apparently dead scales may contain viable eggs.

Some countries, e.g. the United States of America and Australia, require the fumigation of scale-infested orchids on importation. The American requirements are 40 oz of methyl bromide/1000 ft^3 for 2 hr at 60 °F (16 °C) whereas in Australia, 32 oz/1000 ft^3 for $1\frac{1}{2}$ hr is required, though no temperature is specified. Certain large orchid nurseries in the U.S.A. have fumigation chambers for the quarantine treatment of acquired plants which are exposed for 2 hr at 80 °F (27 °C) to 2 lb methyl bromide/1000 ft^3. A high relative humidity and good circulation is required. Fumigation with methyl bromide may damage the young growth of some orchids and these may take a year or so to recover although, normally, healthy plants are usually uninjured. *Cymbidiums* and, to a lesser extent, *Cypripediums* are most susceptible. Work by Griffin *et al.* (1956) showed that, in a R.H. of 70–90%, damage was more severe at 77 °F (25 °C) than at 60 °F (16 °C).

The chemical control of scale insects is made difficult by the protection of the eggs by the mother scale and, in some species, by the impermeability of the hard covering of scale stages. Mealy bugs, too, are difficult to wet with sprays owing to their covering of extruded wax filaments. Insecticides are best used as H.V. sprays, preferably of emulsifiable concentrate formulations, and must be carefully applied so that the fluid penetrates into the crevices of leaf axils, leaf and flower sheaths, etc., which are favoured habitats. Aerosols and smokes of lethal chemicals require frequent applications to achieve satisfactory control. Although soil treatment with an effective systemic compound would offer a simple solution, particularly with pot-plants, no satisfactory material is known.

Provided the host plant is tolerant, sprays of diazinon are, in general, most satisfactory for the control of scale insects and mealy bugs. Unfortunately, diazinon is phytotoxic to many plants such as orchids, cacti and succulents on which coccids are likely to occur. Phytotoxicity may be influenced by different solvents in proprietary formulations and, unless the host is known to be tolerant, a product should first be tried on a few plants. To control mealy bugs and species having soft scales the concentration of diazinon should be 0·02%, whereas to control species having hard, protective coverings 0·04% is recommended.

Parathion at 0·01% is also effective against most species, but where the host plant is tolerant to diazinon, this is preferable owing to its greater safety to operatives. *Coccus hesperidum* is reported as being tolerant to parathion but susceptible to malathion, one application of which at 0·1% is often sufficient. Although, at this concentration, malathion will give adequate control of many other species after several applications, it is generally inferior to diazinon and parathion. For species with hard scales a concentration of 0·15–0·2% malathion may be required to obtain rapid control. Recent work on the control of scales on outdoor, shrubby plants has shown that admixture of a petroleum oil to diazinon or malathion is advantageous, but so far no information is available regarding the value of such mixtures on glasshouse plants. When several applications were given, a mixture of nicotine (0·05%) and a glasshouse-grade petroleum emulsion at 1% gave good control of scales on tolerant plants, including vines, and orchids (other than cattleyas which are susceptible to damage), but this has been largely superseded by previously mentioned insecticides. Carbaryl at 0·1% is recommended in America for use against scales and mealy bugs on woody subjects such as camellia; sprays of sulfotep and trithion are also said to be effective.

It is advisable to prevent the retention of considerable amounts of spray within the hollow crowns or leaf crotches of certain ornamental plants (see p. 35).

Peaches, nectarines, apricots, roses and vines can be sprayed with a tar oil winter wash at 4% when fully dormant. To remove fibrous bark beneath which mealy bugs may be protected from the spray the rods of grape vines should be scraped down before the tar oil application. To avoid marring the appearance of grapes by removing the 'bloom' it is necessary to control mealy bugs and scales after 'thinning' before the berries commence swelling. Grapes not treated when dormant should therefore be sprayed as soon as the shoots have started to extend, and a further one or two applications given at 14-day intervals.

As some ferns are susceptible to organophosphorus insecticides it is advisable to make a preliminary trial with diazinon, malathion or parathion on one or two plants of each species or variety, or to use a nicotine spray (0·05%). Maidenhair fern is particularly susceptible to injury by malathion and diazinon.

Coccids breed continuously in heated glasshouses, consequently the basis of their successful control is several applications of insecticide at intervals of 14 days where temperatures exceed 60 °F (15 °C), and 21 days at lower temperatures.

Coccids can be eradicated from small numbers of plants by periodically removing adult scales, or colonies of mealy bugs and their egg sacs, with a wooden, or bone, scraper and then wetting infested areas with a soft brush either dipped in a petroleum-nicotine mixture, as used for spraying, or in a nicotine solution (0·05%) containing a wetting agent. A malathion emulsifiable concentrate diluted to contain 0·1% malathion should also be suitable, but the odour can be objectionable. Gloves should be worn to avoid contaminating the hands. In an *ad hoc* test, an unidentified mealy bug at the roots of cacti was controlled by watering with 0·01% parathion. Suggested alternatives to parathion for root-infesting species are malathion at 0·05%, dimethoate 0·04%, nicotine at 0·05% or a lindane wettable powder at 0·01% gamma-BHC.

REFERENCES (General to Coccoidea)

BALACHOWSKY, A. S. (1948–50). Les Cochenilles de France, du Nord de l'Afrique et du bassin meditérranéen. *Actual. scient. ind.*, **1054**, **1087**, **1127**, **1202**, 929 pp.

DE CLERCQ, R. & HEUGENS, A. (1964). Voorkomen van *Rhizoecus cacticans* Hambleton in de Sierplantenteelt. *Meded. LandbHoogesch. Opzoekstns Gent*, **29**, 1097–109.

DOUTT, L. (1952). Biological Control of *Planococcus citri* on commercial greenhouse *Stephanotis*. *J. econ. Ent.*, **45**, 343–4.

GRIFFIN, D. E., LUBATTI, O. F. & HARRISON, I. R. (1956). Effect of methyl bromide fumigation on orchids. *Pl. Path.*, **5**, 53–6.

MANICHOTE, P. & MIDDLEKAUFF, W. W. (1967). Life-history of the Cactus Mealybug (*Spilococcus cactearum* McK.). *Hilgardia*, **37**, 639–59.

MORRISON, H. & RENK, A. V. (1957). A selected bibliography of the *Coccoidea*. *Misc. Publs. U.S. Dep. Agric.*, **No. 734**, 222 pp.

MORRISON, H. & MORRISON, E. R. (1965). First supplement to above. *Ibid.*, **No. 987**, 44 pp.

PRITCHARD, E. A. (1959). Orchid pests and their control. *The Orchids: a scientific survey*. Ed. by C. L. Withner. Ronald Press, New York. pp. 459–75.

SAYED, M. T. (1962). Further contributions to the biology of *Planococcus vitis* Nied. and *Planococcus citri* Rosso with special reference to their number of generations. *Bull. Soc. ent. Égypte*, **46**, 459–65.

SAYED, M. T., SOLIMAN, A. A. & SALAMA, H. S. (1962). On the developmental stages of *Planococcus vitis* Nied & *P. citri* Rosso (*Homoptera: Coccoidea–Pseudococcidae*). *Bull. Soc. ent. Égypte*, **46**, 449–57.

SCHMUTTERER, H. (1959). Schildläuse oder *Coccoidea* I. Deckelschildlause oder *Diaspididae*. *Die Tierwelt Deutschlands*, **45**, 1–260.

SNETSINGER, R. (1966). Biology and control of a root-feeding mealybug on Saintpaulia. *J. econ. Ent.*, **59**, 1277–8.

WILLIAMS, D. J. (1962). The British Pseudococcidae (Hom.: Coccoidea). *Bull. Br. Mus. nat. Hist.*, **12**, 1–79.

Family APHIDIDAE: Greenflies

Aphids possess, on the sixth abdominal segment, a pair of tubes called *cornicles* or *siphunculi*, between which the abdomen terminates in a posterior projection known as the *cauda*. Colonies of these insects are commonly seen on the stems and lower leaf surfaces of infested plants. Their ability to transmit viruses from plant to plant may be more injurious than any direct damage caused by their feeding.

Many species have complex life cycles with an annual alternation between woody, winter hosts and herbaceous summer hosts. On the latter, reproduction is parthenogenetic (without fertilization by males) and viviparous (production of living young). When the aphids become overcrowded, winged forms (alatae) may be produced by the parthenogenetic females and disperse to more suitable feeding sites. In late summer, in response to shorter days and lower temperatures, most species produce females called *sexuparae* which migrate to the winter host and produce both males and females (*sexuales*). The latter, after mating, lay eggs in October. These overwinter and hatch in the spring to initiate successive generations of *fundatrigeniae* whose alate forms (*migrantes*)

FIG. 5.23. Nymphs, wingless females and alatae of *Macrosiphum rosae* L. (×5). (J. Hesling.)

fly back to the secondary herbaceous hosts. Glasshouse species usually reproduce continuously on herbaceous hosts throughout the year; there is normally no sexual phase, the aphids being either adult, or nymphal, viviparous females (Fig. 5.23).

The adult, parthenogenetic females contain embryos at all stages of development, the most mature of which already contain embryos themselves. Thus, an adult female is nurturing both daughter and granddaughter embryos. During the first 2 weeks of reproductive life, up to 5 or 6 daughter larvae are produced daily, but, during the following week, only about 2 larvae are born. Reproduction then ceases and is followed by a post-reproductive period of almost 3 weeks during which feeding continues slowly. At this time excretion virtually ceases and the crop becomes distended with insoluble polysaccharides. The main solid matter component in plant sap is sucrose but most of that imbibed is excreted as it is not needed for the synthesis of the proteins required for growth and reproduction. The fecundity of apterous aphids depends on the *quality* of the nutrients ingested rather than on the *quantity*. The conditions under which host plants are cultivated therefore profoundly affects aphid reproductive rate. Aphids were originally believed to suck the sap of plants but it is now known that, once the stylets have penetrated to sieve-tube cells within the phloem, sap is forced through the gut of the insect by the turgor pressure within the plant. The rate at which aphids feed rises rapidly to a maximum and then declines with age. These changes are reflected in the excretion rate of honeydew droplets. Banks (1965) reported that, three days after maturation, adult *Aphis fabae* excrete 30–40 droplets per day, but in older individuals the rate slowly declines.

The multiplication rate of aphids, in the absence of natural enemies and unsuitable weather, is enormous. Each female aphid produces at least 50 daughters during her life-span and, under average glasshouse conditions, each matures in about a week. The influence of temperature is, however, great. Imms (1947) calculated that, at 71 °F (22 °C), 1300 individuals could arise from a single female in 14 days but, if the temperature was 58 °F (14 °C), there would be less than 500 progeny. Assuming unhampered reproduction and survival, the progeny of a single aphid could amount to 12×10^{18} individuals in 4 months.

Flight of alatae is usually confined to the first few days of adult life during which embryo growth is arrested by the inhibition of feeding. Outdoors, in the appropriate environmental conditions, alatae take flight and may then be carried upwards by vertical air movements. While aphid density in the upper air can most conveniently be expressed in terms of 'particle diffusion', the actual number of aphids flying is governed by the number in the correct physiological condition for flight. This condition is influenced by the population density and a diurnal rhythm in nymphal maturation. Evidence from trapping suggests that most alates are only airborne for a few hours.

When air circulation movements permit, aphids return to surface levels and fly close to the ground, often resisting gentle air currents before settling. The longer the flight has been the more intensively the aphid probes its host plant. Feeding begins, embryo development is resumed, and young larvae are produced. If the host is unsuitable these reactions are reversible but, once the aphid is permanently settled, autolysis of the wing muscles precludes further flight.

Some aphids have a special relationship with ants. Most species are merely visited by ants which seem to stimulate their multiplication rate. The association ensures that, while the ants obtain honeydew, the aphids are protected from enemies.

As mentioned at the outset, aphids are of economic importance as vectors of plant viruses. Viruses are transported from infected to healthy hosts in two ways. The most common involves carriage inside the labial groove of the stylets and is referred to as 'stylet-borne'. When the virus is ingested and passes via the haemolymph to the salivary glands, transport is known as *circulative.* It is also possible to classify aphid-borne viruses in terms of the time the virus is retained by the insect. Transmission efficiency within one group, the *non-persistent* viruses (i.e. those which can only be transmitted within a few hours of acquisition) is affected by several factors. These include the duration of the feed (the optimal time being 30 sec), and the degree of hunger experienced by the insect both before, and after, its acquisition feed. These limitations do not apply to *persistent* viruses (which multiply within the insect host before being transmitted) but the virus/vector relationships within this group are more specific and there is commonly a delay, or *latent period,* before the aphid itself becomes infective after feeding on an infected host.

Myzus persicae Sulzer (the peach-potato aphid). This extremely common and polyphagous species is undoubtedly the most important glasshouse aphid. It is variable in colour, ranging from different shades of pink to yellow, or grey-green, and is usually about 2 mm long. The cornicles are long and thin, though slightly swollen in the middle; the extreme tip is characteristically darker than the rest and bears a distinct flange. The cauda is triangular, rather less than half the length of the cornicles, somewhat pale in the apterae (wingless forms) but darker in alatae. The ground colour of the latter is similar to that of apterae but there are characteristic dark markings. All forms possess, between the antennae, well-developed, convergent and roughened lateral frontal prominences, typical of the genus *Myzus.*

Although, under outdoor conditions, *M. persicae* overwinters as eggs on peach, nectarine and some ornamental species of *Prunus,* reproduction under glass is normally continuously parthenogenetic on herbaceous hosts. A vast range of hosts is affected including cactus, cineraria, *Hibiscus* sp., hyacinth, lettuce, orchid, rose, tomato, tulip and violet, but *M. persicae* is of greatest importance on carnations and chrysanthemums. Its reproduction is rapid, the optimum temperature being 79 °F (26 °C); above 82 °F (28 °C) conditions become detrimental. The development period from new-born nymph to adult is 7 days at 75 °F (24 °C), increasing to 43 days at 43 °F (6 °C).

Similarly the interval between the last moult and the birth of the first nymph increases from less than 1 day at 77 °F (25 °C) to 4 or 5 days at 50 °F (10 °C). The fecundity is variable but individual females normally produce between 30 and 50 nymphs.

M. persicae, unlike many other aphids, does not usually form dense colonies on the stems of its host plants; it is a restless species whose populations are in a continuous state of movement. The feeding of large numbers of aphids on infested leaves, buds and flowers causes distortion and occasionally, through injury to the terminal shoot, the growth may be stunted. On carnations pale areas often tend to develop around the individual feeding punctures, giving leaves a mottled appearance. A similar condition has occasionally been seen on tomato fruits. On commercial flower crops this species assumes great importance because even individual insects on the open blooms tend to lower the market value. Studies of aphid behaviour on different varieties of year-round chrysanthemums have shown that, in general, more insects occur in the basal and apical regions of the plant than on the intermediate leaves. Once the flower buds are initiated all the aphids tend to migrate upwards and, if uncontrolled, will cause distorted blooms. Among the more important viruses known to be transmitted by *M. persicae*, and which affect horticultural crops, are 'carnation latent'; 'carnation mosaic'; 'chrysanthemum Good News mosaic'; 'chrysanthemum virus B'; 'dahlia mosaic'; 'lettuce mosaic'; 'orchid mosaic'; 'tomato aspermy'; 'tomato yellow net b'; 'tomato yellow top', and 'tulip-breaking'.

Largely due to the development of insecticide resistance, commercial control of *M. persicae* under glass has become increasingly difficult in recent years, especially in dense beds of carnations and chrysanthemums which are difficult to spray effectively. Resistance to organophosphorus compounds has become quite widespread and even isolated instances of resistance to DDT, BHC and nicotine have also been demonstrated. Undoubtedly a contributory factor to the resistance problem on chrysanthemums and carnations is the fact that most cuttings are supplied by a few large firms. These adopt intensive and efficient pest control programmes to maintain clean plant produce but, when resistant survivors occur, these are subsequently dispersed throughout the country. Development of resistance is, at first sight, difficult to explain with entirely parthenogenetic reproduction, where each population constitutes a 'clone' (namely, the progeny have the same genetic constitution as the parents), but it has recently been shown that a unique type of division, called endomeiosis, occurs within the egg cell of aphids which permits some genetic recombination within clones. The pattern of resistance to different insecticides exhibited by a single strain can be quite complex and often poses serious control problems. There are commonly two colour forms of *M. persicae* on chrysanthemums: a dark-green form which seems susceptible to all insecticides, and a pale-green form which may possess these complex patterns of resistance.

Although the flow of new insecticides has so far kept pace with the resistance potential of the aphid, there is a growing interest in the possibility of reducing the number of insecticide applications by utilizing some measure of biological control. One of the most successful parasites of *M. persicae* is a braconid wasp *Aphidius matricariae*. The adult parasites lay eggs within the bodies of young aphid nymphs and, when fully developed, the parasite larvae completely devour the body contents of the host. These larvae pupate within the aphid's body which assumes a globular shape with a characteristic parchment-like skin. The newly mature parasite emerges through a circular hole cut in the dorsal wall of the aphid abdomen. *A. matricariae* has completely eliminated a heavy aphid population on 3-week-old chrysanthemum cuttings and produced a crop of perfectly clean flowers without use of insecticides. It is possible that successful *integrated* control programmes will be developed utilizing this, or other, parasites, together with certain predatory insects such as ladybird (Coccinellid) larvae, hover fly (Syrphid) larvae, bugs or lacewing larvae. The chemical control of aphids is similar for all species and is dealt with separately on p. 115.

Aphis gossypii Glover (melon and cotton aphid) is a well-known glasshouse aphid which can be troublesome on *Cucurbitaceae* especially when the intensity of the spray programme is drastically reduced, e.g. as when red spider mite is being controlled biologically (p. 196).

A. gossypii varies greatly in colour, ranging from yellow to almost black, although specimens on cucumbers are normally dark green. The antennae are at least half the length of the body; the frontal lobes are almost absent so that the outline of the front of the head, between the antennae, is gently sinuate. The cornicles taper slightly from the base and have a definite, yet inconspicuous, apical flange. The dark green or black cauda is narrow with three hairs on each side.

A. gossypii is a vector of cucumber mosaic and many other viruses. On cucumbers, however, its economic damage is normally restricted to distortion of the infested shoots and to the unsightly, growth-restricting, development of sooty moulds in the warm, damp environment. Yields are reduced where numbers exceed 7/cm^2 of leaf-surface. Occasionally, however, complete defoliation may occur.

Aulacorthum solani Kalt. (the glasshouse-potato aphid), while normally associated with sprouting potato tubers, is found on many glasshouse plants including *Anemone coronaria*, arums, calceolaria, carnation, cineraria, dahlia, gloxinia, lettuce, nasturtium and *Pelargonium peltatum*. It is a vector of chrysanthemum 'virus B' and 'Good News mosaic' virus as well as tomato aspermy virus.

Slightly larger than *Myzus persicae*, *A. solani* (2–2·5 mm long) is a pale- to dull-green species with a darker head and thorax; the apterae are unmarked but the abdomens of the alatae are distinctly banded. The area immediately at the base of the cornicles of apterae and alatae is a deeper green than the remainder of the body. The frontal prominences are outstanding but nonconvergent. The cornicles have almost black and widely-flanged apices. Unlike most other aphids *A. solani* can overwinter as eggs on a number of widely different hosts.

Aulacorthum circumflexum Buckton (the mottled arum aphid) is widespread in all parts of Britain. The apterae (1·5–1·8 mm) are shining green with a characteristic horseshoe-shaped, blackish mark on the dorsum of the abdomen; this mark may be obscure or entirely absent in the younger nymphs (Fig. 5.24). Under glass, reproduction is entirely parthenogenetic. *A. circumflexum* commonly infests both flowers and foliage of arums and cyclamen. It has also been recorded from amaryllis, anthurium, *Asparagus plumosus*, azalea, begonia, calceolaria, chrysanthemum, cineraria, cyclamen, freesia, *Iris tingitana*, lettuce, *Lilium sp.*, *Oncidium* sp., rose, *Schizanthus* sp., and tulip. This species is an important vector of virus diseases including 'dahlia mosaic', 'primula mosaic' and 'tulip-break.'

Macrosiphum rosae L. (the rose aphid). All the members of this genus typically use *Rosaceae* as primary hosts and migrate to various herbaceous hosts in the families *Dipsaceae*, *Umbelliferae* and *Valerianaceae*. *M. rosae* is a large species, often almost 3 mm long, commonly found on young roses under glass during the first year after the plants are budded. Individuals may range from pink to green; they have prominent frontal tubercles but may be positively identified by the long, slender and entirely black cornicles (Fig. 5.23). Under glass reproduction may be viviparous throughout the year but, outdoors, small reddish females, which lay eggs on the young shoots, are produced in the autumn.

Macrosiphum euphorbiae Thomas (=*solanifolii* (Ashm.)) (the potato aphid) migrates from Rosaceae to a wide range of herbaceous plants including potatoes and, under glass, to abutilon, carnation, cineraria, *Cyphomandra* sp., lettuce, tomato and tulip. It is a large (4·0 mm) green aphid with brownish lobes, often

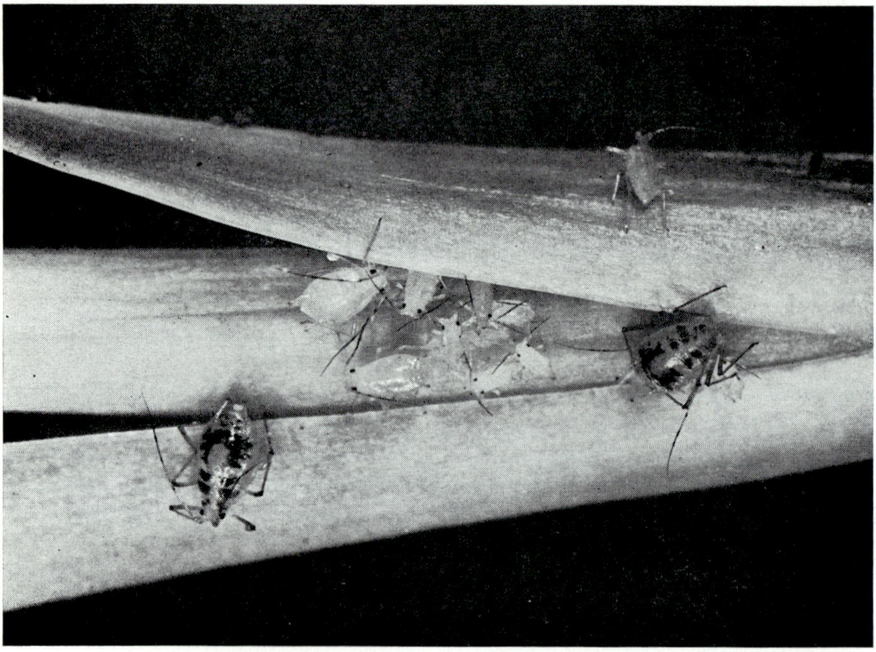

FIG. 5.24. *Aulacorthum circumflexum* Buckton: mottled arum aphid (×10). (A Shell photograph.)

covered with a fine white pubescence, on the thoracic disc. The antennae are longer than the body. The pink cauda is half the length of the cornicles which are long, yellowish, slender and apically reticulate. There are sometimes two colour forms—a green form on carnations which, after prolonged breeding on that host, becomes short-lived and sterile, and a red form on lettuce and *Hyoscyamus niger*. Aphids can transfer from carnation to other hosts without difficulty but establishment on carnation itself is poor. *M. euphorbiae* is a particularly active, restless species and is the vector of many virus diseases including 'anemone mosaic', 'chrysanthemum virus B', 'cucumber mosaic', 'freesia mosaic', 'iris mosaic', 'narcissus mosaic', 'tomato aspermy' and 'tulip-break'.

Macrosiphoniella oblonga (Mordvilko) occurs outdoors on species of *Anthemidae* but is frequently introduced into glasshouses on autumn-flowering chrysanthemums. It is an extremely large green aphid (5 mm long) with dark dorsal marks. The long, green cornicles are reticulate for more than one third of their length but have no apical flange. The blunt cauda is almost as long as the cornicles. Reddish-brown, active, apterous males are commonly found on chrysanthemums in the autumn. They have relatively long legs and unusually small bodies and, if even slightly disturbed, fall to the ground.

Coloradoa rufomaculata (Wilson) is another predominantly green aphid with very long cornicles, exceeding even the antennae in length, which was introduced into Britain from America early in the 1930s. It breeds on the new shoots growing from established chrysanthemum stools and on yellowing foliage low down on the plants.

Nasonovia ribisnigri (Mosley) (the lettuce aphid) (2–2·5 mm long) is light green in colour with a pattern of dark spots on the abdomen. It is distinguished from *Aulacorthum solani* by the W-shaped mark on the forehead between the antennae. The cornicles are long and yellow with blackish apices and are devoid of reticulation. The blunt, finger-shaped cauda is also yellow. *N. ribisnigri* swarms to lettuce in early summer from the overwintering hosts, black and red currant. Reproduction continues in the hearts of lettuce and may cause stunted growth and even prevent 'hearting'. Crop weight is almost inevitably decreased and the appearance of the leaves spoilt. Outdoors, the aphids return to currants between September and November. However, individual apterae often continue to breed on lettuce throughout the winter, and it is at this time that most damage occurs under glass.

Macrosiphoniella sanborni (Gillette) (the chrysanthemum aphid) is very common and widespread on outdoor chrysanthemums and is commonly introduced into glasshouses when summer-grown chrysanthemums are brought in for flowering in the autumn.

M. sanborni is a small (2–2·5 mm), dark, shining, mahogany-brown aphid in which the cauda is considerably longer than the cornicles. The latter are stout and black, enlarged at the base and with a coarsely reticulated pattern on the apical two-thirds of their length. In late summer and early autumn, stems harbour dense colonies which may distort the developing buds and flowers. The chrysanthemum aphid is a confirmed vector of two important viruses, 'vein mottle', and 'virus B'.

Brachycaudus helichrysi (Kaltenbach) (the leaf-curling plum aphid) in common with other species in this genus migrates from a primary host in the *Prunoidea* (*Prunus* sp.) (on which the black eggs overwinter and on which the first generations of aphids cause the leaves to curl characteristically) to secondary Composite hosts (including *Achillea* sp., chrysanthemum, and *Senecio* sp.). It is a small (1·3–2·0 mm) pale-green to orange aphid. Occasionally the dorsum of the abdomen may be pale brown but it is never pigmented with black. The cauda is short, hardly longer than broad, and semicircular with a slight basal constriction. The pale cornicles are short and distinctly flanged.

B. helichrysi attacks both shoots and flowers of chrysanthemums leaving dense deposits of crystalline honeydew.

Dactynotus tanaceti L. is another species occasionally found on chrysanthemums under glass. The dark-brown to wine-red females breed on the undersides of the leaves. The cornicles are long and black. All the hairs on the body and head are knobbed.

Appelia schwartzi Börner (=*amygdali* Buckton). The wingless females of the peach or almond aphid, which are about 2 mm long, vary from yellow-green to brownish-red, though some individuals may be almost black. They are distinctly globular, shiny, aphids with dark heads and dark transverse bands on both head and thorax. The antennae and the cornicles are short with the cauda even shorter than the latter. Overwintering eggs are laid on peach and nectarine, and the nymphs, hatching in April and May, cause curling and distortion of the leaves. Both old and new leaves are attacked and in severe cases may fall from the plants. In late June and early July alatae are produced which fly off to the secondary hosts (*Carduus* spp.). In autumn sexuparae return to peach and produce sexual, oviparous, forms. Under glass, wingless generations are often produced on peach throughout the summer.

Dysaphis tulipae (Boyer de Fonseclombe) (the tulip-bulb aphid). Masses of this wingless, greyish-to-puce-coloured aphid may be found below the dry outer skins of tulip bulbs and gladiolus or iris corms. These aphids have dark heads and a characteristically banded abdomen with scattered rows of darker spots. The cornicles and cauda are dark, the latter being extremely short. Winged forms produced under glass are yellowish-white, sometimes pinkish, with a conspicuous dark area on the dorsum. When the bulbs or corms are set out for forcing under glass the aphids increase rapidly and spread upwards from the bulbs to the leaves and flowering shoots. Infested bulbs often produce stunted and distorted growth. It is a confirmed vector of both 'lily-symptomless' and 'tulip-break' viruses.

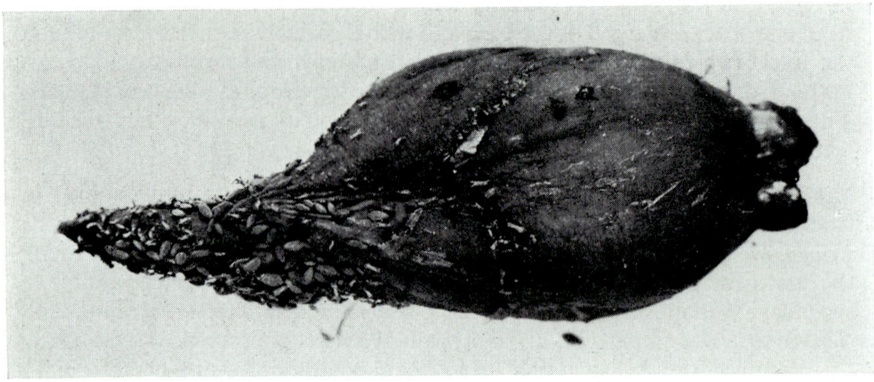

FIG. 5.25. *Rhopalosiphoninus latysiphon* on tulip bulb. (Crown Copyright by permission of Controller of HMSO.)

Two other species, *Rhopalosiphoninus staphyleae* (Koch) (=*tulipaellus* Theobald), which is olive-brown with a yellowish-green head and dark transverse bands centrally on thorax and abdomen, and *R. latysiphon* (Davidson), which is uniformly dark pigmented, may also contaminate stored tulip bulbs (Fig. 5.25). These species, both with swollen cornicles, may be differentiated by the head and first two antennal segments which are strongly spinulose in *R. staphyleae*. During bulb forcing they multiply rapidly and may attack the leaves which turn brown and often shrivel at the tips. Both species complete their life cycles below ground and are unusual in being unattended by ants.

Idiopterus nephrelepidis Davis (the fern aphid), deep-green to almost black in colour, is a destructive pest of many greenhouse ferns including species of *Nephrolepis*, *Polypodium* and *Pteris*, but has also been recorded on cyclamen and *Streptocarpus* sp. The long antennae are almost white, the legs white with black tips, and cornicles white but black basally. The wings of the alatae are mottled with dark patches. Infested fern fronds curl and turn black.

Cerataphis orchidearum (Westw.) (the orchid aphid) strongly resembles a scale insect since the apterae, which are reddish-brown, scale-like creatures with a fringe of short waxen strands, remain practically stationary throughout life. The short legs and antennae remain hidden below the 'scale'. The short cornicles are very broad and the cauda knob-shaped.

C. orchidearum attacks palms and orchids. The leaves are commonly affected and dense colonies also form between the sheaths and stems of orchids. Species of *Cattleya*, *Cymbidium*, *Cypripedium*, *Dendrobium* and *Odontoglossum* are particularly susceptible.

Many other aphid species may be found from time to time under glass but their identification is a matter for the specialist. However, for the purpose of control, it is normally unnecessary to establish the species involved.

Control of aphids

Although, with few exceptions, the aphids which attack glasshouse plants are common outdoors and so readily enter glasshouses, good cultural and hygienic practices are important in preventing attacks arising from infestations on weeds or neglected plants within the houses. Weeds should be removed and, before bringing clean plants into a glasshouse, measures taken to eliminate aphids on any plants to be retained.

Plants which are to be taken into glasshouses and placed with other, clean plants, should be carefully examined and, when necessary, the aphids killed. Attacks on young plants and seedlings, e.g. lettuces (p. 120), can be most expeditiously and economically controlled before planting. The roots of pot-plants such as primulas should be examined for root-infesting species which, if present, will be visible on roots in contact with the pot. Bulbs of tulip, iris and lily, and gladiolus corms, should be examined for tulip aphids which, unless killed before planting (p. 120), will cause considerable damage.

It is important to control aphids on flowering crops before the buds open, for some species may invade the blossoms, marking or distorting the petals and rendering them susceptible to flower diseases. Furthermore, it is difficult to control aphids within the flowers where they are protected from contact with sprays or aerosol droplets. Systemic compounds may not be translocated to flowers in lethal amounts. The susceptibility of flowers to chemical injury further restricts the choice of aphicides.

Apart from *Aulacorthum circumflexum*, which is naturally resistant to BHC, and certain strains of *Myzus persicae* commonly occurring in chrysanthemum houses, aphids are susceptible to gamma-BHC and to most organophosphorus compounds, nicotine and pyrethrum. The latter, although rarely used commercially, is suitable for amateurs' glasshouses.

Except for resistant strains of *M. persicae*, the choice of the most suitable aphicide depends on safety to the consumers of edible produce, to workers and to the plants, together with factors such as cost and convenience of application. The concurrent infestation of plants by other pests may also need to be taken into account. The special problems of controlling aphids on year-round chrysanthemums and on a number of other hosts are described later.

Systemic aphicides

Systemic organophosphorus compounds usually give the most effective control of aphids on those ornamental crops to which they can be applied without causing damage. They are particularly valuable where the aphids infest the tips of shoots, e.g. carnations, or are otherwise protected from contact insecticides. Their use on edible crops is restricted by the necessity of ensuring the absence of potentially harmful residues, but, except with lettuces, aphids on edible crops are sufficiently exposed for control to be obtained with sprays, aerosols and smokes of less toxic or less persistent, non-systemic compounds.

Considerations of safety to workers and to host plants, together with lack of information regarding the aphicidal efficiency and phytotoxicity of some of the more recently introduced compounds have limited the number of systemic compounds used in glasshouses. The following are effective against aphids when applied as H.V. sprays—demeton-S-methyl (0·02%), dimethoate (0·03%), formothion (0·04%), morphothion (0·03%), oxydemeton-methyl (0·02%), phosphamidon (0·02%) and schradan (0·3%). Of these, demeton-S-methyl, oxydemeton-methyl and schradan appear to be the least phytotoxic.

In spite of the considerably greater cost, it may sometimes be advantageous to apply demeton-S-methyl (0·012%), dimethoate (0·03%), formothion (0·04%), or oxydemeton-methyl (0·012%) as soil drenches, using the diluent at 3 fl. oz/6 in. pot and *pro rata* for other size pots. When applied as soil drenches, systemic compounds usually give more lasting protection than when applied as sprays but, under some conditions, e.g. with some very slow growing or mature plants, they may not be translocated in sufficient amounts to kill aphids.

Complete control is usually achieved by a single application of any of these systemic compounds, though a second application may be necessary after an interval of 4 to 6 weeks. Some ornamental plants are susceptible to injury by systemic organophosphorus insecticides (p. 382).

Non-systemic aphicidal sprays

High volume sprays of some non-systemic organophosphorus compounds may also be used to control aphids. On account of their safety to workers and plants, malathion and diazinon are most widely used. The former, at 0·1%, will also control whitefly, thrips and leafhoppers; the latter at 0·02% will, in addition to these pests, also control some strains of red spider mite and leaf-miners. Diazinon is, however, more persistent than malathion and, on edible crops, a minimum of 2 weeks must elapse between its application in H.V. sprays and harvesting, whereas only 1 clear day is necessary with malathion—although to ensure the produce has no 'off' flavour, it is advisable to allow 3 or 4 days. Sprays containing 0·02% phosphamidon have been used successfully on carnations. Good control of aphids may also be obtained by H.V. sprays of gamma-BHC (0·01%) and dichlorvos (0·1%) or, provided the temperature exceeds 65 °F (19 °C), nicotine (0·06%), either of which may enable aphids to be dealt with on plants susceptible to damage by organophosphorus compounds. Nicotine and gamma-BHC may be effective against strains of *Myzus persicae* which are resistant to organophosphorus compounds.

Aerosols and smokes

The high labour cost of H.V. spraying may sometimes be avoided by applying aphicides as aerosols or smokes. Their effectiveness depends on the amount of protection afforded to the aphids by dense foliage, etc., and probably upon the extent of the insects' movements over the foliage. On many hosts gamma-BHC aerosols (0·7 g/1000 ft^3) are both effective (except against *Aulacorthum circumflexum*) and inexpensive. Aerosols of malathion (2·4 g/1000 ft^3), diazinon (1·8 g/1000 ft^3) and dichlorvos (1·0 g/1000 ft^3) are also effective except against resistant strains. Combinations of gamma-BHC or malathion with DDT are available for aerosol production. Where aerosol-producing equipment is not available, smokes of gamma-BHC (0·7 g/1000 ft^3), nicotine from 'shreds', etc. (4 g/1000 ft^3), parathion (0·25 g/1000 ft^3), and sulfotep (1·0 g/1000 ft^3) as well as the vapour of naled (Ortho-Dibrom), generated by volatilization from the heating pipes (pipe temperature exceeding 120 °F) at the rate of 1 fl. oz/10,000 ft^3, provide a convenient method of control. It is important to ensure that the flowers and foliage of tender plants remain dry during fumigation with nicotine or naled. At least 65 °F (18 °C) is required to obtain the best results with nicotine and, as a rule, heating of the houses to maintain this temperature overnight will prevent condensation, as will the heat evolved from piping sufficiently hot to vaporize naled.

Smokes combining gamma-BHC and DDT are available and these, like malathion with DDT aerosols, are useful where control of caterpillars is also required. It may be necessary to repeat applications of aerosols or smokes several times at intervals of 3–4 weeks to eradicate aphids.

More detailed recommendations follow for those crops where special difficulties necessitate specific treatments.

Aphids on chrysanthemums

The control of *Myzus persicae* on direct-planted chrysanthemums is difficult owing to the widespread occurrence of strains having varying patterns of resistance to insecticides, and the difficulty of obtaining effective coverage of the dense foliage in the closely-planted crop. Furthermore, there are marked differences in the tolerances of the various chrysanthemum cultivars to pesticides. Apart from strains of *M. persicae* which may be resistant to many aphicides, and *Aulacorthum circumflexum* which is not susceptible to gamma-BHC, the aphids which occur on chrysanthemums are killed by the range of aphicides already mentioned. Systemic compounds have definite advantages however, since they largely overcome the difficulty of obtaining adequate cover with H.V. sprays as well as the inability of aerosol droplets and, to a lesser extent, of gamma-BHC smoke particles to penetrate the dense foliage of closely planted beds.

Unfortunately many systemic organophosphorus insecticides are highly phytotoxic to chrysanthemums. Demeton-S-methyl, oxydemeton-methyl, morphothion, schradan and thionazin may be used with safety on most varieties. Only the first two, however, are effective against many strains of *M. persicae*. These two insecticides may also control some strains of red spider mite. The high costs of thionazin, which is most effective when applied as a soil drench, makes its use as an aphicide uneconomic but its use as a nematicide (p. 281) will kill aphids, apart from some strains of *M. persicae*, on rooted cuttings.

The main problem with year-round and direct-planted chrysanthemums is controlling *M. persicae*. For this purpose the use of sprays of demeton-S-methyl (0·02%) or oxydemeton-methyl (0·02%) are recommended. The first application may be required about 3 weeks after planting, by which time the plants are established and so less liable to be damaged by the chemicals. A second application may be necessary after a few weeks. It is most important to effect control before the buds show colour and it may be necessary to make a further application at this stage. There is some evidence that demeton-S-methyl is the more effective systemic compound during this phase of flower development, whereas oxydemeton-methyl is more effective earlier.

Demeton-S-methyl and oxydemeton-methyl may also be applied at 0·013% in soil drenches at the rate of $1\frac{1}{2}$–2 gal/yd^2. Although this method is less liable to cause damage than H.V. spraying, it uses more chemical without giving commensurate advantage from a longer period of protection from aphid reinfestation. The use of drenches is, therefore, seldom justified except on certain cultivars which are damaged by sprays. Demeton-S-methyl has shown some superiority over oxydemeton-methyl.

Some strains of *M. persicae* remain highly susceptible to nicotine and as the vapour of this compound, produced by smoking with nicotine shreds or other impregnated materials, is better able to penetrate dense foliage than particulate smokes, good results are obtained by the equivalent of 4 g nicotine/1000 ft^3.

Aphids, other than many strains of *M. persicae* and *Aulacorthum circumflexum* which are not susceptible to gamma-BHC, can also be controlled with morphothion sprays (0·025%) or, on widely spaced plants, by the non-systemic aphicides mentioned earlier (p. 116). Sprays of gamma-BHC (0·01%) and diazinon (0·02%) should be used where leafminer is also a pest. In the absence of leafminer, aerosols or smokes of BHC, malathion, diazinon, dichlorvos and nicotine or naled vapour may be more convenient. Several applications, at

10- to 14-day intervals, may, however, be required to give a degree of control which will ensure that invasion of the flowers does not occur.

Because aphids tend to migrate to the heads of plants as the buds develop, they become more easily controlled by sprays, aerosols and smokes of non-systemic compounds even where chrysanthemums are closely planted. The plants should be carefully examined at this stage so that final control measures can be taken before the flowers open. Where neglect results in the flowers becoming infested the best procedure is to fumigate with nicotine, or gamma-BHC smokes, taking care that the petals remain dry throughout the fumigation. Fumigation with naled may prove equally effective.

The introduction of new chrysanthemum cultivars, together with the large number already grown, makes it impossible to give a comprehensive list of those on which demeton-S-methyl, oxydemeton-methyl and naled can be safely used. Advice should be sought from the National Agricultural Advisory Service, the pesticide manufacturer or the cutting supplier where there is doubt concerning the tolerance of a cultivar to these, or other products.

Recent trials (Worthing, 1967) suggest that some experimental systemic compounds, in particular those at present designated aldicarb and methomyl are very effective against *M. persicae* and also against other pests including red spider mite and leafminer. As they are carbamate insecticides they may be of value against aphids which become resistant to systemic organophosphorus compounds.

Aphids on carnations

Myzus persicae infests the tips of carnation shoots where they may be protected from contact with H.V. sprays, aerosols and smokes within the unexpanded leaves. Consequently their eradication with non-systemic insecticides is rarely achieved except by repeated applications. Although shortage of water causes the leaves at the tips of the shoots to open somewhat and expose the aphids, the procedure is undesirable both culturally and also because it may make the plants more susceptible to chemical injury.

Control is best achieved by H.V. sprays of dimethoate (0·03%), formothion (0·04%), oxydemeton-methyl (0·02%) or phosphamidon (0·02%). To promote wetting of the waxy foliage additional wetting agent is required with proprietary formulations of these insecticides. Demeton-S-methyl sprays are also effective though they impart an objectionable odour to flowers cut within four days. Demeton-S-methyl (0·013%) and dimethoate (0·03%) may also be applied as soil drenches, at the rate of 2 gal/yd^2, to first-year plants and at 3 gal/yd^2 to older plants. They are most effective when plant growth is vigorous and so may fail to give complete control, or only act slowly, if applied to flowering plants in the winter.

Control can be obtained by repeated applications of all the non-systemic aphicides mentioned earlier (p. 116), except where resistant strains occur. Gamma-BHC aerosols and sprays should not be used on the 'Ashington' cultivars during the late autumn and winter or scorching of the leaf tips may occur. The soft apical leaves of young plants may suffer similar injury from nicotine sprays or smokes, particularly if the soil is dry. Marketable blooms should always be cut before applying insecticides.

Aphids on cucumbers

Young cucumbers are susceptible to injury by many insecticides. Seedlings or young plants infested with aphids should be sprayed with parathion (0·01%), demeton-S-methyl (0·02%), dimethoate (0·03%), oxydemeton-methyl (0·02%) or schradan (0·1%). Only one application of schradan is permitted, and it is essential that the following intervals elapse between the application and cutting cucumbers: parathion, 4 weeks; demeton-S-methyl and oxydemeton-methyl, 3 weeks; dimethoate 7 days; and schradan 6 weeks, or 4 weeks from April to July.

Parathion or sulfotep smokes may be used after planting, while malathion sprays or aerosols, diazinon aerosols or sprays and smokes of nicotine may be used safely on established plants after mid-May, provided that, if petroleum emulsions are used to control red spider mite, there is an interval of 4 days between malathion and petroleum applications.

Aphids on lettuces

In some cases aphid infestations are most expeditiously controlled before planting. Attacks on winter-grown lettuces in glasshouses usually result from planting infested seedlings and so can be prevented by dipping these for about 3 min. in nicotine (0·06%) or malathion (0·1%). In the case of plants raised in soil blocks, spraying with nicotine or malathion at these concentrations will give control. Rubber gloves and an apron or overalls should be worn when dipping plants. Where attacks occur after planting, particularly when the leaves begin to fold in, the best control is obtained by spraying with the systemic compounds, demeton-S-methyl (0·02%), dimethoate (0·03%) formothion (0·04%), morphothion (0·025%), oxydemeton-methyl (0·02%), or schradan (0·06%). Earlier, control may be obtained with efficiently applied sprays of malathion (0·1%), diazinon (0·02%), gamma-BHC (0·01%), nicotine (0·06%) or TEPP (0·01%). It is essential that the required interval is allowed between application and the cutting of lettuces. To obviate taint in the case of malathion, this interval should be at least 4 days.

Aphids on bulbs and roots

Tulip and iris bulbs and freesia and gladiolus corms on which tulip-bulb or other aphids are present may be disinfested by immersion, for 15–20 min, in gamma-BHC (0·01%) or nicotine (0·06%). They should then be boxed or, if treated in advance of planting, slowly dried. Alternative treatments are to thoroughly dust with gamma-BHC before planting or to fumigate in a closed container for 5 days with paradichlorobenzene (4 oz/ft^3). To prevent direct contact with the bulbs this chemical should be enclosed in hessian or other open material. The bags of chemical should be distributed among large quantities of bulbs. Failure to control aphids on the dry bulbs will result in damage to the foliage.

Root-infesting aphids are unlikely to occur in commercial glasshouses except where plants, particularly *Primula* spp., are temporarily grown outdoors. They may, however, become a pest in houses where ornamental perennial plants are grown. Watering the soil in the pots, with parathion (0·01%), malathion (0·1%), nicotine (0·06%—no wetting agent required) or diazinon (0·02%) will control root aphids. Parathion and nicotine are, generally, the least phytotoxic of these compounds and should be used unless the others are known to be safe. Two applications at an interval of 14–21 days may be required.

REFERENCES

AUCLAIR, J. L. (1963). Aphid feeding and nutrition. *A. Rev. Ent.*, **8**, 439.

BANKS, C. J. (1965). Aphid nutrition and reproduction. *Rep. Rothamsted exp. Stn*, **1964**, 299–309.

BROADBENT, L. (1959). Insect vector behaviour and spread of plant viruses in the field. In *Plant pathology: problems and progress*. Ed. by C. S. Holton *et al.* University of Wisconsin Press, Madison. p. 539.

CARTER, W. (1961). Ecological aspects of plant virus transmission. *A. Rev. Ent.*, **6**, 347.

HILLE RIS LAMBERS, D. (1938–53). Contributions to a Monograph of the *Aphididae* of Europe. *Temminckia*, **3**, 1–44; **4**, 1–134; **7**, 180–328; **8**, 182–323; **9**, 1–176.

IMMS, A. D. (1947). *Insect Natural History*. Collins, London. 317 pp.

KENNEDY, J. S., DAY, M. F. & EASTOP, V. F. (1962). A conspectus of aphids as vectors of plants viruses. London, Comm. Inst. Ent. 114 pp.

KENNEDY, J. S. and STROYAN, H. L. G. (1959). Biology of aphids. *A. Rev. Ent.*, **4**, 139.

ROBINSON, A. G. (1966). Review of the Fern Aphids in North America with description of a new species and a new genus. *Can. Ent.*, **98**, 1252–9.

STROYAN, H. L. G. (1952). The identification of aphids of economic importance. *Pl. Path.*, **1**, 9–14; 42–8; 92–9; 123–29.

THEOBALD, F. V. (1926). *The Plant Lice or Aphididae of Great Britain*. Vols I–III. Headley Bros., London. 372, 411 and 364 pp.

WORTHING, C. R. (1967). Observations on the chemical control of pests on year-round chrysanthemums. *Proc. 4th Brit. Insect. & Fung. Conf.*, **I**, 63–75.

Suborder Heteroptera

Family MIRIDAE (=CAPSIDAE)

These oval, flattened, plant bugs, commonly called capsids, injure a wide range of flowering and ornamental plants.

Lygus rugulipennis Popp. (the tarnished plant bug or bishop bug) is the mirid found most frequently under glass. Serious injury has been caused to chrysanthemums although sweet peas, asters, dahlias, carnations and zinnias are also attacked. When fullgrown, *L. rugulipennis* is 6·0–7·3 mm long and varies from pale greyish-green to yellowish-green. Though the pronotum has darker markings it is consistently yellower than the ground colour of the forewings. The basal half of the latter lack extensive dark markings or, where such markings are present on the scutellum (the chitinized triangle between the bases of the forewings), they are never in the form of a 'W'. The antennae and legs are pale coloured and long, the eyes dark and very prominent.

Two other colour forms of this species occur in southern England. One is light reddish in both sexes, without dark markings on the pronotum. Males of the other form have deeper red forewings with the pronotum and scutellum yellow; the females are green.

Adults overwinter in dead leaves and long herbage, emerging from hibernation from late March to May. They are active and take flight at the slightest disturbance. The creamy-white, flask-shaped eggs ($1{\cdot}0 \times 0{\cdot}25$ mm) are inserted in stems, petioles or flower buds of young plants in April and May. The bug passes through five nymphal instars. The active nymphs are usually yellow-green to mottled brown; wing pads become visible as they mature. The first adults appear from July onwards.

FIG. 5.26. Damage to chrysanthemum flowers by capsids. (Crown Copyright by permission of Controller of HMSO.)

Outdoors, in Britain, *L. rugulipennis* normally completes two generations annually, the second of these, resulting from eggs laid in late summer, subsequently causes trouble when moved into glasshouses with chrysanthemums.

The nymphs always feed near the shoot apices on the youngest and most succulent tissues causing the stems to become marked with brown calluses. These calluses result from the toxic action of salivary material injected into the plant cells by the bug's stylets. On chrysanthemums, especially after disbudding, injured shoots become stunted, twisted, and occasionally blind, and nodes of the stem become swollen. Capsids may also feed on the leaves, causing them to become puckered and malformed: occasionally even the petals of open flowers may be distorted by feeding (Fig. 5.26).

Lygocoris pabulinus Linn. (the common green capsid (Fig. 5.27)), is abundant throughout Britain and is occasionally, like *Lygus rugulipennis*, introduced into

FIG. 5.27. *Lygocoris pabulinus* L.: an adult capsid (×10). (A Shell photograph.)

glasshouses on chrysanthemums. It completes two generations a year on two groups of host plants; eggs overwinter, and young nymphs of the first generation develop on woody hosts including hawthorn, apple, currant and lime, while the maturing nymphs and second generation are associated with herbaceous hosts including groundsel, potato, nettle, rose and raspberry. Under glass species of *Ipomoea* and *Salvia* may be attacked.

Lygocoris pabulinus (6 mm long) may be distinguished from *Lygus rugulipennis* by the green head and thorax compared with the largely brown, or black, coloration in the latter species.

Curved, creamy-white eggs (1·3 mm × 0·37 mm), rounded at one end, and extended to a neck with a pale-green cap, are laid in June, July and again in later summer. They are inserted in the plant stem, with the cap more or less level with the surface of the bark. Summer eggs hatch in 19–21 days. Feeding by the developing nymphs causes russet speckling towards the bases of the leaves and, in severe cases, the foliage may be twisted and distorted. Nymphs, especially as they mature and before becoming capable of flight, readily fall from host plants and spread from plant to plant over the ground.

Control

On chrysanthemums atomized solutions of DDT (3 g/1000 ft^3) or gamma-BHC (3 g/1000 ft^3) are suitable for control, as are smokes yielding gamma-BHC 0·7 g/1000 ft^3 or DDT 3·3 g/1000 ft^3. DDT sprays at 0·1% and gamma-BHC sprays at 0·01% are also satisfactory.

Plant bugs may be introduced into orchid houses on imported plants and, should they persist and cause damage, can be controlled readily by DDT smokes or by spraying with DDT at 0·05%. DDT-malathion mixtures can be used safely on most orchids.

REFERENCES

AWATI, P. R. (1914). The mechanism of suction in the Potato Capsid Bug (*Lygus pabulinus*). *Proc. zool. Soc. Lond.*, **87**, 685–733.

BECH, R. (1967). Zur bedeutung der *Lygus*-arten als Pflanzenschadlinge. *Biol. Zbl.*, **86**, 205–32.

BUTLER, E. A. (1923). *A Biology of the British Hemiptera-Heteroptera.* Witherby, London. 682 pp.

CARVALHO, J. C. M. & LESTON, D. (1952). The classification of the British Miridae (Hem.) with keys to the genera. *Entomologist's mon. Mag.*, **88**, 231–51.

PETHERBRIDGE, F. R. & THORPE, W. H. (1928). The common green Capsid Bug. *Ann. appl. Biol.*, **15**, 446–72.

SOUTHWOOD, T. R. E. & LESTON, D. (1961). *Land & Water Bugs of the British Isles.* Warne, London, 436 pp.

STEWART, R. K. (1969). The biology of *Lygus rugulipennis* Popp. in Scotland. *Trans. R. ent. Soc. Lond.* **120**, 437–53.

Order LEPIDOPTERA: Moths

Many moths and a few butterflies are pests but, following the widespread use of insecticides in recent years, economic damage from their caterpillars in glasshouses is rare. However, wider use of biological control methods may decrease insecticide applications and so permit some species to retain their earlier importance as pests.

Family NOCTUIDAE (=PHILAENIDAE=AGROTIDAE)

The Noctuid moths have stout bodies, medium-size ($1\frac{1}{2}$ in. wingspread) and sombre colouring. Only the larvae (usually known as cutworms from their habit of eating through stems at the soil surface) cause damage. Normally feeding at night, these larvae frequently remain coiled and concealed below the soil surface or plant debris during the day. When fully grown they are nearly 2 in. long and inconspicuously coloured in shades of brown and grey.

Laconobia oleracea (L.) (the tomato moth) first became a pest in Britain in 1890 and, by 1920, was a serious problem in several glasshouse areas. Both sexes have a wing expanse of $1\frac{1}{4}$–$1\frac{3}{4}$ in. The forewings are dark purplish-brown with a characteristic, jagged, sub-terminal line. The 'stigmata' (reniform and orbicular marks) of the forewings are faintly outlined with white. The hind-wings are grey with darker shading around the hind margins. The moths do not fly by day (when they hide in soil crevices) and are often only observed when disturbed, (e.g. by watering). The males are polygamous (there are usually twice as many females) and mating takes place, after emergence, as soon as the wings are fully expanded. The female begins to oviposit 2–3 days later.

About 1000 eggs are laid by each female in batches of between 60 and 300 on the underside of tomato leaves. Each egg (0·75 mm diam.) is hemispherical, with the dome slightly depressed; the transparent shell, except for the smooth floor, is strongly ribbed and reticulate. At first the eggs are greenish but soon become pale yellow or almost white. Under average glasshouse temperatures they hatch in 8–10 days.

There are five to seven larval instars which are completed in 33–39 days. In all stages of growth transverse rows of dark setae arise from black, dot-like sclerites. The spiracles on the prothorax and first eight abdominal segments become conspicuous in the third instar, the white peritreme being encircled in black. When full-grown the larva is $1\frac{1}{2}$–$1\frac{3}{4}$ in. long. The head, pale green in the first two instars, becomes white to greyish-brown with marbled areas. The general body colouration is yellowish- or reddish-brown with three dark-grey dorsal stripes partially obscured by underlying greyish-black stripes. An additional grey stripe runs along the upper edge of the lateral yellow stripes. The prothorax has a single spot, while the meso- and metathorax as well as the abdominal segments I–VII each have two setigerous black spots. The eighth and ninth segments have a single black dot posterio-laterally, but the apical segment is devoid of any such spots or dots.

At first, the larvae feed gregariously and skeletonize the leaf surface but, after the second moult, they disperse and, because they have difficulty in climbing up the stems of new host plants, their numbers are often severely depleted. Older larvae not only devour large areas of leaf but may also attack stems (see Fig. 5.28), so predisposing them to fracture; they also eat large holes in both green and ripening fruit. Two or three half-grown larvae are capable, within a single night, of completely stripping the leaflets from several leaves, leaving only the bare petioles. When about to pupate, the caterpillars become active and search out crevices in dry situations such as walls, woodwork and debris. A light, silken cocoon is spun and, 2 or 3 days later, a brown pupa ($\frac{3}{4}$ in. long) is formed which gradually assumes a shining black. There are apparently two types of pupae with different durations of development. A short-period form, most commonly found in May and June, emerges in

16–50 days, but the long-period type, which is found in increasing numbers between July and late autumn, emerges in 100–300 days.

L. oleracea is widely distributed in Britain, and outdoors feeds on brassicas, other *Cruciferae*, and *Chenopodiaceae*. Under glass it may also occasionally skeletonize cucumber leaves, and sever the leaves, or eat the calyx and petals, of carnations.

No really effective predators or parasites of the tomato moth have been found amongst the twenty so far recorded. The most common parasite is the large

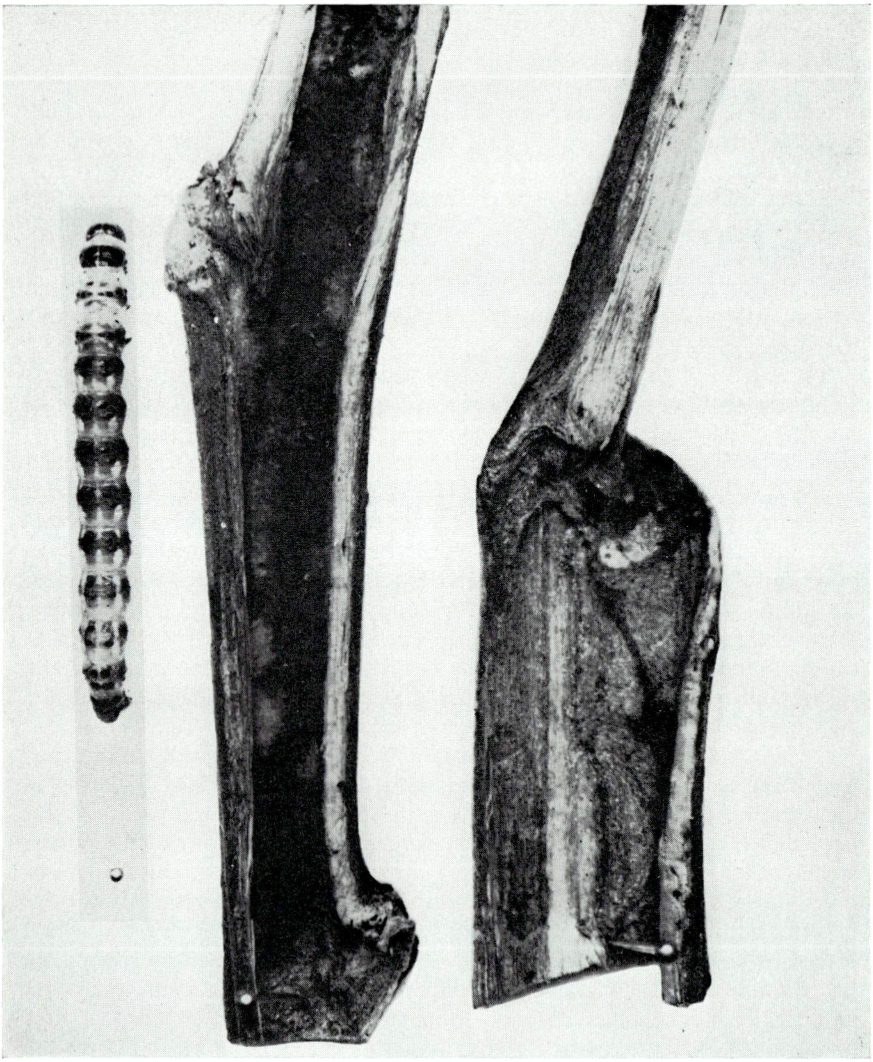

FIG. 5.28. Atypical injury by tomato moth. (Courtesy of Glasshouse Crops Research Institute.)

reddish-brown Ichneumonid, *Pimpla instigator* F., which attacks the chrysalis, but its seasonal emergence is not well co-ordinated with that of the host under glasshouse conditions.

REFERENCES

LLOYD, LL. (1920). The habits of the Glasshouse Tomato Moth, *Hadena* (*Polia*) *oleracea* and its control. *Ann. appl. Biol.*, **7**, 66–102.
SPEYER, E. R. & PARR, W. J. (1948). The Tomato Moth. *Rep. exp. Res. Stn Cheshunt*, **1947**, 41–62.

Phlogophora meticulosa (L.) (the angle shades moth) with a 2-in. wing-span has fawn forewings patterned with olive-green and rose-pink markings. The yellowish-white hindwings, crossed by two narrow, dark lines, are shaded

FIG. 5.29. *Phlogophora meticulosa* L.: angle-shades moth, adult (× 2) (Crown Copyright by permission of Controller of HMSO.)

brown at the margin (Fig. 5.29). When at rest, the wings are folded so that the insect resembles a crumpled leaf. Outdoors, the moths are on the wing from May to October. At one time caterpillars were found under glass only in the autumn or winter, having hatched from eggs laid on chrysanthemums before they are brought into the houses for flowering. The modern practice of lighting year-round varieties may, however, attract the adults to glasshouses earlier in the year.

The radially sculptured, pale greenish-yellow eggs which are laid in small groups on the leaves hatch in about 10 days (Fig. 5.30). There are normally six instars. The mature larva ($1\frac{1}{2}$ in. long) is smooth and velvety, it is dark green to brown with a single, broken-white, dorsal stripe and V-shaped grey markings as well as dark-grey stripes laterally. The prominent, white, oval spiracles are ringed with black. Young larvae skeletonize the upper leaf surface and, when disturbed, hang from the foliage on threads. Second, and later, instars may eat large holes in the leaves and nibble the margins of blossom-buds, so distorting the flower owing to inhibition of growth at the point of injury. When open flowers are attacked the caterpillars work their way from the margin to the centre and trim the petals evenly. All feeding occurs at night; in daytime the larvae hide in curled leaves or amongst petals. When fully grown (after about 7 weeks) the caterpillars burrow about 1 in. into the soil, spin a loose cocoon into which soil particles are woven, and form a chestnut-brown cocoon $\frac{3}{4}$ in. long. The moths emerge after 3 weeks.

REFERENCES

PARR, W. J. & SPEYER, E. R. (1941). Some observations on the life-history of the Angleshades Moth (*Brotolomia meticulosa* L.). *Ann. appl. Biol.*, **28**, 29–33.

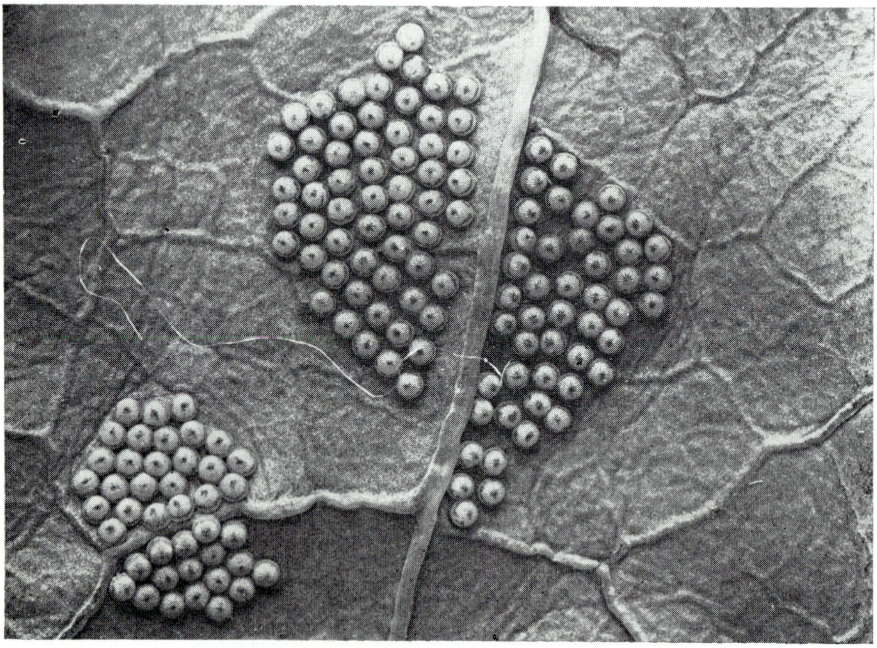

FIG. 5.30. Egg mass of *Phlogophora meticulosa* (×5). (Crown Copyright by permission of Controller of HMSO.)

Phalaena typica L. (the gothic moth) is also greyish-brown, and has two irregular transverse, whitish lines, and faint 'stigmata' on the forewings (wingspread $1\frac{1}{2}$ in.). The dull, grey-brown hindwings are devoid of markings. The moths fly in late summer and lay clusters of eggs on the leaves. The larvae, which hatch in about 14 days, feed gregariously for 2 or 3 weeks. The caterpillars are a brownish-grey dorsally but, as they mature, a paler lateral band appears above the prolegs. The body bears darker mid-dorsal and lateral stripes as well as indistinct lateral markings which produce a diamond-shaped

FIG. 5.31. *Mamestra brassicae* L.: cabbage moth, adult ($\times 4$). (A Shell photograph.)

pattern along the back. In the autumn the larvae can cause severe damage to the flowers and foliage of chrysanthemum before pupating in the soil and emerging as moths in late spring.

Peridroma saucia (Huebner) (the pearly underwing), with a wing-span of 2 in., is at least $\frac{1}{4}$ in. larger than the tomato moth (p. 125). It has reddish-brown forewings which are covered with irregular dusky markings but there is no marginal white line as in *Laconobia oleracea*. The distinctive hindwings, on which the venation is clearly demarcated, are clear pearly white. The pearly underwing, though an irregular pest of tomatoes under glass, has similar habits to the tomato moth. Some authorities have claimed that outbreaks may arise from immigrant individuals from the Continent.

The caterpillars, when fully grown, are 2 in. long and vary from green to brown; they are marked with dark dashes on each body segment. On the penultimate segment is a conspicuous W-shaped mark and a black V, or

H-shaped, mark on the head. The spiracles are black and not, as in *L. oleracea*, white ringed with black.

Mamestra brassicae (L.) (the cabbage moth), which is a widespread pest of brassicas, occasionally damages carnations and chrysanthemums under glass. It is a large, grey-brown to black moth (wing-span 1½ in.) with an irregular, white, transverse line near the margin of the forewing and a white outline to the orbicular mark (Fig. 5.31). The hindwings are brown with a central spot. Outdoors the adults appear in May and June, and lay ribbed, reticulated eggs in dispersed groups on the leaves of brassicas, lettuce, peas, onions and even

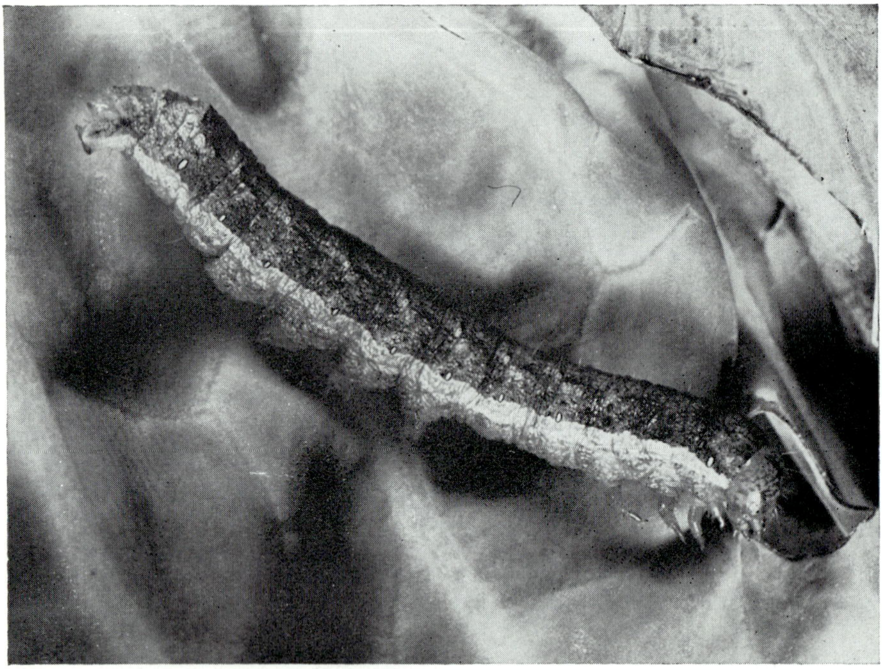

FIG. 5.32. Larva of cabbage moth (×2). (A Shell photograph.)

tomatoes. The caterpillars, green when young, later turn grey-brown with a dusky central line speckled with white and a pale yellow lateral stripe (Fig. 5.32). When full grown they are 1½–2 in. long. They feed voraciously for about 3 weeks before pupating in the soil in August. Some moths may emerge from these pupae and give rise to a second generation which matures in late autumn to overwinter as pupae.

Plusia gamma (L.) (the silver Y moth) is known to cause extensive damage to crops of French beans and chrysanthemums under glass (see Fig. 5.33). It does not apparently survive the winter in Britain but arrives here as an annual migrant. The forewings (1½ in. wing-span) vary from silver-grey to almost black with a distinctive pale yellow, or silvery-white Y-shaped, mark centrally.

FIG. 5.33. Noctuid damage to chrysanthemums. (By kind permission of the Plant Pathology Laboratory, Ministry of Agriculture, Fisheries and Food.)

The hindwings are light brown with a dark border. Unlike most moths it is active during the day. The females normally lay up to 500 whitish, reticulate eggs in June which hatch in about 12 days to greenish caterpillars known as 'false loopers' because they have only three pairs of prolegs, instead of the normal five. The brownish head has black markings and there is a dark-green dorsal stripe edged with white. The larvae mature and pupate in July. On year-round chrysanthemums this moth may continue to breed until the autumn. The caterpillars are difficult to find but their presence is betrayed by the large frass granules which accumulate on the leaves below the site of damage.

Spodoptera littoralis Boisd. (the Mediterranean climbing cutworm) occurs in some of the warm countries chosen as especially suitable for the winter production of chrysanthemum cuttings. It can be transported on the cuttings; in the latter part of 1963 it appeared in over twenty chrysanthemum nurseries in different parts of Britain. At that time it was known as *Prodenia litura* F., and there was some concern as the older larvae, unlike other cutworms, proved to be insensitive to DDT. The adult moth has brown forewings (span $1\frac{1}{5}$–$1\frac{4}{5}$ in.) with whitish, irregular markings, marginal lines and clearly defined white 'stigmata'. The hindwings are whitish with brown margins (Fig. 5.34). Eggs are laid in clusters of 20–500 completely covered with pale-brown hairs

FIG. 5.34. *Spodoptera littoralis* Boisd.: Mediterranean climbing cutworm, adult (×2). (By kind permission of the Plant Pathology Laboratory, Ministry of Agriculture, Fisheries and Food.)

detached from the abdomen of the female (Fig. 5.35). On year-round chrysanthemums they hatch in 8 days, and the six larval instars are completed in 33 days. At first the caterpillars are greenish but later become very dark brown with conspicuous black and yellow markings on each body segment. There are two pairs of yellow spots on the first two segments behind the head (Fig. 5.36). When fully grown the caterpillar is $1\frac{1}{2}$ in. long. The reddish-brown pupa, 1 in. long (Fig. 5.37), is found in the soil and the moths emerge after about 46 days.

The caterpillars feed nocturnally and, when newly hatched, feed only on the superficial tissues of the lower leaf surfaces, but older individuals eat entire leaves and may severely damage buds and flowers (Fig. 5.33).

Experiments have shown that *S. littoralis* cannot survive outdoors in British winter conditions. The eggs fail to hatch after 10 days exposure at 35 °F (2 °C) so the storage of infested cuttings under these conditions (e.g. during transit) effectively prevents any further outbreaks.

FIG. 5.35. Egg mass of *Spodoptera littoralis* (×2). (By kind permission of the Plant Pathology Laboratory, Ministry of Agriculture, Fisheries and Food.)

Control of Noctuidae

The use of DDT has literally eradicated tomato moth from commercial houses. Reinfestation by moths entering from outside occurs very infrequently; usually on such a small scale that picking off leaves bearing young, undispersed, caterpillars, or localized insecticidal treatment of infested plants are the only necessary measures.

If newly planted tomatoes are infested a very light application of a 5% DDT dust at $\frac{1}{2}$–$\frac{3}{4}$ oz/100 ft^2) may be applied. With larger plants DDT dusts must be applied with discretion as an interval of at least 2 weeks must elapse between the last application and picking. Dusts should be blown into the air above the plants and allowed to settle, and, unless an efficient dusting appliance is available, their use on plants with well-developed fruit is inadvisable. Sprays of DDT, either H.V. at 0·1% or L.V. at 3·0% (2–5 gal/acre), can be used. To avoid local overdosing with L.V. application it is best to direct the spray above the plants. Wettable powder formulations should be used in sprays on young tomatoes until the first truss is in flower. The interval between spraying and picking must be at least 2 weeks.

If an attack is widespread on fruiting plants, DDT as smokes (3·5 g/1000 ft^2) or aerosols (3·5 g/1000 ft^2) should be used; a minimum interval of only 2 days is then required.

FIG. 5.36. Larva of *Spodoptera littoralis* (×2). (By kind permission of the Plant Pathology Laboratory, Ministry of Agriculture, Fisheries and Food.)

Although tomato moth caterpillars have been known to cause damage to cucumbers grown after infested tomato crops, they, like other noctuid caterpillars, occur very rarely nowadays. The skeletonizing of the leaves by newly emerged caterpillars is obvious, and careful removal of caterpillars provides the simplest control. Spraying with derris or malathion, or aerosol applications of azinphos-methyl or malathion will probably prove effective against young noctuid caterpillars.

Cutworms (other than of *Spodoptera littoralis*) attacking ornamentals may also be readily controlled with DDT. For those which feed at soil level a 5% DDT dust should be applied to the surface of the soil at 1 oz/yd^2. The most efficient method of applying DDT to control foliage- or flower-eating caterpillars is best decided by the grower, who should bear in mind that deposits from aerosols and smokes are smaller, and less persistent, than those from sprays. Dusts, if badly applied, may detract from the appearance of foliage and flowers. High volume spraying of the opening flowers of chrysanthemums or carnations during the late autumn and winter is undesirable, especially when the plants are being grown at low temperatures to retard maturity, as fungal diseases, such as *Botrytis cinerea*, may be encouraged.

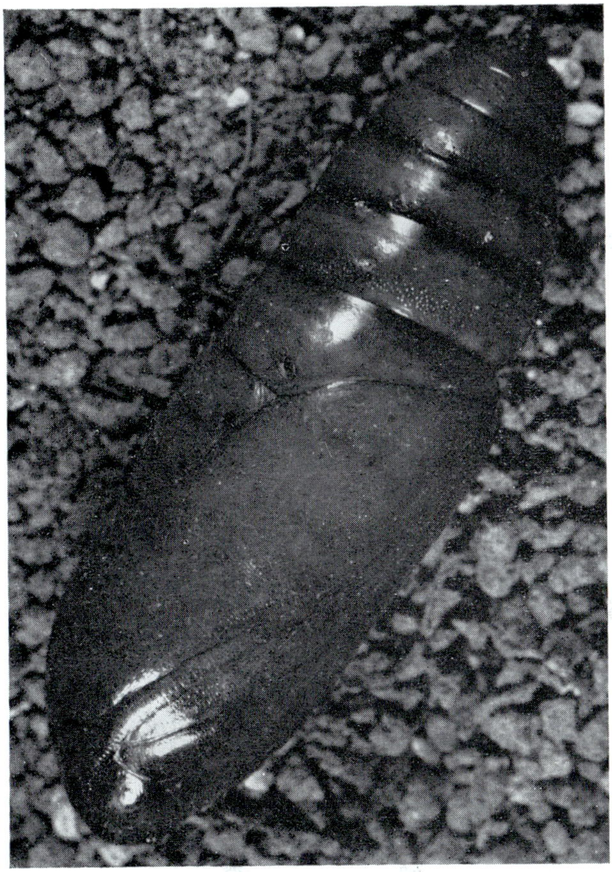

FIG. 5.37. Noctuid pupa (*Spodoptera littoralis*) (×4). (A Shell photograph.)

It is frequently necessary to control diseases and other pests in addition to caterpillars. DDT may then be combined with fungicides and other insecticides in specialized 'chrysanthemum' sprays, and with malathion or gamma-BHC in aerosols or smokes.

The caterpillars of *S. littoralis*, up to the third instar, are not susceptible to DDT; they can be killed by H.V. sprays of dichlorvos (0·05%), trichlorphon (0·1%) or TDE (0·075%), or by dichlorvos aerosols (1 g/1000 ft^3). Repeated treatments at weekly intervals until flowering are advised. If the caterpillars are larger than the third instar, spraying with 0·04% dieldrin may be necessary to kill them. Some chrysanthemum cultivars (varieties) are damaged by dichlorvos (p. 384).

Family TORTRICIDAE

Cacoecimorpha pronubana (Huebner) (the carnation tortrix), an important pest in Southern Europe, has gradually extended its range northwards and is now widely established in southern England, France and parts of Germany.

FIG. 5.38. Carnation tortrix (*Cacoecimorpha pronubana*), adult (×3). (A Shell photograph.)

The relatively inactive moths may immediately be recognized by the bright orange hindwings. However, when at rest, the wings are held roof-wise over the body and, as the forewings are greyish-brown, present a sombre appearance (Fig. 5.38). Wingspread is 18–22 mm though the males are slightly smaller. The forewings are reticulated with darker brown lines and there is a dull, reddish-brown patch which broadens from the middle of the front edge to the anal angle of each wing. The hindwings are bright orange bordered with black.

Following mating, shortly after emergence, oviposition lasts 12–14 days. Up to eight egg-masses, each containing between 10 and 200 eggs, are produced by each female. Each mass, which is covered with mucilage, consists of light-green, reticulate, oval eggs overlapping like shingles on a roof. The average incubation period is 16–17 days increasing to 24 days at 55 °F (13 °C), and decreasing to 12 days at 70 °F (21 °C).

The caterpillars, which develop through seven instars, are yellowish- to olive-green with sparsely distributed green tubercles each bearing a long seta. When full grown the larva is $\frac{3}{4}$ in. long with a light-brown head bearing dark markings and a yellowish-green prothoracic shield with four dark posterior

FIG. 5.39. Pupa of carnation tortrix (Courtesy of Glasshouse Crops Research Institute.)

markings. There are normally two broods annually, but prolonged larval life causes much overlapping and caterpillars of all ages may be found throughout the year. Larval development usually takes from 2 to 3 months. Newly hatched caterpillars are active and disperse from the egg masses soon after hatching. They feed in leaf axils, within buds and opening flowers, or in shoot tips. If disturbed, they drop on silken strands on which they may be carried about the nursery. The youngest larvae construct a small silken shelter on the upper surface of the leaves or even on the flower petals. By the third instar they eat entire leaves and spin silk at the growing points in a characteristic fashion, finally they bore into the base of flower buds (Fig. 5.39). The dark brown to black pupa is enclosed within a dense web of silk at the final feeding site. After 14–45 days, depending on season, the adult moth emerges and lives for up to 3 weeks. Carnations are the primary host but *C. pronubana* also breeds on species of *Acacia*, *Chorizema*, *Coronilla*, *Grevillea* and *Nandina*.

REFERENCES

FISHER, R. C. (1924). The life-history and habits of *Tortrix pronubana* Huebner with special reference to the larval and pupal stages. *Ann. appl. Biol.*, **11**, 395–447.

HERFS, W. (1963). Laboratory investigations on the bionomics of *Tortrix pronubana* Hb. I. *Z. angew. Ent.*, **51**, 42–54; II. *ibid.*, **52**, 1–28; III. *Z. pflkrankh.*, **70**, 193–8.

SPEYER, E. R. (1938). Pests of carnation and their control. *Sci. Hort.*, **6**, 72–84.

Epichorista ionephela Meyr (the South African carnation worm). In recent years many carnation cuttings raised in Kenya and South Africa have been imported into Europe and, on this material, *E. ionephela* has been introduced to Denmark and Norway and has also been intercepted by the quarantine authorities in Britain.

In South Africa the carnation worm is a widespread pest of outdoor carnations and pears. Other hosts include chrysanthemum, lucerne, lupin, *Oxalis acetosella*, roses and *Rumex acetosella*.

The female moth has a wingspread of $\frac{3}{4}$ in., the male being slightly smaller; the forewings are reddish-brown to sand-coloured with diffusely scattered, small dark-brown spots and a larger dark area on the hind margin. The hindwings are light grey. Rarely more than two egg clusters, each of about 25 eggs, are laid on a single plant either at the base, or apex, of a leaf. The incubation period is about 9–12 days. The young larvae, which are yellowish with dark-brown heads, nibble both the edges of the leaves and the superficial tissues. When full grown, after 4 to 8 weeks, the head becomes light brown and the body green with dark dorsal and paler, lateral stripes. The brown pupa is formed within a silken cocoon made either in a feeding tunnel within the plant or between spun leaves. Moths emerge in 9–26 days. As many as five overlapping generations are produced in South Africa.

The larva burrows into the shoots either apically, so destroying the growing point, or through the side, usually at, or below, a leaf sheath. The entrance holes, of which there may be several on an attacked plant, are made obvious by accumulations of frass. This hollowing of the stem inevitably destroys the plant. Occasionally larvae attack the buds.

Control

The vigilance of the persons implementing the Importation of Plants and Plant Products Order has so far prevented this pest from being introduced into Britain, and it is essential that these officials should be given every assistance.

The detection of eggs or larvae on a vast number of cuttings is difficult and there is a possibility that the pest may enter the country. Nursery workers handling cuttings or young plants should report any suspicious symptoms. Cut blooms imported from South Africa should never be taken into the vicinity of carnation houses. In Denmark fumigation of imported cuttings with parathion or lindane (4·5 g active ingredient per m^3) or low temperature storage are being used as preventive measures.

Immediate control measures are required if the pest is discovered. Infested plants should be removed and burnt, and the remainder sprayed at weekly intervals with 0·1% DDT, 0·015% azinphos-methyl or 0·1% carbaryl until the pest is eradicated. The burrowing habit of the larvae makes efficient control difficult.

REFERENCES

FJELDDALEN, J. (1965). Attacks of the carnation tortrix *Epichorista ionephela* Meyr found in Norway. *Gartneryrket*, **55**, 340–2.

THYGESEN, TH. & BJERGGARD, A. (1966). Experiments on control of South African Tortrix Moth by fumigation in vacuum chambers. *Tidsskr. PlAvl*, **70**, 224–31.

OTHER SPECIES OF TORTRICIDAE

Several other species of tortricid moths have been recorded under glass but their precise identification is a matter for the specialist:

Species	Host	Caterpillars
Cacoecia oporana L.	Vines, cucumbers	Greenish with dark dorsal stripe and black head
C. rosana L.	Rose	Head and prothorax shining brownish-black, body dull green, darkest dorsally
Tortrix diversana Hb.	Peaches, apricots	Green with dark dorsal and lateral stripes, head and prothorax black
T. costana Schiffermueller	Ferns	Dark grey-green with white spots on a black head
Peronea comariana Zell.	Strawberries, azaleas	Head and prothorax yellow brown, body pallid green
Ditula angustiorana Haworth	Vines, apricot, azalea	Dark green. Head and prothorax yellow-brown, the latter with black flecks

Control

Larvae of most, if not all, tortricids which attack glasshouse plants can be controlled by light applications of a 5% DDT dust or by DDT aerosols and smokes (3·5 g/1000 ft^3). They are also controlled both by the sprays recommended for *Epichorista ionephela* and by spraying with 0·075% TDE. If fruit of grapes and peaches is to be picked within 4 weeks, aerosols or smokes of DDT should be used.

Family TINEIDAE

Caloptilia azaleella (Brants) (the azalea leafminer), which is a native of Japan, has been distributed to both Europe and America by the trade in plants.

The forewings are elongate with both the basal portion and apical fringe mainly black and a wide yellow blotch along three-quarters of the front margin. The hindwings are grey with long hair fringes.

FIG. 5.40. Leaf-rolling by azalea leafminer. (Crown Copyright by permission of Controller of HMSO.)

The elongate, oval, flattened eggs, one to five per leaf, are laid singly on the underside of the leaves, usually alongside a vein. After 4 days they hatch to release minute, legless larvae which have very powerful mouthparts. They tunnel into the leaf immediately below the egg and construct an elongate mine within the tissues. By the third moult the larva is greenish-yellow and vacates the mine to curl back the tip of a leaf with silk (Fig. 5.40). Within the cavity so formed it lives and feeds. The pupa is formed within a cocoon often constructed on an undamaged leaf.

Infested bushes are often defoliated and even damaged leaves appear severely scorched. *Azalea indica*, which is commonly attacked, is usually grown cold until the onset of flowering when the plants are brought into heated houses.

Under these conditions the adults emerge in numbers, oviposition takes place freely and may be followed by severe defoliation.

Control

Plants should be sprayed with either gamma-BHC (0·012%) or parathion (0·01%) shortly before taking the plants into the glasshouse. Such sprays should be repeated two or three times at 14-day intervals after the plants are housed. As sprays may damage the flowers it is important to control the pest before the flower buds open.

REFERENCES

HODSON, W. E. H. (1927). The Azalea Leaf Miner, *Gracillaria azaleella* Brants. *Jl. R. hort. Soc.*, **52**, 54–9.

GENERAL REFERENCES TO LEPIDOPTERA

BUCKLER, W. (1885–9). The larvae of *British Butterflies & Moths*. 9 Vols. *Ray Soc. Publs*, 1508 pp.

FORD, L. T. (1949). A guide to the smaller British *Lepidoptera*. *S. Lond. ent. & N.H. Soc.* 230 pp.+Suppl. 15 pp. (1958).

MEYRICK, E. (1929). *A Revised Handbook of British Lepidoptera*. Watkins & Doncaster, London. 914 pp.

SOUTH, R. (1961). *Moths of the British Isles*. Vols. I and II. Warne, London. 427 and 379 pp.

Order COLEOPTERA: Beetles

Family SCARABAEIDAE: Cockchafers

Larvae of several species may injure the first crops grown in newly erected glasshouses. The grubs are white, fleshy creatures with characteristically curled, wrinkled bodies, and three pairs of strongly developed thoracic legs (Fig. 5.41). They hatch from eggs laid in the soil and, depending on the species, take from 1 to 5 years to mature. Chafer larvae are entirely subterranean, rising to the superficial soil layers at night and descending deeper by day. They feed on the roots of plants which, when severely damaged, show retarded growth and flag easily in bright sunshine. The problem decreases within a year or so as the population matures, for the beetles normally lay eggs amongst grass or cereals. Occasionally larvae enter pots containing chrysanthemums or other plants which have stood outdoors, causing them to wilt. Although the adults may occasionally damage the petals of roses, narcissi and other blooms they are never sufficiently numerous to be a glasshouse pest.

Control

A gamma-BHC dust, as advised for the control of wireworms (p. 144), or a 5% DDT dust at $\frac{1}{2}$ oz/yd^2, well mixed in to a depth of 2–3 in., kills cockchafer larvae in glasshouse borders. If cockchafer larvae are present below wilting plants they should be watered with 0·01% gamma-BHC or with 0·04% DDT. With pot-plants the insecticide should be applied when the soil is moist, at about 4 fl. oz to 8-in. pots or $2\frac{1}{2}$ fl. oz to 6-in. pots.

REFERENCES

MILNE, A. & LAUGHLIN, R. (1956–60). Biology and ecology of the garden chafer, *Phyllopertha horticola* L. *Bull. ent. Res.*, **47,** 7–22; **49,** 685–99; **50,** 39–52, 467–8; **51,** 353–78.

VAN EMDEN, F. I. (1941). Larvae of British Beetles II. A key to British lamellicorn larvae. *Entomologist's mon. Mag.*, **77**, 181–92.

FIG. 5.41. Cockchafer larva (*Melontha vulgaris* F.) (×2). (A Shell photograph.)

Family ELATERIDAE

Click beetles and wireworms are occasionally troublesome where glasshouses are newly erected on grass or cultivated land which has become weed-infested. Wireworms are the larvae of click beetles, so-called because, when on their backs, they are capable of leaping into the air with a distinct clicking sound. The leap is made possible by a spine on the ventral side of the prothorax which fits into a hole in the mesothorax. If the body is bent this spine springs out and flings the insect from the ground. The beetles are mostly dull with parallel-sided bodies tapering to a point apically. The head, concealed by a helmet-shaped pronotum, carries serrate antennae. In spring, batches of 40–100 eggs are laid within 3 in. of the soil surface. They hatch in 5 to 6 weeks into elongate, yellowish, cylindrical wireworms with short legs and a remarkably tough outer skin (Fig. 5.42). For the next 4 years these larvae live within the soil, approaching the surface between March and May and again from July–September. At other times they may be as deep as 2 ft. They pupate within an oval, earthen cell.

As, during the first year, wireworms mostly feed on the turf interred by soil culture, glasshouse crops are usually unaffected. But, during the next 3 years, they may feed on the roots of almost all crops. Tomatoes, lettuce, carnations and chrysanthemums are all subject to severe damage. Bulbs, corms and rhizomes may also be infested. While crop loss may be severe, wireworms eat only a small amount of plant tissue. They make small pits in the main root system so causing the wilting and death of both seedlings and young plants. Tomatoes and beans are normally attacked below ground level; the wireworm tunnels up within the stem and in larger plants, may reach as much as 6 in. up the main haulm. The larvae of the four species most likely to be encountered under glass may be identified in the keys provided by van Emden (1945).

Control

When glasshouses are erected on soil likely to have a considerable population of wireworms and leatherjackets it may also contain cockchafer and

FIG. 5.42. Wireworms (×4). (By kind permission of the Plant Pathology Laboratory, Ministry of Agriculture, Fisheries and Food.)

nuctuid moth larvae. These can cause irreparable damage to plants within a few days of planting and consequently preventive measures should be taken unless expert examination of the soil has shown it to be free from pests.

Wireworms, and other larvae associated with grassland, are easily controlled by working a gamma-BHC dust into the top 2 or 3 in. of soil 2 or 3 weeks before planting. Dusts specially prepared for soil application are more economic than dusts developed for application to foliage. A dust containing 0·4% gamma-BHC should be applied at $\frac{1}{2}$ oz/yd^2 (1$\frac{1}{2}$ cwt/acre). To avoid tainting, which is likely to occur with BHC, root crops such as radishes should not be grown for at least 18 months after treatment.

Where wireworms enter older glasshouses from adjoining infested land, this treatment can be applied to a 2–3 ft strip at the sides. The best preventive measure, however, is to maintain a weed-free strip alongside the houses.

If plants are being attacked by wireworms they should be watered with 0·01% parathion or, where the plants are tolerant and there is no danger of tainting, with 0·01% gamma-BHC. The rate of application depends to some extent on the plants; with tomatoes about 1 pint of diluted parathion or $\frac{1}{2}$ pint of gamma-BHC is required, the parathion treatment being repeated after 3 weeks.

Where newly planted lettuce are attacked the soil surface can be dusted with gamma-BHC and this washed into the soil by heavy damping.

REFERENCE

VAN EMDEN, F. I. (1945). Larvae of British Beetles. V. Elateridae. *Entomologist's mon. Mag.*, **81**, 13–37.

Family CURCULIONIDAE

Otiorrhynchus sulcatus (F.) (the black vine weevil), is sometimes referred to as the 'cyclamen grub', as it is a destructive pest on this crop. The adult has the head prolonged in the form of a broad rostrum and the antennae sharply elbowed. It is 9–10 mm long, basically black, but patches of yellow hairs on the elytra give newly emerged specimens a mottled appearance (Fig. 5.43). The elytra are joined so that the beetle is flightless and so spread is by walking. Vine weevils are common outdoors in late spring and wander into many glasshouses through low ventilators and under doors. They feed at night, causing a distinctive injury by eating out irregular notches from the margins of the leaves of vines, peaches, rhododendrons and other plants. During the day they hide in crevices or the soil.

Reproduction is parthenogenetic, no males being known, and begins about 10 weeks after maturation from the pupa. Under glass, weevils mature in greatest numbers in the late autumn. As individuals oviposit for at least 4 months, producing about 1000 eggs at an average rate of 7 per day, eggs will generally be found in spring and early summer. They are deposited near plants in the surface soil. Eggs are spherical (0·7 mm in diameter), white at first but becoming pale brown within a few days. They hatch in about 3 weeks to legless, yellowish-white grubs with markedly wrinkled bodies and brown heads. When disturbed in soil they lie in a characteristically curved manner (Fig. 5.44). They feed for about 80 days throughout the spring and early summer before pupating in earthen cells about 12 mm in diameter within an inch or so of the soil surface. Hence eggs laid in mid-winter give rise to full-grown larvae which may cause extensive damage the following autumn. The adults emerge about 7 weeks later. Though there is only one generation per year there is much overlapping as outdoor populations lay eggs in early autumn so that, at any time, all stages may be found. Severe damage is done to cyclamen roots (Fig. 5.45): young

FIG. 5.43. Adult *Otiorrhynchus sulcatus*: vine weevil ($\times 7$). (A Shell photograph.)

larvae attack rootlets but older grubs may destroy the entire root system and make cavities in the corms. *Adiantum* and other ferns, begonia, carnation, gloxinia, orchids, primulas, sanseveria, saxifrages and lettuce are all liable to attack. Smith (1932) lists 77 potential hosts, and recently Mason (1960) recorded that, in addition, *Cissus antarctica* Vent. was attacked by both adults and larvae but *Hedera canariensis* Willd., *Spathiphyllum* sp. and *Fatshedera lizei* Guillaum were damaged only by the adults. When damage does occur to pot-plants they may wilt and die almost overnight.

Control

Where vine weevils, or their larvae, occur in glasshouses treatment of the potting compost with an insecticide is the best way of preventing further attacks. The admixture of aldrin dust at 2 oz $1\frac{1}{4}\%$ per bushel of compost should be highly effective and nonphytotoxic. The use of heptachlor, as recommended in the U.S.A., is unacceptable in Britain.

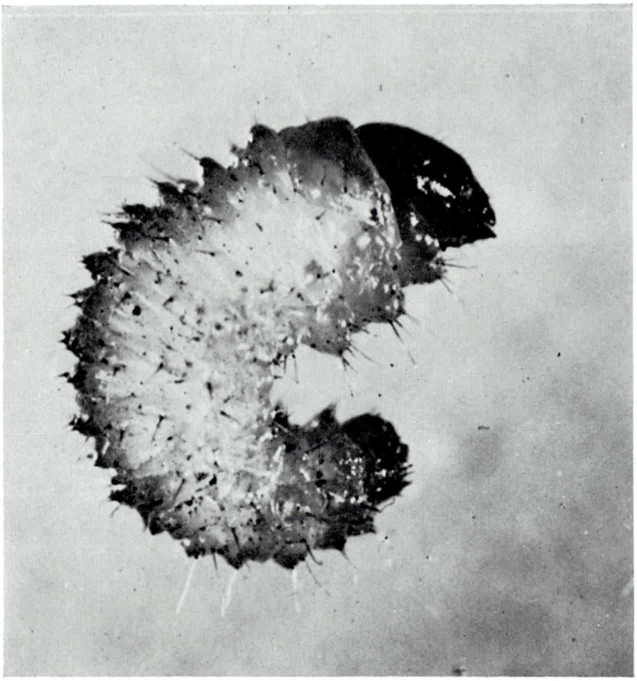

FIG. 5.44. Larva of vine weevil ($\times 5$). (A Shell photograph.)

Other acceptable substitutes are gamma-BHC and DDT. Although the former is more active against the larvae, DDT, in view of its lower phytotoxicity, is recommended at the rate of $\frac{1}{2}$ lb of a 5% dust per bushel of compost. For plants known to be tolerant to soil applications of BHC, $\frac{1}{2}$ lb of a 3% dust per bushel can be used. As an alternative the application of 20 to 25% DDT wettable powder as a dust to the surface of the soil, especially near collars of plants, will destroy larvae hatching from eggs over a period of several weeks.

To destroy larvae attacking plants established in pots or borders, a 0·02% gamma-BHC emulsifiable concentrate should be applied in sufficient amount to soak the soil to a depth beyond that at which the young grubs are feeding.

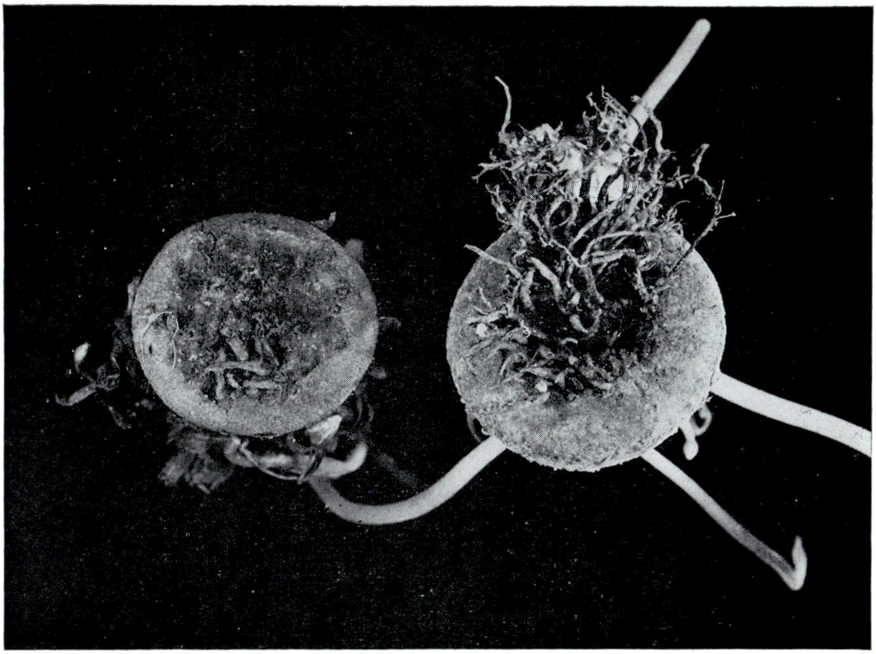

FIG. 5.45. Vine weevil damage to cyclamen roots. (A Shell photograph.)

Spraying plants with 0·1% DDT or 0·02% BHC, or smoking in the late evening with parathion or BHC will kill adults feeding on the foliage. Such applications should be repeated when fresh damage is caused by later emerging adults.

As this pest often occurs on nurseries growing a wide range of ornamental plants, the phytotoxicity hazard of any chemical must be considered.

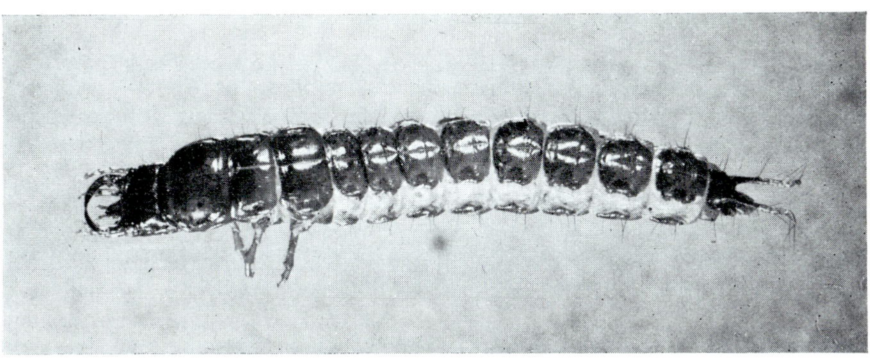

FIG. 5.46. Predatory larva of a ground beetle (Carabidae) (×3). (A Shell photograph.)

REFERENCES

MASON, C. (1960). Observations on the life-history and control of the vine weevil on cyclamen and foliage plants. *Pl. Path.*, **9**, 29–33.

SMITH, F. F. (1932). Biology and control of the Black Vine Weevil. *Tech. Bull. U.S.D.A.*, No. **325**, 45 pp.

Family CARABIDAE: Ground Beetles

Passing reference should be made to this large family of predacious beetles, members of which are occasionally found in soil and compost. The adults are very active, usually large insects with well-developed head, thorax and abdomen and long thread-like antennae. The active larvae are characterized by a pair of prominent cerci at the hind-end of the heavily chitinized body and sharp sickle-shaped mandibles (Fig. 5.46). They are valuable predators of soil insects.

Order DIPTERA: Flies

Suborder Nematocera

Family TIPULIDAE: Crane Flies or 'Daddy Long Legs'

These flies have a prolonged head, almost in the form of a proboscis, a thin body and long, fragile legs. They abound in moist grassland throughout Britain. The adults are not injurious but the larvae, or 'leatherjackets', sometimes cause damage in newly erected glasshouses. They normally feed on turf, but when deprived of this, will attack many plants. Leatherjackets may be introduced into glasshouses either in turf or, occasionally, in soil on the roots of chrysanthemums when these are brought into glasshouses for flowering. Bulbs, corms, tubers, rhizomes and underground stems of all kinds may be superficially damaged or tunnelled and any such plants brought into glasshouses for forcing and flowering may introduce the commoner species, *Tipula oleracea* Linn., *T. paludosa* Meig. and *Nephrotoma maculata* Meig. These species may be separated quite readily by the colour of the body. *T. paludosa* is grey with a distinct reddish-brown tinge and indistinct stripes on the thorax, *T. oleracea* is dusty-grey with pale stripes on the thorax, while *N. maculata* is yellow and black. Other species may be identified by reference to Coe *et al.* (1950).

In early autumn black, shiny, elliptical eggs ($0{\cdot}3 \times 1{\cdot}0$ mm) are laid in grassland on, or near, the soil surface either in small groups of 5–6 by the flies probing with their ovipositor, or singly when dropped freely during flight. They hatch in 10–14 days (as long as 3 months in *N. maculata*) into brownish-grey leatherjackets (Fig. 5.47) with small black heads and a rough, wrinkled body which is somewhat rubbery to the touch. The eggs and newly hatched larvae are sensitive to desiccation so that survival is highest when cool, wet weather prevails in late summer. The larvae, which move through the soil by means of rows of short bristles, feed during the autumn and again the following spring. When full grown, leatherjackets are about an inch long and feed voraciously, causing wilting and dying in attacked plants. The pupae (Fig. 5.48) which are formed in the soil, move to, and project from, the surface to liberate the adult within 14 days. Each female lays between 200–500 eggs.

Control

Where glasshouses have been newly built on grassland, weedy arable land, or where older houses have been re-soiled with soil from such sites, the ground should be examined to see if leatherjackets or other soil pests are present. If the soil is cultivated in the autumn and allowed to dry out thoroughly before being used, leatherjackets will probably be destroyed by desiccation. After the removal of 'lifted' chrysanthemums from glasshouses in districts where crane flies are numerous, the border soil should be examined for leatherjackets.

Where leatherjackets are the only pest a 5% DDT, a 0·4% gamma-BHC dust, or a 3% BHC dust should be applied at $\frac{1}{4}$ oz/yd², raked in lightly and the

FIG. 5.47. Leatherjacket (cranefly larva) (×3). (A Shell photograph.)

surface of the ground moistened. Alternatively, the soil can be sprayed with 0·05% DDT, or 0·02% gamma-BHC, using 1 gal/100 ft².

Should wireworms also be present, the treatment with BHC to control these will destroy leatherjackets.

Control can also be obtained by baits. To 28 lb of coarse bran add an amount of a wettable powder formulation of gamma-BHC or DDT equivalent to 4 oz of active ingredient—1 lb of Paris green powder (a scheduled poison) may be substituted for these chemicals. Add about $1\frac{1}{2}$ gal of water to make the bait moist, yet friable. Mix thoroughly and spread evenly at a rate of 1 lb/100 yd². Rubber gloves should be worn when mixing and distributing the bait, and care must be taken to avoid inhalation of the insecticidal dusts. Dry soil surfaces should be dampened before distributing the bait.

REFERENCES

BRINDLE, A. (1960). The larvae and pupae of the British Tipulinae (Diptera, Tipulidae). *Trans. Soc. Br. Ent.*, **14**, 63–114.

COE, R. L., FREEMAN, P. & MATTINGLY, P. F. (1950). Handbooks for the Identification of British Insects. IX (2), Diptera 2. Nematocera. *R. ent. Soc. Lond.* 216 pp.

LAUGHLIN, R. (1960). Biology of *Tipula oleracea* L. I. Growth of larvae. *Entomologia exp. appl.*, **3**, 85–97.

FIG. 5.48. Pupa of cranefly (×4). (By kind permission of the Plant Pathology Laboratory, Ministry of Agriculture, Fisheries and Food.)

Family MYCETOPHILIDAE: Fungus Gnats

The larvae of several species of these delicate, dark grey or black flies are important pests of glasshouse plants. In most species the larvae are saprophagous, feeding on decaying plant tissues, but a few are primary pests. It is not possible to identify the larvae; the adults are also difficult but Tuomikoski (1960) has clarified their taxonomy. Those species likely to be encountered within glasshouses may be distinguished in the following key:

KEY TO GLASSHOUSE SCIARIDS

1. Posterior portion of pronotum hairy..........................
   Termitosciara (Plastosciara) perniciosa Edwards
2. Posterior portion of pronotum hairless........................ 3(4)
3. Female wingless: Maxillary Palp 1 jointed....................
   Pnyxia scabiei (Hopkins)
4. Female winged; Maxillary Palp 3 jointed...................... 5(6)
5. Male genitalia with basal papilla or tuft of hairs, clasper tapering to distinct apical claw (see Fig. 9.13)............................
   Lycoriella solani (Winn.) (=*fenestralis* Zett)
6. Male genitalia without basal papilla or hair tuft. Clasper ovate with apical claw hidden by tuft of hairs (see Fig. 9.14)................
   Bradysia paupera Tuomikoski

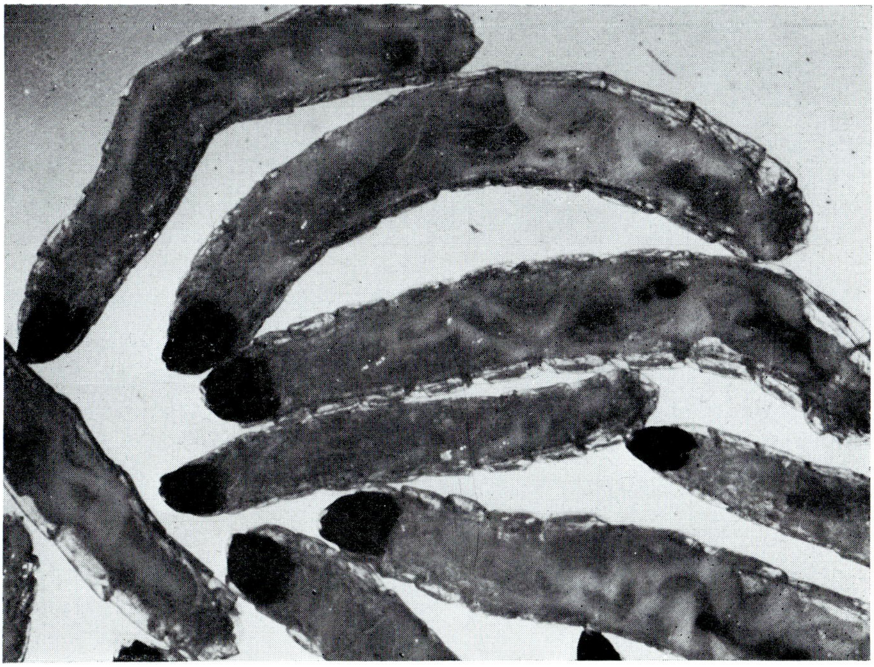

FIG. 5.49. Sciarid larvae (×15). (J. Hesling.)

Larvae of *Termitosciara perniciosa* and *Pnyxia scabiei* are slender and white with black heads (Fig. 5.49), they cause severe damage to cucumbers, especially if they are short of water, by attacking the tap roots and stem cortex near soil level, so causing collapse of the plants. Many observers have claimed that the larvae only attack growing plants when these are allowed to dry out but, while no experiments have been made to demonstrate this, there have been numerous cases where plants appeared to have been correctly watered right up to the time of the attack. Each female *P. scabiei* lays about 100 oval, white, shiny eggs (0·15 × 0·09 mm) in clusters within soil or compost (Fig. 5.50). Hatching occurs within 5 days. The larvae mature after 14 days into white 'free' pupae each within a dark, silky cocoon. The adult emerges about 8 days later. The females are wingless but the males are of two types, with long, or short, wings.

The common species *Bradysia paupera* is becoming an important pest of seedlings, rooted cuttings, and young plants. Adult females (Fig. 9.9), which are 3·0–3·5 mm long and black with a greyish-brown abdomen, are often seen

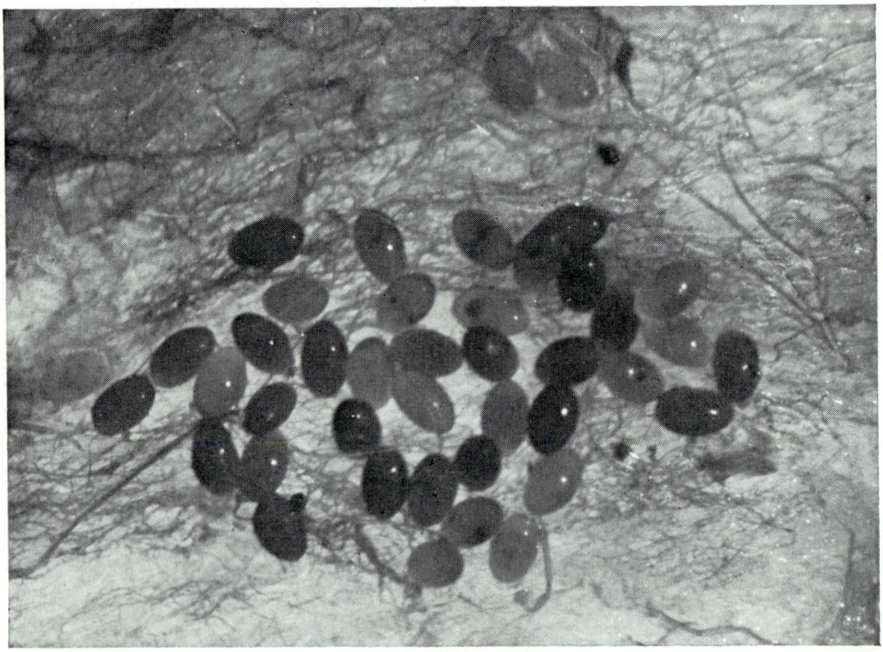

FIG. 5.50. Eggs of sciarid (*Bradysia paupera* Tuom.) (×50). (J. Hesling.)

hovering at the base of large plants, or running over the soil. They are attracted to steamed soil and also to dried blood used as a fertilizer on pot-plants. Mating occurs a few hours after emergence; 3 days later each female lays over 100 elongate eggs (0·2 × 0·1 mm) which hatch in 9–12 days at 60 °F (15 °C), or only 4 days at 72 °F (22 °C). The larvae pass through four instars in 24–27 days. When mature, some construct a cocoon of silken threads into which particles of soil are woven, but others form entirely unprotected pupae in the soil (Fig. 5.51). The light-brown pupae are 2 mm long. At 58 °F (15 °C) emergence occurs within 8 days; at 72 °F (22 °C) within 4 days.

The larvae feed on the root hairs of growing seedlings which may merely grow more slowly or actually collapse. This species has caused damage to

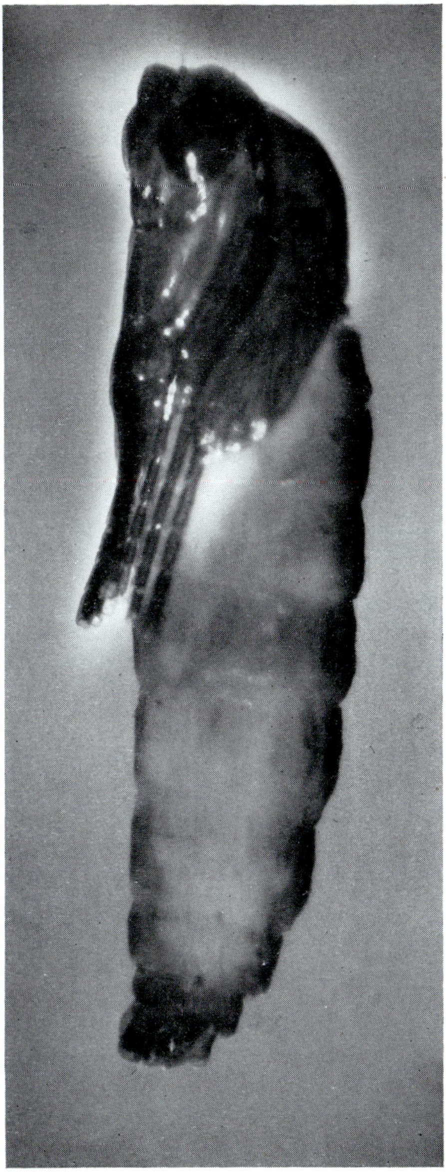

FIG. 5.51. Pupa of *Bradysia paupera* (×65). (J. Hesling.)

carnation cuttings in Germany, and is an important pest of many plants including cyclamen, sweet peas, orchids, primulas, freesia seedlings, and chrysanthemum cuttings. An unusual form of damage has been caused to the leaves of cucumbers on which the females laid eggs. The larvae skeletonized the leaves in a similar manner to young lepidopterous caterpillars. The pupae were anchored by silken threads to the leaf surface. The successful development by hundreds of flies on apparently healthy tissues confirms that this species is primarily phytophagous. *Lycoriella solani* (Winn.) is a widespread insect which has frequently been indicted as a primary pest. This is still conjectural, for it is often saprophagous and its life history is discussed in this light under Mushrooms on p. 320. Although it has caused damage to the roots of azalea and *Mesembryanthemum* sp. it normally feeds on rotting plant tissues and is frequently found in damaged bulbs, slug-eaten celery roots, potato tubers, and on tomato roots affected by brown root rot.

Control

Where sciarids are a recurrent pest, soil which has been sterilized by steaming or baking should be protected by covering with sheets, or by storage in closed bins.

Parathion at 0·01% when applied to pots, boxes or borders, is very effective against larvae and non-injurious to plants and seedlings. Sufficient should be applied so as to wet cucumber beds to a depth of 6 in.; this requires about 1 gal per linear yard of the normal-sized bed, and care must be taken to thoroughly wet the soil around the tap root. One application usually suffices to eliminate the pest unless infested dung is added when mulching.

Pots and boxes should be watered sufficiently to soak their contents, taking care to avoid wetting the foliage of plants such as cyclamen and crassulas, which are susceptible to damage. It may be necessary to repeat the application to pot-plants if they are exposed to further infestation. When diluting and applying parathion the protective clothing stipulated for Part II substances (see p. 38) must be worn.

Owing to its highly poisonous nature parathion is only obtainable by professional growers, and so malathion, at 0·01%, should be used by amateurs. Incorporation of diazinon in potting compost at 50 p.p.m. also shows promise of being effective against the young larvae without being phytotoxic to seedlings.

REFERENCES

GERBACHEVSKAYA, A. A. (1963). Leaf-midges (Diptera; Lycoriidae) injurious to vegetables and common mushrooms in hothouses in the Leningrad Region. *Ent. Obozr.*, **42**, 496–511.

MADWAR, S. (1934). The biology and morphology of *Pnyxia scabiei* Hopkins (Dipt. Nematocera). *Bull. Soc. ent. Égypte*, **17**, 136–149.

PARR, W. J., CROCKER, C. & SPEYER, E. R. (1954). A sciarid fly injurious to seedlings. *Rep. exp. Res. Stn Cheshunt*, **1953**, 36–9.

SPEYER, E. R. (1923). Mycetophilid flies as pests of the Cucumber Plant in Glasshouses. *Bull. ent. Res.*, **13**, 255–9.

TUOMIKOSKI, R. (1960). Zur Kenntnis der Sciariden (Dipt.) Finnlands. *Ann. zool. Soc. 'Vanamo'* **21**, 1–164.

Family CECIDOMYIIDAE

These minute and delicate flies have long antennae decorated by conspicuous whorls of hairs, and hairy wings with only a few veins. They develop from larvae attacking a wide range of ornamental plants (Barnes, 1948). The larvae are characterized by a minute head and a *sternal spatula*, or anchor-shaped process, on the ventral side of the thorax. Midges are a difficult group for the non-specialist to identify on taxonomic characters alone, and are, for practical purposes, best identified by the characteristic damage caused by the larvae. Harris (1966) has, however, produced a useful key to the economically important species.

Diarthronomyia chrysanthemi Ahlberg (the chrysanthemum midge) first appeared in the U.S.A. but, in 1927 and again in 1936, was introduced into England on chrysanthemum cuttings imported from America. It spread from these sites of introduction and was widely established on commercial, autumn flowering varieties. While primarily a greenhouse pest it can persist and breed outdoors, though it does not breed on weeds or summer chrysanthemums. Under glass it breeds throughout the year producing about eight overlapping generations.

The adult fly, about 2·5 mm long, is reddish-brown—the female tending to be more reddish than the male. The flies emerge on sunny mornings, the males living for only about 6 hr. Even when fertilized, the females rarely live longer than 1 day. There is usually a preponderance of females in the ratio of 3:2. Mating occurs within a few minutes of maturation and in light, warm conditions oviposition is often completed by noon on the day of emergence. The bright-red eggs are laid singly on new growth, irrespective of whether this is terminal or axillary. They are placed either within the folds of young leaves, or amongst the bracts of flower buds. Individual females lay up to 100 eggs, the average being about 30. The incubation period varies between 3–14 days, dependent on temperature, humidity and season. Development is favoured by moist, warm conditions. Within 24 hr the minute, legless, pale-orange larvae burrow within the leaf tissues. About 7 days after they have begun to feed, their presence is shown by minute, light patches within the leaves. An oval, or cone-shaped, gall 2 mm long projecting obliquely from leaves, stems, buds or calyxes becomes visible between 16 and 100 days from hatching of the eggs (Fig. 5.52). The duration of the larval stage varies between 2 weeks and several months. The pupal phase likewise lasts from 6 days to several weeks, according to the environmental conditions. Five generations may occur annually in unheated glasshouses, and as many as eight with continuous heat. Prior to emergence the pupa projects from the gall to allow the adult to escape. In slight attacks the galls are scattered over the upper leaf surface but, in more severe cases, the plant may be so heavily galled that the leaves become distorted, the stems twisted and swollen and the flowers misshapen. Untreated infested cuttings grow into weak plants.

Control

Since infestations of chrysanthemum midge usually result from bringing infested cuttings or plants into glasshouses, such material, unless it has been obtained from a reliable source, should be examined for the presence of galls. The occurrence of the pest in Jersey must be notified, and affected plants destroyed in accordance with the Destructive Insect and Pest (Jersey) Law, 1960. Destruction of infested plants is not required in Britain, but measures must be taken to control the pest.

Spraying with 0·01% parathion, 0·01% diazinon or 0·01% gamma-BHC gives good control of all stages, and will eradicate an infestation provided several applications are made at intervals of about 10 days in summer, or 21 days under cooler conditions in autumn and winter. Emulsifiable concentrate formulations of these insecticides should be used and care taken to thoroughly wet both the flower stem and young foliage with the insecticide.

FIG. 5.52. Galls of chrysanthemum midge. (By kind permission of the Plant Pathology Laboratory, Ministry of Agriculture, Fisheries and Food.)

Particular care must be taken to 'clean-up' infested stock plants. The young growth, and any older growth, bearing galls should be removed before the first spraying, and drenching sprays applied to thoroughly wet developing 'cutting' shoots.

Dusting with 5% DDT, or 0·3–0·6% gamma-BHC, at 5-day intervals over a period of about 3 months will control the pest by killing adults, but will cause undesirable disfigurement of the foliage by deposits.

Applications of thionazin, at 0·02%, to rooted cuttings for the control of chrysanthemum eelworm will probably control all larval stages of the midge, but may fail to prevent the emergence of adults from pupae formed before treatment.

REFERENCES

BARNES, H. F. (1939). The Chrysanthemum Midge. *Jl R. hort. Soc.*, **64**, 503–6.

— (1948). Gall Midges of Economic Importance. *IV. Ornamental Plants and Shrubs*. Crosby Lockwood, London. 165 pp.

HARRIS, K. M. (1966). Gall midge genera of economic importance (Diptera: Cecidomyiidae) Pt I. *Trans. R. ent. Soc. Lond.*, **118**, 313–58.

Dasyneura rhodophaga (Coquillet) (the rose bud midge) has only doubtfully been recorded in Britain but it is a most important pest of outdoor and glasshouse roses in the U.S.A. and Canada. The flower and leaf buds of infested roses become distorted, turn brown and often die. No flowers are produced and young shoots are frequently malformed.

The adult, which is yellowish-brown and less than 1 mm in length, oviposits on the sepals of flower buds, in the axils of young leaves or within leaf buds. The yellowish eggs hatch in 2 days and the white larvae feed on the young tissues causing infested buds to shrivel; distorted growth results when feeding occurs on the upper surface of young leaves and petioles. As many as 30 larvae have been found in a single rose bud. The larvae turn orange as they approach maturity about a week later, and drop to the soil where they pupate in cocoons. The adult midges emerge from these cocoons in about a week. The normal period of activity is from May to November, the winter being passed in the soil as pupae.

Control

The pesticides advised for the control of this pest in America are DDT, malathion, parathion and sulfotep. It is probable that other organophosphorus insecticides, in particular diazinon, would also prove effective. The choice may be influenced by the presence of additional pests.

The persistence of DDT is advantageous where overlapping generations are possible. A wettable powder, or emulsion formulation, diluted to 0·1% DDT and applied at high volume both to the bushes and the soil surface beneath, should be effective. However, H.V. application to roses is considered undesirable by many growers, consequently the best control measures are the application of DDT, diazinon or malathion aerosol concentrates at rates equivalent to 3·3 g; 1·8 g and 2·4 g/1000 ft^3 respectively, or, on a small scale, the use of either a 5% DDT dust or smokes of DDT, parathion or sulfotep. As it is necessary to eradicate the pest to prevent serious loss of marketable blooms, the application should be repeated at least twice at 10-day intervals.

Dasyneura affinis (Kieffer) (the violet leaf midge) is an important pest of wild and cultivated violets grown both outdoors and in frames. The females may live up to 1 week, but, as in other midges, the males are short-lived. Each female can lay up to 100 eggs within the rolled-up leaf margins and, sometimes, the open flowers. Oviposition generally takes place in early afternoon. The gall which forms 8–11 days after oviposition often causes a bluish colouration of the distorted leaves. In severe cases all the central leaves are malformed and the whole plant is dwarfed with almost no flowers (Fig. 5.53). Continuous infestation for 3 years may cause death. Where flowers are attacked the sepals and petals are swollen and may be rolled upwards. As many as 20 larvae may be found in a single gall 4–8 days after oviposition. Dependent on the species

FIG. 5.53. Damage caused by violet leaf midge. (Courtesy of Glasshouse Crops Research Institute.)

of *Viola* the galls may be bare or hairy. The larvae, white at first, become pale orange as they mature. They pass through four moults and, after maturing in 6 weeks, pupate in cocoons within the galls. The pupal stage lasts 10–12 days and about four generations are completed annually outdoors, but under glass an almost continuous sequence of at least five generations is possible. In cold frames and outdoors the winter is passed as a larva within the gall.

Control

In view of the difficulty in controlling this pest, care should be taken to select only clean plants for frames or glasshouses. Although good control may be possible by hand-picking and destruction of affected leaves with small batches of plants, complete control requires periodic examination of the plants and is often impractical.

The usual method is based on killing the egg-laying adults. With plants in the open this is accomplished by spraying with 0·1% DDT or dusting with 5% DDT before the peak egg-laying periods, *viz* early May, early July, mid-August and mid-October. The disadvantage is that attacks of spider mites are likely to be severe following destruction of predators by DDT. As climatic conditions in glasshouses are likely to modify the life cycle of the midge, routine applications of DDT sprays or dusts at 10-day intervals over a period of 6 weeks, commencing when galls are first observed, are advised.

The systemic compounds demeton (0·02%) and schradan are effective in controlling the larvae within the leaves though demeton is not currently approved for use in glasshouses in Britain.

Other systemic compounds likely to give effective control if a single application is given immediately galls are observed, include demeton-S-methyl, dimethoate, disulfoton, oxydemeton-methyl and thionazin.

REFERENCES

COUDER, J. (1933). Étude monographique de la cecidomyie de la violette. *Bull. Soc. Hist. nat. Toulouse*. **65**, 193–279.

GODAN, D. (1956). Contributions to the ecology of *Dasyneura affinis* Kieff. *Z. angew. Ent.*, **39**, 1–19.

LEATHERDALE, D. (1965). The Violet-leaf midge. *Gdnrs' Chron.*, **158**, 120.

Parallelodiplosis cattleya (Molliard) (the cattleya midge) causes both spherical and elongate galls, about the size of a pea, on the aerial roots of orchids (*Cattleya* spp. and *Laelia* spp.). Large numbers of these galls may reduce the growth of infested plants. Each gall contains several deep-orange larvae in individual cells. The larvae pupate within their cells but the pupae protrude from the gall to permit the adult to emerge. Several generations of the orange-coloured midges are produced each year.

Control

The application of insecticides to control more common pests prevents attacks by this midge. An attack can usually be controlled by cutting off and burning galled roots but if galls are numerous and their removal too laborious the roots on the outside of the compost should be sprayed with 0·01% gamma-BHC as a wettable powder. Several applications, at intervals of 14 days, may be required. It is inadvisable to use an emulsifiable concentrate unless the orchid is known to be non-susceptible.

Suborder Cyclorrhapha

Family SYRPHIDAE: Hover flies

Hover flies are readily recognized by their habit of hovering, with rapidly vibrating wings, before darting to another 'hover station' a few feet away. They are usually brightly coloured, often with yellow stripes on a black or blue-black ground; indeed, many bear a superficial resemblance to wasps, while some of the hairy species may be mistaken for bees. The larvae have diverse habits, feeding within bulbous tissues and fungi or living as saphrophages in decaying organic material, in the rotting wood of trees, or in the nests of ants and bees. To the horticulturist the most important species are those whose larvae feed predaciously on aphids and other plant lice (Fig. 5.54). These predatory larvae, which are commonly found in association with large aphid infestations, are usually flattened, somewhat slug-like creatures, tapering anteriorly to the small head. The tough cuticle is predominantly green or brown, and is covered with regular rows of transverse corrugations; at the blunt, hind-end of the body are a pair of short respiratory horns. Outdoors, syrphid larvae are often responsible for eliminating dense aphid colonies, but they do

FIG. 5.54. Syrphid larva predatory on aphids. (A Shell photograph.)

not normally achieve economically useful control. There is, however, considerable interest in these insects because of their potential in integrated control programmes under glass, and, in France, a successful technique for controlling *Myzus persicae* on lettuce has been demonstrated. Difficulties of mass breeding aphidophagous species may, however, limit this potential.

Merodon equestris Fab (the large narcissus fly) is an important pest of bulbs. The adult is a large fly 12 mm long with a stout, densely hairy body of variable colour. The ground colour is black but the hairs may give the insect an almost entirely tawny appearance, or it may be red, orange or grey, so giving a strong resemblance to bumble bees. Syrphids are, of course, distinguished from the latter by possessing only one pair of wings. *Merodon* may be separated from drone flies, which it superficially resembles, by the broad tooth-like process on the underside of the hind femora.

M. equestris has one generation a year, adults appearing between late April and the end of June. Flies may emerge in forcing houses as early as February, but are then unlikely to find suitable conditions for oviposition. Outdoors, the

FIG. 5.55. Damage caused by large narcissus fly. (Courtesy of Glasshouse Crops Research Institute.)

females live for about 17 days and during bright, warm, windless days, 40 small, oval, white eggs are laid singly on the soil surface near, or on, bulbs or foliage. Ideally, the fly crawls down into the soil cavity formed above the nose of the bulb by the dying foliage laying only one egg on, or near, each bulb. The eggs are susceptible to both desiccation and excessive damp. Small, active, hairy larvae hatch in 10–15 days and burrow down through the soil alongside the bulb, eventually tunnelling into it through the undersurface, or side, of the base plate. The entrance hole resembles a rusty pin-prick. At first, the larva feeds only at the base of the bulb, but later it enters the scales and, within a few weeks, excavates the whole centre of the bulb (Fig. 5.55). As the legless larva grows it becomes yellowish and cylindrical in shape, with a slight taper anteriorly; at the hind end is a horny, black respiratory process, with, on either side, another minute projection. Full grown larvae are 20 mm long and occasionally, as food becomes depleted, they may migrate from one bulb to another. The winter is passed as a larva within the bulb, and in early spring this is vacated either via the base or neck, and the larva pupates

FIG. 5.56. Larvae of *Eumerus tuberculatus* Rond. in narcissus bulb. (J. Hesling.)

at the soil surface. The puparium is roughly pear-shaped, tough and leathery, with the posterior projection plainly visible. The adult fly emerges about 6 weeks later.

While narcissus is the principal host, *Merodon* larvae have been found in species of *Amaryllis*, *Cyrtanthus*, *Galanthus*, *Galtonia*, *Habranthus*, *Hyacinthus*, *Iris*, *Leucojum*, *Scilla* and *Vallota*. While early attack is difficult to detect, infested bulbs, examined in the autumn, are softer than usual, especially in their 'neck' region.

Eumerus (=***Paragopsis***) spp. (the small narcissus flies) *E. tuberculatus* Rond. and *E. strigatus* Fall. are also pests of bulbous plants, the former principally attacks narcissus and onion, the latter more frequently attacks hyacinth, onion and parsnip.

Both flies are about 6 mm long, generally black, with three pairs of white crescent-shaped marks on the abdomen, and white lateral lines on the thorax. The species may be distinguished by the presence, in male *E. tuberculatus*, of a small projection on the ventral surface of the hind femora.

The first generation appears from late April to the end of June, while the smaller, second generation is on the wing from July until October. The female, which usually selects damaged bulbs for oviposition, begins to lay within 5 days of emergence. The oval eggs (2·5 mm long) are laid in groups of 5–40 on the bulb collar, as well as in folds of the coat, and on the ground in the vicinity of bulbs. They hatch in 2–12 days (50–77 °F; 10–24 °C). The larvae have a rather transparent cuticle and, when fully grown, are 8 mm long. The posterior respiratory horn is red; in addition there are a pair of small single tubercles and an even smaller pair of 'twin' tubercles. The larvae are found principally in the neck region, though the whole bulb may be reduced to a rotting mass (Fig. 5.56). The first generation larvae pupate after 2–4 weeks, and the flies emerge 9–20 days later. *Eumerus* sp. are more closely associated with decay than *Merodon equestris* and show a preference for damaged tissues, especially that caused by the eelworm *Ditylenchus dipsaci*; indeed, they may be purely saprophagous.

The second generation of flies emerges from July onwards. Some progeny of this generation mature the same year, but most overwinter as larvae, to emerge in the spring of the following year. It is this generation which is commonly found emerging harmlessly from stored and forced bulbs.

Control

The control of narcissus flies is more the responsibility of the bulb producer than of the person purchasing such bulbs for forcing under glass. Freedom from both narcissus fly and eelworm should be stipulated by the nurseryman. It is possible to prevent oviposition either by covering the plants with hessian, or by earthing-up the bulbs so as to fill spaces in the soil left by dying leaves.

Consignments of bulbs for forcing should be examined with care and any which are soft, particularly in the neck region, should be rejected. Bulbs which have been partially destroyed by larvae will either fail to grow, produce only weak and distorted main leaves, or small, 'grassy' foliage. In bulb-producing areas where narcissus flies occur, measures should be taken to prevent infestation before lifting. Dipping the bulbs in diluted aldrin or dieldrin emulsifiable concentrates will afford protection for 2 years. The use of aldrin or dieldrin is acceptable in Britain, provided the concentration does not exceed 6 oz active ingredient per 20 gal (equivalent to 0·18%). If the basal plates of the bulbs have commenced to swell this treatment should not be applied, but alternatively, the drills may be treated by dusting with aldrin immediately before planting. For this purpose the active ingredient should not exceed 7 lb/acre, i.e. 5 cwt of a $1\frac{1}{4}$% dust/acre. Half this quantity is sufficient to give protection. Where the bulbs have not been protected against flies, their immersion for 3 hr in a bath containing a gamma-BHC emulsifiable concentrate diluted to 0·05% gamma-BHC, and containing sufficient additional wetting agent to promote immediate wetting of the bulbs, will kill the larvae. Treatments must be applied as soon as possible after lifting bulbs for forcing.

Hot-water treatment, primarily directed at eelworm and scale mite control, will also destroy larvae within the bulbs.

REFERENCES

ANON. (1958). Narcissus Pests. *Minist. Ag. Fish. Food. Bull.*, **51**, H.M.S.O.

FORSYTH, J. (1967). Prevention of damage to bulbs by Narcissus fly larvae. *Daffodil Tulip Yb.*, **1967**, 104–7.

HODSON, W. E. H. (1927). The bionomics of the lesser bulb-flies, *Eumerus strigatus* Fall. & *Eumerus tuberculatus* Rond. in South-west England. *Bull. ent. Res.*, **17**. 373–84.

ZABIROV, SH. M. (1963). On the ecology of the Onion-bulb fly *Eumerus strigatus* Fall. (Diptera Syrphidae). *Ent. Obozr.*, **42**, 736–43.

FIG. 5.57. Adult *Phytomyza syngenesiae* Hardy: chrysanthemum leafminer (×10). (A Shell photograph.)

Family AGROMYZIDAE: Leafminers

The larvae of these small flies 'mine' in the stems or leaves of a wide range of host plants, and several species often become established in glasshouses.

Phytomyza syngenesiae Hardy (the chrysanthemum leafminer), a widely distributed polyphagous species, is an important pest of chrysanthemums. The adult fly (Fig. 5.57), 2–2·5 mm long, is black with a light-grey bloom; the lower portions of the head and abdomen are yellow. The wings extend beyond the body when at rest. The flies are relatively inactive and make only short flights. Adult females feed on the plant sap exuding from cone-shaped cavities made by inserting the ovipositor just below the epidermis. After a short time the feeding punctures, which are far more numerous than those made for oviposition, become obvious because small bleached areas form followed by characteristic calluses (Fig. 5.58). At 65 °F (19 °C) the adults live for about 18 days and, where the mean temperature is reduced to 41 °F (5 °C), for as long as 24 days. At the higher temperature each female, while laying her complement of 50–75 eggs, produces about 4 eggs per day. The smooth, translucent-white eggs (0·4 × 1·5 mm) are inserted into the undersides of the leaves. They hatch in 3–6 days into smooth, white larvae, tapering anteriorly to the head which has prominent mouth hooks. They feed in the palisade tissues on the ventral leaf surface. At first the mines are narrow, but as the larvae grow, they widen and become more tortuous. The mines usually begin at the oviposition site, within 10 mm of the leaf margin, and work gradually towards the mid-rib. The removal of the palisade tissue makes the mines white and conspicuous (Fig. 5.59), as are the granules of frass distributed at intervals along the tunnel. There is evidence that there are several strains within the species each of which deposits its frass in a distinct pattern (Hussey & Gurney, 1962). In light attacks mines occur only singly but, in severe cases, there may be 20–30 larvae in a single leaf, which then becomes completely blanched.

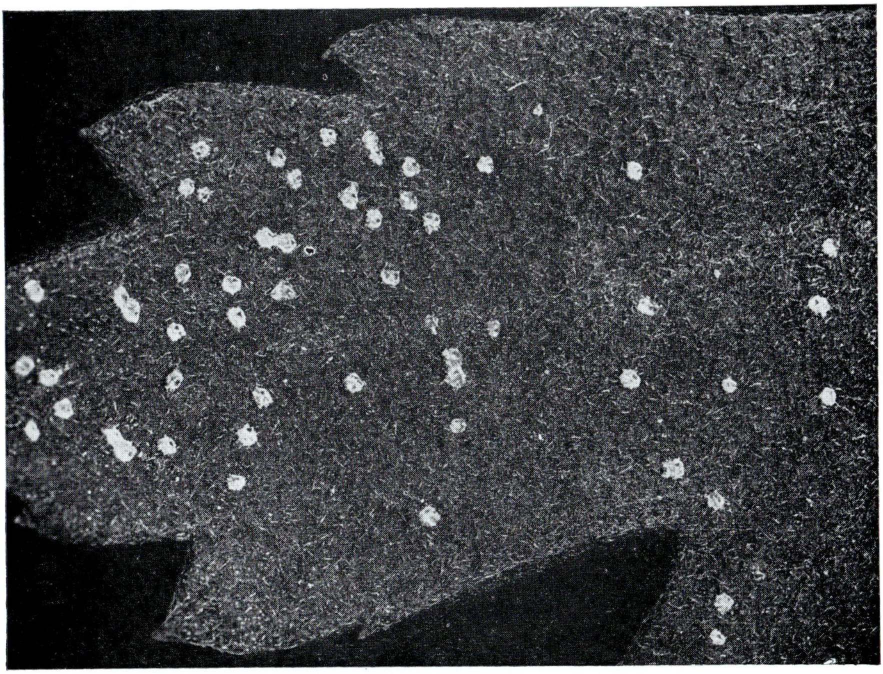

FIG. 5.58. 'Feeding marks' of adult *Phytomyza syngenesiae*. (Courtesy of Glasshouse Crops Research Institute.)

Pupation takes place in the mines after 5–10 days' larval feeding. When first formed, the pupal skin is yellowish but soon becomes dark brown. At 70 °F (21 °C) the adult flies escape, through a circular hole at one end of the puparium, after 9 days; at 60 °F (16 °C) after 12 days, or as long as 55 days at 40 °F (5 °C). Mating occurs a day or so later.

Outdoors, the first generation of flies appears in May, and the second in late July. The latter generation overwinters as pupae.

While almost all chrysanthemum varieties may, from time to time, be found infested, there are considerable differences in susceptibility. Cultivars Galaxy, Vibrant, Long Island Beauty, Tuneful and Delightful, are particularly vulnerable, while others such as Golden Delightful, J. R. Johnson, Pink Pride, Milestone, Yellow Shasta, Yellow Iceberg, Cerise Shoesmith and Dark Delmar, are relatively resistant.

The foliage of many other cultivated *Compositae* especially cinerarias may be disfigured by the tunnels and, under glass, where several generations may develop annually, light attacks must be controlled.

FIG. 5.59. Disfigurement of foliage by chrysanthemum leafminer. (Courtesy of Glasshouse Crops Research Institute.)

Control

The occurrence of adult feeding punctures necessitates prompt measures to prevent disfigurement by mined leaves.

The usual chemicals employed are gamma-BHC or diazinon. Parathion, though very poisonous, is also effective but should only be used where required to control other pests. Although not officially stipulated, it is strongly recommended that there shall be an interval of at least 4 days between parathion sprays and disbudding, cutting or packing flowers.

Two or three applications, at high volume, of 0·01% diazinon or 0·01% BHC at 14-day intervals give good control. In spite of increasing the frequency, and number, of applications some growers may find it more convenient to apply these chemicals as aerosols by using atomisable concentrates, or as smokes at 7-day intervals. Reports of difficulty in controlling leafminer on year-round chrysanthemums have revealed that strains tolerant to BHC and diazinon have been selected. However, many poor results have resulted from inadequate application. Systemic compounds have certain advantages, and evidence that chrysanthemum leafminer can be controlled by these materials has been gathered from incidental observations of trials in which their aphicidal properties were investigated. Sprays and soil drenches of thionazin killed larvae within the mines. There is also field evidence that leafminer rarely occurs where oxydemeton-methyl is regularly applied for aphid control.

REFERENCES

COHEN, M. (1936). The Biology of the Chrysanthemum Leaf Miner *Phytomyza atricornis* Meig. *Ann. appl. Biol.*, **23**, 612–32.

GRIFFITHS, G. C. D. (1967). Revision of the *Phytomyza syngenesiae* group (Dipt. Agromyzidae) including species hitherto known as *P. atricornis* Mg. *Stuttgarter Beitr. z. Naturh.* **177,** 1–28.

HUSSEY, N. W. & GURNEY, B. (1962). Host selection by the polyphagous species *Phytomyza atricornis* Meig. (Dipt. Agromyzidae). *Entomologist's mon. Mag.*, **98**, 42–7.

Liriomyza solani Hering (the tomato leafminer) has occurred in SE. England, where it attacks both tomato and, more rarely, cucumbers since 1926. Where mines are formed on established plants little damage is caused, but serious injury may occur, both in the cotyledons and terminal shoots, when eggs are inserted in tomato seedlings. If the cotyledons are large, the larvae may restrict tunnelling to these primary leaves but, where they are smaller, the larvae tunnel up and kill the terminal shoot.

The adults are 2 mm long, the head is black above and yellow below, the thorax black with a yellow spot dorsally and the abdomen banded black and yellow. The flies live for about 12 days and cut 'feeding pits', with their ovipositor, mostly on the upper surface of the leaves. One individual fly has been found to cut 71 of these pits within 4 hr. Mating occurs during the first day after emergence, and oviposition begins a day later; 50–100 eggs are laid at a rate of about 7 per day. The eggs are white (0·25 × 0·15 mm) and are inserted in slits (about 0·35 mm deep) cut in the superficial leaf tissue. Hatching occurs after 6 days and the larvae pass through three instars in 10 days; in this time tortuous mines are made between the upper and lower leaf epidermis (Fig. 5.60). When full-grown the larva is about 8 mm long, pale orange with a dark ring round the posterior end of the body; it finally tunnels to the surface of the stem, or leaf, and emerges through a slit, cut by the mouth hooks, before falling to the ground. The external larval skin hardens to form a puparium which rests about 5 mm below the soil surface. The pupal period lasts from 36–65 days.

Control

Since the damage caused to tomato seedlings at the cotyledon stage in the late winter and early spring arises from eggs laid by flies emerging from puparia in the soil, prevention of adult emergence is most important. If miners were present in the late autumn, it is inadvisable to propagate tomatoes unless the soil has been treated to destroy puparia.

FIG. 5.60. Mines of *Liriomyza solani* in tomato. (Courtesy of Glasshouse Crops Research Institute.)

This can be achieved effectively by steaming the soil. No information is available regarding the effect of chemical sterilants on leafminer puparia, but it is anticipated that most soil fumigants used to control nematodes or the fungi causing brown root rot of tomatoes, will be effective provided the surface is well sealed during the sterilization process. Formaldehyde treatment is, however, not effective.

Should an attack occur during the propagation of tomatoes, immediate action is necessary directly adult feeding marks become visible.

Tomato seedlings, especially when reared under poor winter light conditions, are susceptible to injury by many pesticides. Spraying with parathion at 0·01%, or with diazinon at 0·02% at 10- to 14-day intervals has proved satisfactory on both seedlings and young plants.

Parathion smokes at 7-day intervals, though tending to be more phytotoxic than sprays, will also control the pest. Atomized solutions should not be used until several true leaves are formed. Plants at, or beyond, the stage when the first truss is clearly visible can be sprayed with 0·01% gamma-BHC at 14-day intervals. If it is more convenient to use aerosols, atomized solutions of gamma-BHC (0·7 g/1000 ft^3) or diazinon (1·8 g/1000 ft^3) are satisfactory if undertaken several times at weekly intervals. Smokes of gamma-BHC or parathion applied weekly are also satisfactory. Light dusting with malathion as soon as the symptoms appear is also effective. A minimum period of 14 days should be allowed between the last application of BHC or diazinon spray and harvesting; but only 2 days if BHC is applied as an aerosol or smoke; and 1 day if parathion smokes are used. Every endeavour should be made to control infestations on tomatoes in the autumn to prevent pupation in the soil.

REFERENCE

SPEYER, E. R. & PARR, W. J. (1949, 50, 51). The Tomato Leaf Miner, *Liriomyza solani* Hering. *Rep. exp. Res. Stn Cheshunt.* **1948**, 43–51; **1949**, 48–56; **1951**, 51–4.

Phytagromyza dianthicola Venturi has recently caused severe damage to carnations in Belgium. The typical larval damage reduces the vigour and production of the plants. Eggs are laid on the leaves, and hatch after 2–4 days into a larva which tunnels for 10–12 days before pupating within the leaf. The adults emerge after a fortnight and produce 6–7 generations annually.

REFERENCE

HEUNGENS, A. (1966). Voorkomen van de mineervlieg *Phytagromyza dianthicola* Venturi in de Anjerteelt. *Meded. RijksFaculteit Landbw. Gent.*, **31**, 107–16.

Family TRYPETIDAE

The fruit flies are distinguished by the marbled, or banded wings, and often cause damage to plants by laying eggs on the fruits within which the larvae develop. The larvae of some species, however, live within leaf tissues causing characteristic 'blotch' mines. At least three species have been found on chrysanthemums.

Urophora (Euribia) zoe (Meigen) (the chrysanthemum blotch miner) disfigures winter-flowering chrysanthemums with large irregular blister-like mines. These are caused by stout, yellowish larvae about 6 mm long, hatching from eggs laid on the leaf surface and feeding within the tissues. When full-grown they emerge to pupate in the soil. There are two generations of flies each year. The adults are striking yellow flies with dark areas on each side of the thorax and large iridescent wings dappled with brown.

Paroxyna misella (Loear), recorded from Middlesex, Kent and Scotland, forms a terminal gall of the apical shoot and associated leaves. As many as eight larvae may occur in a single gall but occasionally the larvae make blotch mines in the leaves instead of tunnelling within the shoot. There are at least two generations per year, the winter being passed in the pupal phase.

Tephrites conjuncta (Loew) distorts the flower heads when the larvae feed at the base of the florets and on the surface of the capitulum. Two larvae are usually found in each infested flower and both pupate therein.

Control

Changes in chrysanthemum culture have reduced the occurrence of blotch miners. They usually commence their attack before the housing of chrysanthemums grown outdoors during the summer.

Spraying with emulsifiable formulations of gamma-BHC (0·01%) destroys larvae within the mines and the methods suggested to control chrysanthemum leafminer (p. 166) are also effective. Prior to the development of modern insecticides nicotine at 0·1% together with a wetting agent was used.

REFERENCES

HENDEL, F. (1927). Trypetidae. Lindner *Flieg. Palearkt.*, **49**, 221 pp.

HODSON, W. E. H. & JARY, S. G. (1939). A new insect pest of chrysanthemums. *Agriculture*, **46**, 54–6.

LECLERCQ, M. (1967). Contribution à l'étude des Trypetidae (Dipt.) palearctique et de leurs relation avec les vegetaux. *Bull. recherches Agron. Gembloux*, **2**, 65–105.

SHAW, M. M. (1958). *Tephrites conjuncta* (Loew) on chrysanthemums in Scotland. *Pl. Path.*, **7**, 76.

Family PSILIDAE

Psila nigricornis Meig. (the chrysanthemum stool miner) has caused widespread attacks since 1933 and it is now especially prevalent in N. England and SE. Scotland.

The adult is a shining, brown-to-black fly with a distinct bluish or greenish sheen. It is 4–5 mm long and may usually be distinguished from the very similar carrot fly *P. rosae* (F.) by having completely black antennal segments—the third segment is partly yellow in the latter species. The presence of seven digitations in the anterior spiracles of *P. nigricornis* larvae but only five in *rosae* is a better character. There are two generations each year; the flies, active from mid-May to early July and again from late August to early October, frequently feed on the flower heads of umbelliferous plants.

Groups of about four white, elongate eggs are laid in the soil around the stems of newly planted chrysanthemums. They hatch in 10–14 days and the first instar larvae feed on the fibrous root system. As larvae grow, they become yellowish-white with a shiny and markedly toughened skin (Fig. 5.61). The body tapers towards the extremities, gradually towards the minute head, but abruptly to the tail. When fully grown (6 mm) the larva forms a yellow-brown, cigar-shaped puparium about 4 mm long, which is flattened, on one side, at

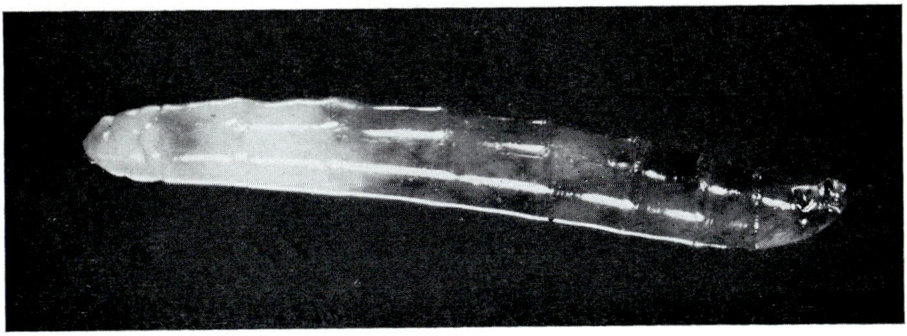

FIG. 5.61. Larva of *Psila nigricornis* Meig.: chrysanthemum stool miner (× 15). (A Shell photograph.)

the head end. Puparia are found either in the soil or in the superficial mines of the attacked plants.

Under glass, larvae of the second generation, feeding in the roots of late-flowering chrysanthemums or in the stools of early-flowering varieties destined for cutting production, may cause extensive damage. Long tunnels are formed in the outer tissues which, above the larval mines, tend to become dry and split, so causing deep wounds in the rootstock (Fig. 5.62). Where damage is

FIG. 5.62. Damage by chrysanthemum stool miner. (Courtesy of P. Osborne, Edinburgh School of Agriculture.)

extensive, no new vegetative shoots are produced, but even a light attack may adversely affect cutting production. Flies may emerge from the boxed stools as early as February or March, but it is not known whether these are capable of reproducing successfully as no larvae have been found on chrysanthemums so early in the year.

Although there is no evidence of attacks on outdoor lettuce, serious losses in crops following autumn chrysanthemums have been recorded in glasshouses. At first, attacked plants wilt as the larvae bore into the root-collar region but, when the tap roots are attacked, many die off completely. Celery is also liable to attack by this species.

Two further species, *P. bicolor* Meigen, and *P. limbatella* (Zett.) have been found on chrysanthemums.

Control

Adults and larvae are easily killed with BHC provided the larvae have not tunnelled into the roots and stem bases. Where they are protected from contact insecticides, a good measure of control can be obtained with the systemic insecticide demeton-S-methyl (0·05%). While no trials have been made to compare the merits of other organophosphorus insecticides the close relationship of both *P. nigrocornis* and *P. bicolor* with the carrot fly suggests that diazinon may be effective.

The bases of chrysanthemum plants, in houses in which the pest has recently occurred, should be thoroughly sprayed with 0·01% gamma BHC or well dusted with a gamma BHC dust in early May and early September. Where planting is done between these dates the treatment should be applied within 7 days, with only one application made to plantings after July.

Provided infested plants have been removed from a glasshouse and the soil has been efficiently steamed, a following crop of chrysanthemums or lettuces should remain clean. Where lettuce has to be planted before November in an unsteamed border vacated by infested chrysanthemums, it is suggested they be drenched, within 7 days of planting, with 0·04% dimethoate applied at 1 gal/yd^2. Treatments applied to damaged plants will not result in recovery so that prompt treatment, at the first sign of damage, is essential.

Contact insecticides have failed to control existing attacks on lettuces and drenches of dimethoate or other systemic insecticides may not be economic except where the attack is severe. If such compounds are tried, it is essential that the recommended intervals between application and cutting should be observed, as their persistence is greater at low temperatures.

Hot-water treatment of chrysanthemum stools is completely lethal to miners in all stages of development.

REFERENCES

COLLIN, J. E. (1944). The British Species of Psilidae. *Entomologist's mon. Mag.*, **80**, 214–24.

OSBORNE, P. (1961). Comparative external morphology of *Psila rosae* (F.) & *P. nigricornis* Mg. (Dipt., Psilidae) third instar larvae and puparia. *Ibid.*, **97**, 124–7.

VERNON, F. D. R. (1962). Observations on the biology and control of the Chrysanthemum Stool Miner. *Pl. Path.*, **11**, 38–41.

Family ANTHOMYIIDAE

Delia brunnescens Zett. (the carnation fly), though not common under glass, is a widespread pest of various species of *Dianthus*, regularly causing 'blotch' mines in the leaves, and tunnels in the roots of carnations, pinks and Sweet William. Another closely related species, D. *fugax* (Meig.), is commonly associated with these attacks but Miles (1957) has established that it is entirely saprophagous. The males of these species may be distinguished by the presence, in *brunnescens*, of a protuberance on the base of the ventral surface of the second tarsal joint of the middle legs.

Adult *D. brunnescens* are 6–7 mm long. In males the head is brownish; the thorax dark brown in front, and yellowish-grey behind; the flat abdomen has a narrow black median stripe and narrow black bands on the front edges of the segments. The females have a yellowish-grey head with a brighter yellow patch above the antennae; the thorax and abdomen are yellowish-grey.

Only one generation is produced each year; although the flies are active throughout the summer from June onwards, eggs are not laid until the autumn (September–November). They are white, elongate-oval, 1·0 mm long and markedly reticulated with strong longitudinal ridges which may be irregular or broken. They are laid on the leaves or in the leaf axils and, being non-adhesive, are easily dislodged. Outdoors, the incubation period is 8–14 days; larvae hatching from eggs laid on the leaves wander over the leaf surface before penetrating the tissues. All three larval instars feed on the mesophyll and excavate almost straight mines along the leaves. Larvae hatching in the leaf axils tunnel directly into the main stem.

When mature, the somewhat sluggish larvae are about 9 mm long and creamy-white. The stout body tapers anteriorly to the small head and is obliquely truncated with a pair of bifid tubercles at the rear. The smooth, shiny body segments are clearly demarcated by bands of rounded denticles along their front edges and on ventral surfaces. Even inside glasshouses larvae live about 6 months and feed throughout the winter, surviving exposure to severe frosts. Most leave the host plant to pupate within a barrel-shaped puparium about 6 mm long. A few puparia may, however, form in the damaged tissues. The flies emerge after about 8 weeks.

Since this species only damages carnations in winter, longitudinal mines in carnations at other times of the year are probably caused by an Agromyzid fly *Dizygomyza flavifrons* (Mg.) which makes conspicuous silvery blisters on the leaves of outdoor carnations. The life history of *D. brunnescens* suggests that the need for control measures may be forecast by examination of the plants in late September.

Control

With the widespread use of pesticides to control other insects, attacks by the carnation fly in commercial nurseries have occurred less frequently. They are more frequent in private glasshouses to which they may be introduced on cuttings of outdoor *dianthi*. Any plant material from outdoors must be carefully scrutinized before, and for some time after, being placed in a glasshouse or frame.

Spraying the plants with 0·01% gamma-BHC, or dusting with a gamma-BHC dust will, if properly timed, prevent egg laying and destroy newly hatched larvae. It is important to ensure that the lower parts of the plants are adequately covered as the most serious damage is caused by larvae entering the main stem.

If glasshouse carnations were attacked the previous autumn or winter, preventive spraying or dusting should commence in July and be continued, at 3-week intervals, until mid-November. This extended period is necessary to allow for any modification of the normal life cycle under glasshouse conditions.

The first indication of damage is usually the appearance of linear mines in the leaves in late September. Immediate application of sprays of 0·01% gamma-BHC or 0·02% diazinon should be made when these are seen, with one or more further applications at 3-week intervals. Carnation foliage is difficult to wet and additional wetting agent should be added to obtain adequate cover of the leaves and stems. It is, however, unlikely that these sprays will kill larvae which have actually entered the stem. Although experimental proof is lacking it is probable that soil applications of the systemic insecticide dimethoate at 0·03%, using 2–3 gal/yd^2, will destroy larvae feeding within the stem. To plants in 6 in. (size 32) pots, 4 fl. oz of the drench should be given and 2 fl. oz to 3 in. (size 60) pots.

Spraying with gamma-BHC or diazinon should also prove effective against *Dizygomyza flavifrons*.

The 'Ashington' cultivars may be susceptible to injury by gamma-BHC during the late autumn and winter, and applications from October to March are inadvisable.

REFERENCES

LEIBER, E. (1965). On the habits of the carnation fly *Phorbia* (*Delia*) *brunnescens* Zett. *Mitt. biol. Bund-Anst. Ld-u. Forstw.*, **115**, 229–32.

MILES, M. (1957). Studies of British Anthomyid Flies. VIII. The Carnation Fly, *Hylemyia brunnescens* (Zett.). *Bull. ent. Res.*, **48**, 219–28.

SÉGUY, E. (1932). Études sur les Anthomyides. 6e note. Notes biologiques et taxonomiques sur les mouches de l'oeillet. *Encycl.ent.* (B), Diptera, **6**, 71–81.

Delia platura (Meig). (the bean seed fly or freesia maggot) has caused severe damage to freesia plants in the Channel Islands. The larvae tunnelled into the stems just below ground level, so killing many seedlings.

The greyish-brown adults, 5 mm long, become active in May. They mate in swarms, at heights up to 30 ft, before laying about 50 eggs over a period of several weeks. The white eggs (0·3 × 1·0 mm) are often laid in newly cultivated soil; Channel Island growers noticed that unshaded crops were more severely affected, suggesting that the flies preferred bright sunlight.

The eggs hatch in about 3 days into larvae which feed for 1–3 weeks. The larvae are typical white maggots with prominent, curved mouth hooks; the terminal segment bears 12 tubercles and the posterior spiracles 8–10 projections. These larvae are primarily saprophagous and mature even when decaying organic matter is the only available food. It follows that infestations are determined mainly by cultural operations and available organic matter, rather than host plants. In the presence of organic matter plants may remain unaffected.

There are normally three or four generations annually in Britain.

Delia antiqua (Meig.) (the onion fly) has typical white, tapering larvae with a blunt posterior bearing eight tubercles. Larvae occasionally infest tulip bulbs and so may be brought into glasshouses. Eggs are laid in the soil near the bulb which may subsequently contain as many as 30 larvae which tunnel up through the base. They normally pupate in the soil.

Control

Watering the soil in pots or borders with gamma-BHC (0·02%.) will control attacks of this insect on freesias. Good results have also been obtained with parathion (0·01%) and diazinon (0·02%), but the greater persistence of BHC is probably advantageous.

These insecticides should also control onion fly.

REFERENCES

MILES, M. (1950). Studies of British Anthomyiid flies. I. Biology and habits of the bean seed flies, *Chortophila cilicrura* (Rond.) and *C. trichodactyla* (Rond.). *Bull. ent. Res.*, **41**, 343–54.

— (1958). Studies of British Anthomyiid flies. IX. The biology of the onion fly. *Bull. ent. Res.*, **49**, 405–14.

MILLER, L. A. & MCCLANAHAN, R. J. (1960). Life-history of the Seed-Corn Maggot *Hylemyia cilicrura* (Rond). and of *H. liturata* (Mg.) (Diptera: Anthomyiidae) in Southwestern Ontario. *Can. Ent.*, **92**, 210–21.

Order HYMENOPTERA: Wasps, Bees and Ants, etc.

Most of these insects are parasites but a few are phytophagous and, like the sawflies and seed-infesting chalcids, are pests. The order also includes bees, wasps and ants, well known as social insects, which occasionally require control under glass.

Family CHALCIDIDAE

Eurytoma orchidearum Westwood (the cattleya fly) is a minute (3 mm), black, slender wasp which is a major pest of orchids in America and has occasionally caused severe damage in this country. The adults have four small wings which, apart from a small club-like vein on the forewing, are almost without veins. The females lay eggs singly within the tissues of the small pseudo- or flower-bulbs, several being deposited in a single bulb. The larvae burrow within the tissues hollowing out and destroying the embryo flowers and occasionally the young leaves and rhizomes. Infested shoots may become dark and swollen; larvae in the leaves and rhizomes induce gall-like swellings. When mature, the larva forms a whitish pupa from which the adult emerges about 2 months after the eggs were laid.

Control

DDT, applied several times at 4- to 6-week intervals, as dusts (5%) or sprays kills the adults. BHC is probably equally effective. Larvae within galls are unlikely to be killed and consequently the galls should be removed and destroyed where this is practical.

Family VESPIDAE: Wasps

Wasps may occasionally be troublesome under glass where they are attracted to ripening grapes, peaches or nectarines. Nests made from partially masticated bark are constructed in cavities within earthen banks. Only the queen wasps survive the winter and, in the spring, they make small nests within which the first workers develop. Large numbers of this caste are produced during the summer but late in the season males and queens are produced to carry over the species through the winter.

Control

Where wasps are troublesome it is advisable to screen ventilators but should nests have to be destroyed, entrance holes should be dusted with about 1 tablespoonful of a 20–25% wettable powder formulation of DDT, 5% BHC dust or 2% derris dust. Carbon tetrachloride ($\frac{1}{4}$–$\frac{1}{2}$ pint) injected into the nest at dusk will also kill the wasps. In an emergency, wasps in peach houses can be killed, not later than 2 days before picking, by smoking with DDT.

Family FORMICIDAE

Several species of ants may be troublesome in glasshouses; like wasps, some attack ripening fruit but the most serious damage is caused when they transport aphids and scale insects to new host plants. The ants are attracted to the honeydew produced by these insects which they 'cultivate' to ensure adequate supplies. Where pot plants are kept for long periods on the benches nests may be constructed in the soil resulting in injury through water shortage. The permanent nests formed in the soil contain several castes of sexual and worker ants, all of which have specialized duties including feeding of the larvae, foraging for food, defence and maintenance of the nest.

Control

Ants nests should be destroyed by soaking with gamma-BHC. In unplanted soil, a dilution of e.c. to 0·03% may be used but in pots this concentration should be reduced to 0·01%. Where nests cannot be destroyed the ground surface over which the ants travel should be liberally and frequently dusted with 20% wettable powder formulation of gamma-BHC.

Two baits have recently been approved for the control of ants, one based on 1% trichlorphon and the other on 1% propoxur. These are very effective when placed on ant trails.

Family APIDAE

Bees, whether domesticated or wild, are familiar producers of honey. Outdoors, they play an important role in cross-pollinating plants but, in cucumber houses, this activity may cause severe losses as fertilized cucumber fruit is swollen, bitter-tasting and commercially unacceptable. Although varieties without male flowers are now being developed, bees must be excluded from most crops and it may be necessary to screen ventilators for this purpose. Gardeners may find removal of male flower buds the most convenient means of preventing cucumber fertilization by bees.

Chapter 6

Arthropod Pests Other Than Insects

Order ISOPODA: Woodlice, Sowbugs or Slaters

The body is segmented with a clearly defined head region bearing a pair of compound eyes, two pairs of antennae, one of which is minute, a pair of mandibles and two pairs of forwardly directed, leaf-like maxillae. The major portion of the animal consists of seven clearly demarcated and laterally expanded thoracic segments. A pair of simple, jointed walking appendages is attached to the ventral surface of each thoracic segment. The narrower and comparatively short abdomen is composed of six, more or less flattened, segments often ending in a pointed tail or 'telson'. Each of the first five abdominal segments has, on its ventral surface, a pair of leaf-like plates containing a system of minute air tubules. In some genera these plates overlap to form a cavity within which the air tubules can 'breathe' in a moisture-saturated atmosphere rather like a gill. Indeed, woodlice are probably the most recently evolved organisms to change from a water to a land habitat. They are, however, incompletely adapted for terrestrial life and, to conserve the water content of their bodies, must remain in damp places.

The daily pattern of activity in these creatures is correlated with the atmospheric humidity. Most feeding takes place at night when they move about rapidly on the soil surface as the humidity increases. This nocturnal activity largely determines the distribution of the different species but the necessity to seek refuge from relative dryness, as well as to hide during the day, modifies the pattern of spread. By day, they seek out damp dark situations in the fabric of the glasshouse, in rotten wood or below bark, stones, decaying vegetation, manure heaps and other permanently damp situations.

Woodlice feed, for the most part, on decomposing cellulose; indeed, all the common species can be reared successfully on a diet of fermenting straw. They are particularly attracted to dried blood and vast numbers used to congregate on the bed surface when this substance was more widely used as a top dressing for tomatoes and cucumbers. They will, however, feed on roots and stems (Fig. 6.1) or eat irregular holes in leaves (Fig. 6.2) even where these do not touch the ground.

About twelve species have been recorded from British glasshouses but only four of these have been consistently associated with damage to plants.

Armadillidium nasutum Budde-Lund belongs to a genus whose members can all roll up into spheres with the tail appendages projecting beyond the head. Its colour varies from slate to dark brown with symmetrical, yellowish mottling on each thoracic segment.

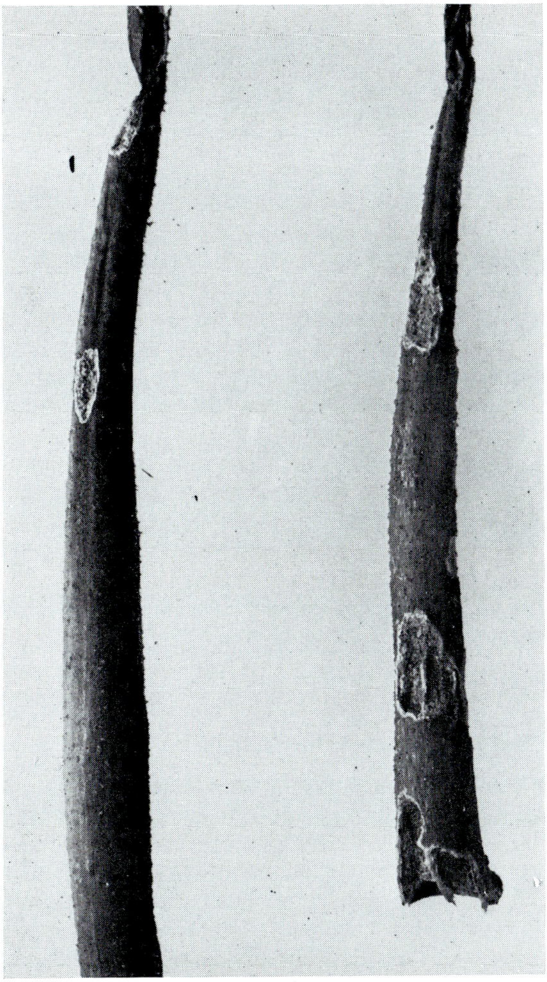

FIG. 6.1. Woodlice damage to stems of cucumber. (Crown Copyright by permission of Controller of HMSO.)

A. nasutum is particularly common where, as in cucumber houses, the temperature is kept fairly high. It is very active, living just below the bed surface by day and feeding at night.

Gravid females have been found from April to September. They may be recognized by the brood pouch, a cavity formed by the overlap of pairs of

FIG. 6.2. Woodlice damage to cucumber leaves. (Courtesy of Glasshouse Crops Research Institute.)

plates growing out from the base of all the legs between the second and fifth thoracic segments. Two broods are produced, up to 50 eggs being carried in the pouch for about 1 month. The young, white woodlice, which have fewer legs than the adults and are about 1 mm long, leave the pouch 1 week after hatching. They attain their mature colouration within a fortnight and after

4 months grow to about 5 mm. They become sexually mature the following year.

A. nasutum causes serious damage to potted cucumbers and young transplants by chewing the stems and pitting leaves with pear-shaped holes. Chewing at ground level may cause premature death of mature plants by partially or completely girdling the stem, thereby interrupting translocation of water. Woodlice have damaged tomato seedlings, either by eating through the stem at the soil surface or even chewing out the growing points. Except where the leaves touch the ground older plants are not normally attacked. Winter lettuce is unaffected even by large numbers of this species.

Where it can burrow into loose material *A. nasutum* frequently overwinters outdoors, heaps of soil from cucumber borders being favoured sites.

Armadillidium vulgare (Lattreille) varies from dark slate to pale yellowish-grey. It may be distinguished from *A. nasutum* in that, when it is rolled into a sphere, its antennae are completely concealed, whereas they remain visible in the latter species. *A. vulgare* produces over 100 young in each brood.

It is normally averse to excessive heat and is therefore the dominant species in cold tomato houses and frames, especially in the first few years after erecting glasshouses on grassland. In winter it burrows deep in the soil, specimens having been recovered as much as 8 ft below the surface. Although inconspicuous, as it feeds largely below ground, it can destroy seedlings by eating through the stems and, by eating the fibrous root system, has been found to cause stunted growth in quite large plants.

Oniscus asellus L. is possibly the most common British woodlouse and is the species found so frequently under rotting bark and beneath rockery stones. It is 12 mm long when full-grown, dark mottled greyish-brown with a marbled pattern along the sides. *O. asellus* cannot roll up when disturbed and produces only 30 young in each brood.

Under glass it chews the aerial roots of orchids and frequently damages young, tender growth of carnations and sweet peas.

Porcellio laevis Lattreille is a large (18 mm), uniformly slate-blue species with a smooth body. It is often introduced into glasshouses in stable manure and causes damage to the roots of cucumbers, *Adiantum* and other ferns.

Porcellio dilatatus Brandt in which the head and body are tuberculate, sometimes occurs in vast numbers in cool houses, especially below seedboxes or flower pots. It is rarely found below foliage during the day. Seventeen young are carried in the brood pouch for 10½ weeks in the first generation but for only 5 weeks in the second.

Control of Woodlice

Strict cleanliness is most important in preventing the infestation of glasshouses, mushroom houses or frames by woodlice. Rubbish and debris should not be tolerated within houses or in their vicinity. If seedboxes, pots and staging cannot be stored under cover, they should be stacked away from weeds and rubbish in as dry a place as possible. To remove hiding places, the structure of glasshouses or frames should be maintained in good condition by pointing brickwork and replacing rotten wood. Manure for making-up cucumber beds or preparing mushroom compost should be stacked on a clean concrete, or hard-earth, base away from rubbish which may harbour woodlice.

Where a vacant glasshouse is known to be infested with woodlice they should be controlled, before bringing in plants or sowing seed, by drenching the soil with a phenolic (tar acid) soil sterilant. Several types of phenolic fluids are available; as ordinary disinfectants may contain creosote-type oils harmful to plants, only those made by a reputable manufacturer should be used. Most should be diluted to 1 in 100 before application to the soil at 5 gal/yd^2. Planting can be done as soon as the phenols have disappeared from a treated soil. In view of differences in commercial products the maker's instructions should be followed but, as a general guide, the period between application and planting varies from 6 weeks, when the soil temperature is 40 °F (5 °C), to 3 weeks when the soil temperature is 55 °F (13 °C). Staging or other structures should be thoroughly sprayed with the diluted phenolic product, taking particular care to drive the spray into crevices in which woodlice may be hiding.

If woodlice are sheltering in the compost of pot-plants, watering with parathion at 0·02% may be the most satisfactory control measure since practically all plants are tolerant to soil application of this pesticide. As parathion is highly poisonous and not available to amateurs, plants may alternatively be watered with 0·1% DDT or 0·02% gamma-BHC.

Cucumber beds, by virtue of their open, organic composition provide ideal conditions for woodlice, but the application of 0·1% DDT or 0·02% parathion at ½–1 gal/linear yard of bed, depending on the size of the bed and the amount of mulch, will destroy them. DDT dusts have proved less effective. Except where larvae of sciarid flies (see p. 151) are also present, DDT is preferred on account of its persistence and lower toxicity. Baits of bran and Paris green may also prove effective.

Caution. Although DDT is not injurious to cucurbits when applied to the soil it is very damaging to the foliage and care must be taken to avoid wetting or splashing the leaves. Prior to the introduction of DDT, baits consisting of 1 part (by weight) of copper aceto-arsenite (Paris green) admixed with 28 parts of bran or 56 parts of dried blood, were used to control woodlice. The baits were distributed on the soil surface, on benches or staging.

Laboratory tests by Vernon & Dennis (1966) showed that bran-Paris green and bran-sodium fluoride baits, sprays of DDT (0·1%), BHC (0·5%) and parathion (0·013%) or 5% DDT dusts were very effective against *O. asellus* when applied to the soil. The baits killed most rapidly, but that containing sodium fluoride lacked persistence. Woodlice may be controlled immediately before planting, either by applying a 5% DDT dust to the soil (1 oz/yd), or by spraying with 0·1% DDT or 0·05% BHC to thoroughly wet the surface of the soil. In the absence of more attractive food material sprinkling bran-Paris green, bran-sodium fluoride or dried blood-Paris green baits on the soil at ½–1 oz/yd is effective.

Dusting the soil surface, etc., with DDT dust or distributing bait is also suitable where woodlice are attacking young plants in the ground or seedlings in boxes. Young seedlings should not, however, be dusted unless they are known to tolerate DDT. Instead, the boxes or pots may be lifted and dusts applied to the staging or border. Baits should not be sprinkled on the leaves of seedlings or be in contact with their stems at soil level as they may cause injury or provoke attack by fungi such as *Botrytis cinerea*.

REFERENCES

EDNEY, E. B. (1954). British Woodlice (Isopoda). *Linn. Soc. Lond. Synopses Br. Fauna*, **9**.

HATCHETT, S. P. (1947). Biology of the Isopoda of Michigan. *Ecol. Monogr.*, **17**, 47–79.

HEELEY, W. (1941). Observations on the life-histories of some terrestrial Iospods. *Proc. zool. Soc. Lond.* (B), **111**, 79–149.

VERNON, J. D. R. & DENNIS, E. B. (1966). Experiments on control of woodlice. *Pl. Path.*, **15**, 116–8.

Class DIPLOPODA: Millepedes (or Thousand-Legged Worms)

These can be distinguished from insect larvae, which they superficially resemble, by the possession of a head, a thoracic region consisting of three segments (the posterior two each bearing a single pair of legs) and an abdominal, or trunk, region of many segments each bearing two pairs of legs (Fig. 6.3). The body is quite hard. They are sluggish creatures and, when disturbed, frequently coil up like a watch-spring.

Millepedes live within soils and humus piles, etc., where they feed mainly on dead plant material such as leaves or decaying twigs though some species eat fungal hyphae or even small dead animals. The use of large quantities of organic matter to improve soil fertility favours the rapid increase of millepedes which are important agents in its breakdown. As with woodlice, moisture is the critical factor influencing their life and habits. They seem unable to find their way to damp sites by sensory perception so their movements are influenced merely by avoidance of dry situations. This negative behaviour permits local wandering in the higher nocturnal humidities but prevents them from entering unsuitable habitats. In dry conditions they often burrow deep into the soil but, to obtain water, they frequently attack growing crops. Where high populations develop, even a temporary shortage of water may result in severe crop damage. Millepedes are attracted to the dilute sugar concentrations found in plants so that, once initiated, an attack on living tissue does not necessarily cease when water again becomes abundant.

Freshly germinating seedlings may be eaten completely through and, although their mouthparts are relatively weak, millepedes may gnaw the surfaces of the roots and stems of older plants. While plants with a thick cortex are unaffected, thinner and more succulent parts may be scarified leading to wound infection by saprophytic fungi and bacteria. Indeed, it is usually difficult to decide whether millepedes initiate or follow the diseased condition of the tissue with which they are commonly associated.

Several species have been recorded in glasshouses but only two are common.

Oxidus gracilis (C. L. Koch) is a tropical species which has become widely distributed in greenhouses in temperature regions. It is flattened with well-developed dorso-lateral keels which, on the hind segments, are produced into backwardly directed teeth. The upper surface varies from chestnut to dark reddish-brown. The lower surfaces and the lateral keels are amber coloured. *O. gracilis* has 19–20 body segments and is about 1 in. long when full-grown.

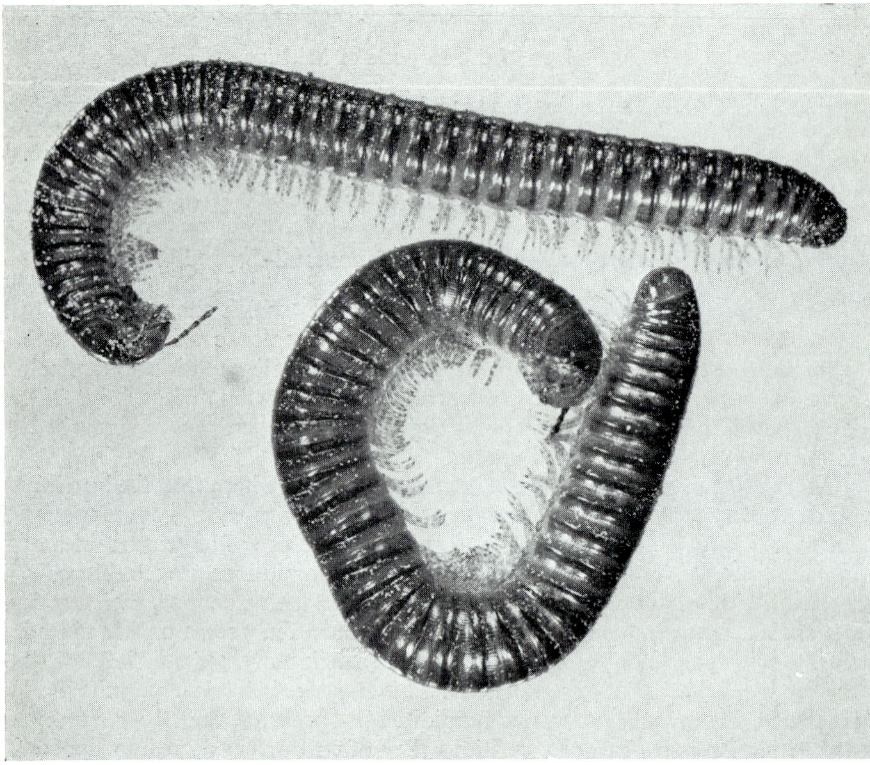

FIG. 6.3. Adult millepede (*Tachypodojulus niger* Leach) (×6). (A Shell photograph.)

Where the temperature is maintained above 70 °F (22 °C) breeding takes place throughout the year. The female produces her eggs in clutches of 20–300 and places them in small cavities in the soil 3–15 mm below the surface. Incubation takes about 10 days. The young larvae are cream coloured and have only three pairs of legs. They pass through seven moults, increasing the number of body segments and legs at each. Moulting takes place in special elliptical chambers the walls of which are made with soil and organic matter by the millepede. These chambers protect the newly moulted, inert and helpless larva until the new skin hardens. Although the larvae are relatively inactive the adults move about freely at night. Development is complete within 5–6 months at about 65 °F (18 °C).

O. gracilis is particularly common in cucumber houses; populations of 250/ft^2 have been found within 2 months of making up the beds. Gnawing of young plants may severely damage the cambial layers of the main stem for up to 2–3 in. above the bed surface. By damaging roots it is widely regarded as playing an important role in providing entrance for fungal pathogens. An unusual type of damage was reported by Rolfe (1964) who found *O. gracilis* eating holes in tomato plant pots made of compressed paper treated with bitumen. The pots disintegrated as they were moved for planting out.

Blaniulus guttulatus. (Bosc.) (the spotted snake millepede) is a slender, cylindrical species about 12 mm long with 30–60 body segments. It is whitish-cream to yellow with a row of red or purple spots along each side of the body. These spots are repugnatorial glands which exude an irritant which probably deters predators. Fifty to one hundred large, elongate eggs are made within a nest of soil particles cemented together with saliva. Hatching occurs within about 2 weeks. *B. guttulatus* is an important pest of many agricultural and garden crops including sugar beet, potatoes and strawberries. Under glass it has caused damage to carnations, chrysanthemums and especially young beans where these are grown as a catch crop. It is commonly found feeding in tulip bulbs and cyclamen corms where it extends damage initiated by slugs and wireworms.

Choneiulus palmatus (Nemec) and ***Nopiulus minutus*** Brandt, closely related to *Blaniulus guttulatus*, have also been found in British glasshouses. They may be distinguished from *Blaniulus* which has no ocelli, or simple eyes, by the possession of six to eight in a single row. *C. palmatus* is grey-brown, *N. minutus* cream coloured. Both species have been found associated with bulbs introduced into glasshouses for forcing.

Control

Attention to cleanliness, mentioned under the control of woodlice, is equally important in avoiding infestations of millepedes. The substitution of composts containing peat, for mixtures containing leaf mould or rotted manure, reduces the risk of introducing millepedes into glasshouses and frames.

Infestations are rarely discovered until the millepedes have caused some damage, but where they are known to be present in empty houses the application of phenolic fluids, as for woodlice control, will eliminate them. As an alternative, gamma-BHC dust can be thoroughly worked into the top 3 in. of soil, a few weeks before planting, at a rate equivalent to 8 oz 0·5% dust to 100 ft^2 or the soil watered with 0·01% gamma-BHC at the rate of 1 gal/yd^2.

To prevent attack on seedlings, in boxes or pot-plants, the surface of the borders or benches should be dusted with gamma-BHC, taking care to avoid application to susceptible seedlings.

Where millepedes are causing damage to the roots, watering with parathion (0·01%), nicotine (0·05%) or mixtures of nicotine (0·05%) and gamma-BHC (0·005%) will effect control if applied in sufficient quantity to soak the infested compost or surface soil. If cucumber beds contain considerable quantities of undecomposed manure they may require up to 1 gal/linear yd. Watering with 0·1% DDT, as for woodlice control, also controls millepedes.

Nicotine soil drenches cause millepedes to come to the surface and, unless resprayed, some may recover; mixtures of nicotine and gamma-BHC, however, overcome this disadvantage. Nicotine alone, at 8 oz of 95–98% nicotine per 100 gal, together with a wetting agent at the rate recommended for normal spraying, should be applied in the late afternoon or evening, and any millepedes lying comatose on the soil surface lightly sprayed with some of the same solution first thing the following morning, using about 1 gal/120 ft^2. Addition of 0·005% gamma-BHC emulsifiable concentrate to the diluted nicotine avoids respraying but such mixtures must be used with care on seedlings and young plants.

Fiedler (1965) suggested that some South African species are less readily killed by these insecticides, concentrations of 0·2% gamma-BHC and 0·05% parathion being required. Against some species 0·2% malathion was effective.

REFERENCES

BLOWER, J. G. (1958). British Millipedes (Diplopoda). *Linn. Soc. Lond. Synopses Br. Fauna*, **11**, 1–74.

BRADE-BIRKS, S. G. (1939). Notes on Myriapods. XXXVI. Sources for description and illustrations of the British fauna. *Jl S-east agric. Coll. Wye*, **44**, 156–79.

CAUSEY, N. B. (1943). Studies in life-history and ecology of the Hothouse Millipede (*Orthomorpha gracilis* Koch). *Am. Midl.*, **29**, 670–82.

CLOUDSLEY-THOMPSON, J. L. (1950). Economics of the 'Spotted Snake Millipede' *Blaniulus guttulatus* (Bosc.). *Ann. Mag. nat. Hist.*, **12**, 1047–57.

EDWARDS, C. A. & GUNN, E. (1961). Control of the Glasshouse Millipede. *Pl. Path.*, **10**, 21–4.

FIEDLER, O. G. H. (1965). Notes on the susceptibility of millipedes to insecticides. *J. ent. Soc. Sth. Afr.*, **27**, 219–25.

LANG, J. (1959). Einige Bemerkungen zur Okologie von *Orthomorpha gracilis* Koch und zu ihren Vorkommen in Prager Glashäusen. *Cesk.Spol. Zool. Vestnik.*, **23**, 97–122.

ROLFE, S. W. (1964). Glasshouse millipede damaging 'tom pots'. *Pl. Path.*, **13**, 45.

Class CHILOPODA: Centipedes

Centipedes have the body divided into segments each bearing a pair of legs. The first segment behind the head has specially adapted limbs, or poison claws, to capture and hold prey while it is killed by venom from glands opening at their tips (Fig. 6.4). The common British species, *Haplophilus subterraneus* Shaw, is light brown and about 2 in. long when fully extended. Centipedes are nocturnal creatures living in damp places under stones or bark or within soil and leaf-litter. They have an interesting reproductive habit involving parental care of the young. The mother remains with the eggs to prevent them becoming desiccated or covered with fungal hyphae. It is believed that, as she 'licks' the eggs, a fungitoxic material is excreted. There is no larval stage, additional segments being added posteriorly as the centipede grows.

These active, predatory creatures are beneficial since they feed on many soil insects and millepedes.

FIG. 6.4. Immature centipede (*Lithobius forficulatus*) (×3). (A Shell photograph.)

REFERENCE

EASON, F. H. (1964). *Centipedes of the British Isles*. F. Warne, London. 294 pp.

Class SYMPHYLA: Symphilids

Symphilids, which abound in light, humic soils, superficially resemble very small centipedes. They possess a distinct head with long antennae consisting of 5–60 segments. The body, which is composed of 14 segments, bears 12 pairs of legs, and is terminated by a pair of cerci arising from the thirteenth segment (Fig. 6.5).

Edwards (1959) revised the British species and showed that five phytophagous and three saprophytic species occur in glasshouses. Of the phytophagous species *Scutigerella lineatus* Edwards, *S. palmonii* Michelbacher and *S. immaculata* Newport are found in soil below many cultivated plants and occur frequently in glasshouses. Two other species *Hanseniella unguiculata* Hansen and *H. calderia* Hansen are tropical species which, as they have weakly developed mandibles, are probably not primary plant pests but may occasionally be found amongst the roots of exotic plants in hothouses.

Scutigerella immaculata (Newport) (glasshouse symphilid or garden centipede). Infestations of this widely distributed species, which is about 6 mm in length and white, may follow introduction of manure or infested pot-plants into glasshouses. It lays clusters of about a dozen distinctive, pearl-white eggs (0·5 mm in diameter), each covered with a network of ridges, in tunnels made by the adults in the soil. The eggs may be near the surface but, if the soil is dry, oviposition occurs much deeper. Depending on the temperature incubation takes 10–23 days and throughout this period the female remains with the eggs. As the eggs mature they become gradually darker, assuming a pale brown colour immediately prior to hatching. Newly hatched larvae have only six pairs of legs and six antennal segments and are relatively inactive. The posterior part of the body is conspicuously enlarged. Within 2 days they moult to a form more similar to that of the adults and possessing seven pairs of legs and 13 antennal segments. They mature, after about 3 months, to light-cream adults which may vary in size from 2–8 mm. Symphilids are unusual in that moulting continues, at approximately monthly intervals, throughout adult life, as many as 40 moults having been recorded in a life span of 4 years or more.

FIG. 6.5. Symphilid *Scutigerella immaculata* Newport, adult (×15). (Courtesy of Rothamsted Experimental Station.)

Symphilids feed voraciously on young and tender roots, individuals eating their own weight in 1 day. While direct damage to the root hairs and rootlets can check growth and predispose the plants to wilt in strong sunlight, the feeding damage may also initiate root-rots by bacteria and fungi. Yeasts, bacteria, fungi and even dead soil animals may also be eaten. Under glass, tomatoes, lettuce, cucumbers, French beans, sweet peas, lilies, African violets, and chrysanthemums may all be severely attacked. Stems of plants such as sweet peas may be hollowed out. Leaves of lettuce, where they touch infested soil, may be riddled with feeding holes similar to those produced by springtails. Corky calluses, which often develop over the superficial lesions on attacked roots, give a gnarled appearance. Affected lettuce make little growth, fail to heart, and often succumb to *Botrytis* attack. Tomatoes are frequently attacked soon after planting out; the older roots are injured and the new rootlets eaten as fast as they develop. The stems of the stunted plants become bluish and, while the lower leaves turn yellow, the upper foliage assumes a characteristic dark green colour. Since symphilids are not uniformly distributed within a house symptoms are patchy but tend to be more frequent along paths, walls and pipes, and around pillars.

Edwards (1958) showed that cultivated soils containing a high proportion of organic matter are particularly suitable for symphilids. These delicate creatures do not tunnel but move about in earthworm burrows and other natural soil crevices. The vertical distribution of the population varies with the season. In bare soil, numbers of *S. immaculata* near the surface rise to a peak in May, decline sharply to a minimum in July, and later increase very slowly during the autumn (Edwards, 1959). Near tomato roots the population increases sharply within a week of planting, presumably due to migration from the bare surrounding soil, and remains at a high level till May. Thereafter the population penetrates deeper than 12 in. into the soil, some individuals have been recorded as far as 5 ft below the surface.

Though reproduction in glasshouses is higher during spring and summer, when total numbers increase most rapidly in response to the warm, moist soil and abundance of food, *S. immaculata* breeds throughout the year. In autumn, high soil temperatures, in excess of the optimum range of 59–68 °F (15–20 °C), tend to force the population deeper into the soil. This movement is probably dictated by the need for 100% relative humidity in the soil air. So long, therefore, as the culture system provides warm, moist soil with an abundance of food, symphilids remain a potential pest. Recent damage to year-round chrysanthemums illustrates how a change of cultural technique may extend the period during which damage is possible.

Control

Once symphilid attacks have developed, it is probable that they will reoccur the following year, as their migration to depths prevents effective or economic treatment with insecticides. Indeed, it is unusual for the entire population to be within the upper foot or so at any period. Sterilization with steam in the late autumn or winter, as usually practised, is normally of little value, as are treatments with metham-sodium, D.D. or chloropicrin. Although better results are usually obtained by sterilizing in early spring, after the removal of a crop (e.g. lettuces), they do not justify the selection of this treatment solely for the control of symphilids.

It is possible to examine soils for symphilids before planting by taking samples to a depth of 3 ft. These are gently stirred in a vessel of water so that the symphilids float to the surface but, as the pest is seldom uniformly distributed, many samples must be examined. In practice, it is usually preferable to wait until the symphilids migrate to the plant roots. Confirmation of suspected attacks can then be obtained by removing plants, with the soil surrounding their roots, and immediately placing them in a vessel of water. The symphilids, released by gently breaking up the ball of soil, can be counted on the surface of the water, taking care not to confuse them with Collembola (springtails).

Where previous attacks have occurred, additional plants can be set out between the rows for removal and examination. Controls should be applied when ten or more symphilids are found at the roots of any one plant.

Parathion has been used successfully since 1949 and remains the most satisfactory pesticide in respect of insecticidal efficiency and absence of phytotoxicity. Diazinon, though rather more phytotoxic, is almost as effective and less dangerous.

The soil around the roots of affected plants should be drenched with 0·01% parathion or diazinon. With tomatoes this application can usually be made to serve as the first 'ball watering' and about 1 pint of diluted insecticide given to each plant. With other plants the quantity should be such that the soil in the root zone is thoroughly wetted.

For tomato plants gamma-BHC wettable powder at 0·02% can be used as a drench instead of parathion or diazinon, neither of which is available to amateurs, provided the soil surrounding the plants is moist and open. The control obtained is normally inferior to that given by the organophosphorus materials, and may slightly harden plants, particularly if they are planted in a soil depression which does not drain readily.

Gamma-BHC can be used more safely as a pre-planting treatment, either by thoroughly mixing a dust into the top 4–5 in. of soil at a rate equivalent to 2 oz of 0·4% dust/yd^2, or by watering the planting holes, in the case of tomatoes or similarly planted subjects, with about $\frac{1}{2}$ pint of 0·01% gamma-BHC. These treatments should be made several days before planting.

Precautions. There is some risk of injury if BHC is applied to soil used for propagating cucumbers. It should not be used for the treatment of soils to be planted with root vegetables which might become tainted by residues.

REFERENCES

EDWARDS, C. G. (1958). Ecology of Symphyla Pt. I. Populations. *Entomologia exp. appl.*, **1**, 308–19.

— (1959a). Ecology of Symphyla. Pt. II. Seasonal Soil Migrations. *Entomologia exp. appl.*, **2**, 257–67.

— (1959b). A revision of the British Symphyla. *Proc. zool. Soc. Lond.*, **132**, 403–39.

— (1961). Ecology of Symphyla. Pt. III. Factors controlling soil distribution. *Entomologia exp. appl.* **4**, 239–56.

MICHELBACHER, G. E. (1938). The biology of the Garden Centipede (*Scutigerella immaculata* Newport. *Hilgardia*, **11**, 55–148.

Order ACARINA: Mites

Mites and ticks are part of the arthropodan class *Arachnida* which also includes the scorpions, spiders and harvestmen. They are characterized by a reduction of the body segmentation to four indistinct regions; indeed, it is usually only possible to distinguish two—the *gnathosoma* bearing the mouth and associated appendages, and the *idiosoma* carrying the legs. While all adult mites have four pairs of legs, the first larval stage normally has only three pairs. Simple eyes may be present but, where absent, the animal is dependent on sensory setae on the body and legs. Mites occur in most habitats from salt water to the organic debris on high mountains. Many species are important medical and agricultural pests.

Family TETRANYCHIDAE: The Spider Mites

Tetranychus urticae Koch and *T. cinnabarinus* Bois., the glasshouse red spider mites, are probably the most serious pests in glasshouses throughout the world, and in certain circumstances, may damage outdoor plants.

There has been much controversy about the identity of these mites. For many years they were known, in Europe, as *T. telarius* (L.), and, in America, as *T. bimaculatus* Harvey. Recent study shows that there is a complex of species and, temporarily, the confusion has been resolved by calling the green two-spotted mite with a winter diapause form *T. urticae* Koch, within which the names *telarius* and *bimaculatus* are sunk as synonyms. A similar reddish-carmine mite, without an overwintering form, is identified as *T. cinnabarinus* Boisduval.

In a recent paper Boudreaux & Dosse (1963) have concluded that, if the rules of the International Code of Zoological Nomenclature are to be strictly applied, the reddish-carmine mite must bear the name *T. telarius* (L.). Nomenclatorial changes in important pests are regrettable because it becomes uncertain to which form different control measures apply.

Both the greenish and carmine-red species occur regularly in glasshouses; the latter has no specially adapted hibernating form and is normally associated with perennial crops.

The soft, pear-shaped and arched body of *T. urticae* is subdivided only by a transverse groove behind the second pair of legs. The palpi extend to the apices of the femora of the first pair of legs and between them the stylet-like chelicerae arise from a mandibular plate which, while puncturing leaf tissue, is moved rapidly in and out by powerful muscles. On either side of the body there are two 'eye-spots' slightly behind the second pair of legs. The abdomen is two-thirds as broad as long and bears sixteen curved, simple setae. The legs are slender, the anterior pair being the longest with the second pair reaching the tips of their tibiae. The other two pairs are directed backwards, the third pair reaching the tip of the abdomen, while the fourth pair project far beyond. When full grown the females are 0·45 mm long. The male, 0·3 mm long, has a slender and rather pointed abdomen. The short, stout, penis has an upwardly directed apical hook bearing a flattened barb. The shape of this barb is used as a diagnostic character.

Females produce about 100 spherical, pearly eggs 0·1 mm in diameter (Fig. 6.6) which are deposited singly, though commonly close together, either on the leaf surface or attached to strands of silk. As the eggs develop they become reddish. The incubation period varies from 3 days at 90 °F (32 °F)., to 28 days at 50 °F (10 °C). The six-legged, flesh-coloured larva emerges and when food supplies are adequate, remains near the oviposition site all its active life. At 70 °F (21 °C) it is active for only 1 day before entering a quiescent, immobile, non-feeding stage for a further day or two. This resting stage is firmly attached to the leaf surface, usually near a vein, by silken fibres. During the resting period the legs are rigid, the front two pairs held straight forward parallel to the palps. The next stages, protonymph and deutonymph, also have active and quiescent stages each lasting 1 to $1\frac{1}{2}$ days. In the late nymphal stages the sexes may be distinguished because the males are more elongate. Adult males are more active than the females, searching out and waiting beside a female deutonymph resting stage to copulate when she completes her final moult. Egg laying begins 24–48 hr after fertilization. Within a day of maturation, unfertilized females lay eggs which develop into males. Fertilized females produce both sexes, with a slight preponderance of females.

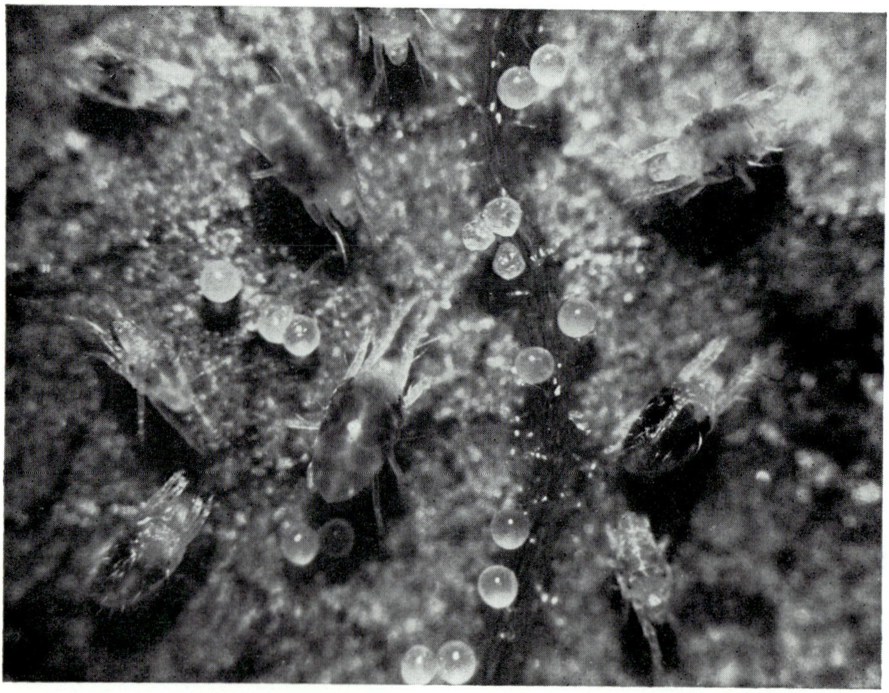

FIG. 6.6. Adults, nymphs, and eggs of glasshouse red spider, *Tetanychus urticae* Koch. (×50). (Courtesy of Glasshouse Crops Research Institute.)

Nearly 200 plant species have been recorded as hosts of *T. urticae*. In glasshouses it is the dominant pest on many plants. Mites are most active on the underside of the leaves; individuals are not obvious but, on most plants, their presence becomes apparent by a fine mottling of the upper leaf surface caused by the puncturing of leaf cells by the mandibular stylets (Fig. 6.7). The cells immediately adjacent to these punctures collapse and dry out so that, as feeding progresses, the leaves become hard and almost parchment-like. The green chlorophyll, within the punctured cells, is partially dissolved by enzymes secreted by the mites' salivary glands and a yellow tint is induced, along with mottling on the upper surface. Where infestations occur on fruiting plants, damage to the fruit ranges from fairly severe, as on tomatoes, to none on cucumbers. As it moves over the leaves, *T. urticae* forms a web of fine silken strands and, when mites are abundant, entire plants may be enveloped in sheets of this material. The silk is produced from glands discharging their secretion

FIG. 6.7. Characteristic leaf damage caused by spider mites. (Courtesy of Glasshouse Crops Research Institute.)

on the upper surface of the stylets. Normally, the silken webbing does not increase the damage but, on flowering plants such as carnations and chrysanthemums, the extensive and unsightly weft of dense webbing formed on the flowers renders blooms unsaleable (Fig. 6.8).

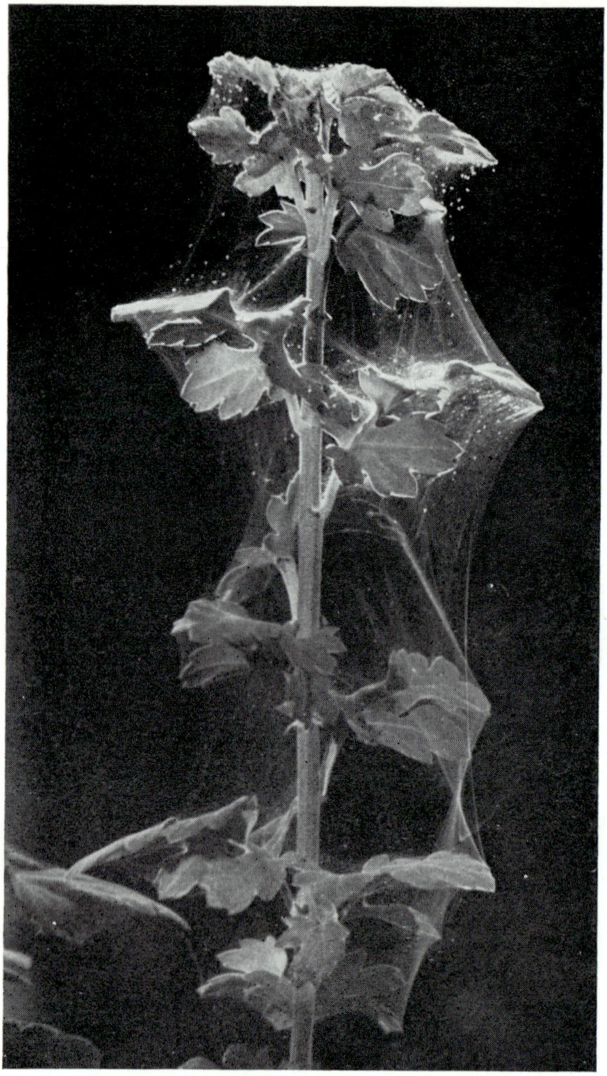

FIG. 6.8. Spider mite webbing on chrysanthemum. (J. Hesling.)

The dispersal mechanisms of *T. urticae* are important features of its behaviour under glass. Newly matured females settle for oviposition after a limited upward dispersal from the point of emergence. So long as the leaf tissues provide adequate food there is no further migration. As the population and competition for food increase, younger females congregate at the apex of the plant. The tip of the apical leaf is bent down by the weight of jostling mites and individuals drop on silken threads about half an inch long. Immediately such a thread is formed others hurry down it, continually thickening the strand and often trapping other mites. Trapped individuals are covered by the silken web from which they seldom escape. The 'rope' extends quite rapidly downwards often at a rate of 2 ft/hr. Outdoors, ropes may be blown some distance away from the host plant. In the glasshouse however, air movement is restricted and the ropes hang vertically but may occasionally be dispersed during cultural operations. Mites drop from the end of the ropes by 'free fall' without producing further silk. Once on the ground they respond to the plane of polarized light from the sky and so move in fixed directions across the floor of the glasshouse. Plants encountered are climbed but most mites fail to find other host stems and climb the glasshouse structure to regain a feeding site on leaves in contact with the glass. Because of these limited vertical movements a glasshouse mite population tends to remain in one position so that acaricide-resistant populations are often encountered in the same portion of affected houses from year to year. This positional stability is reinforced by the behaviour of hibernating mites in the autumn.

The factors inducing *T. urticae* to enter winter diapause, or hibernation, are day length, temperature and nutrition; these factors act in a complex way on the developmental stages. The critical day length below which diapause is induced depends on latitude, and in Britain is about 12½ hr. The effect of nutrition is difficult to define, and even in mid-winter when the short day length encourages diapause, mites on vigorously growing plants remain unaffected. In practice, if any two of the factors tend to be diapause-inducing most of the mites hibernate. The inducing factors act in succession on the larvae, proto- and deutonymphs with an intensity which is correlated with the size of the young mite. No change is induced during nymphal development but the resulting adult females become bright red, do not feed and, unlike active feeding forms, are negatively photokinetic. This leads to their hiding in cracks and crevices of the glasshouse. Hibernating mites do not resume activity until they have experienced several hours below 42 °F (6 °C). Most mites hide in the apex of the roof, particularly in the ridge plate and capping, of older wooden houses. Being far from heat sources they are influenced by the external, rather than the internal, glasshouse environment. Spring emergence is, therefore, governed chiefly by the outside weather. Consequently, if it is warm early in the year, newly planted cucumbers may be infested by mites descending from hibernation. If it is cold, infestation may be delayed until the cucumbers have reached the tops of the wires. Females, newly emerged from winter diapause, feed for 2 or 3 days before laying up to 10 eggs. Subsequently development is similar to that already outlined for summer forms. Occasionally, when the first 'normal' generation is produced in mid-winter, the short day length may reinduce diapause in January or February. The scale of an infestation is largely governed by the size of the overwintering population. Because day length tends to be the dominant factor affecting diapause there is a case for artificially increasing the day length to about 16 hr from early September. When 150 watt lamps, spaced 20 ft apart, were used to extend the day length or provide a 2-hr 'break' in the night, the diapausing population was drastically reduced. In pre-war days, growers tried to kill overwintering mites by burning sulphur and washing down the house structure with cresylic acid. These methods largely ceased when efficient acaricides became available and the continued red spider problem is largely due to the survival of overwintering mites.

Colour variations in the *Tetranychus* complex cause confusion. Starvation alone produces bright orange-red females indistinguishable from hibernating forms. Another form with red colouration found under glass is initiated in mid-summer, apparently by high temperatures and host plant desiccation. These conditions induce a resting, or aestivating, stage enabling the species to survive heat and desiccation in dry, tropical areas. These resting forms resume activity when monsoon rains encourage fresh vegetative growth.

Another orange-red mite is that produced by hybridization between *T. urticae* and *T. cinnabarinus*. These hybrids are short-lived and almost infertile but are not uncommon where both species occur.

The damage caused by spider mites varies with the plant attacked but on beans and cucumbers, the leaves may be shrivelled and the plants killed. Early infestations, uncontrolled over 5 months, decrease cucumber yield by 40%. On chrysanthemums little leaf damage is caused but the value of the flowers may be severely reduced by a covering of silk. Because females oviposit mainly on the youngest, upper leaves the largest mite populations tend to occur on the older leaves where the progeny mature. On cucumbers, especially when cordon-grown, the build-up of mites may be reduced by pruning off infested, older leaves before the mites have matured and migrated upwards.

Biological control

An interesting development in the control of *Tetranychus* spp. on glasshouse crops (such as cucumbers and orchids), is the use of the Chilean predatory mite *Phytoseiulus persimilis* A.-H. Over a wide temperature range (optimum of 76–86 °F (25–30 °C)) this mite develops twice as fast as its prey and, under the best conditions, female *P. persimilis* lay about 4 eggs per day and consume 30 red spider eggs or 24 immature stages. Two predators liberated on a cucumber plant so increased and spread that they were present on every infested leaf within 19 days and completely eliminated the *T. urticae* within 33 days. This pattern of control has been obtained consistently both in small and large houses.

To exploit a biological control technique it is essential to determine the maximum pest density which can be tolerated without decreasing yield—this density is called the economic threshold. Assessments of mite damage on cucumber leaves showed that the crop is unaffected unless the leaf injury exceeds a 'critical level', roughly equivalent to a 30% loss of the phytosynthetic area (Hussey & Parr, 1963a). Further, the leaf damage increases and decreases at predictable rates before and after control. So long as the predator is liberated when leaf damage is slight so that there is time for control before the diminishing number of spider mites causes damage equivalent to the critical level, the operation will be successful (Hussey *et al.*, 1965). During the 28- to 33-day interval between a correctly-timed predator introduction and elimination of the prey on cucumbers, the damage level cannot approach the economic threshold, even with the most rapid rates of increase in leaf damage yet recorded.

The most important factor in achieving predictable fluctuations in leaf damage during control by the predator is that, initially, almost every plant must be infested with spider mites. This may occur naturally during propagation but, if not, plants must be artificially infested at planting out.

A critical feature of the biology of *Phytoseiulus* is the period for which it can persist after the host mite has been eliminated. Preliminary experiments suggest that it survives for up to 4 weeks, although this period is unusually short as many other phytoseiid mites can subsist on honeydew and pollen for longer periods. If there is no further infestation, therefore, a systematic introduction of the predator gives commercial control in 4 weeks and the survivors maintain it for at least another month. So two introductions should provide control throughout the usual (20-week) cucumber crop. To maintain enough predators, without making two introductions, it may be possible to use a proportion of the cucumber plants (about 1 per 100) to rear *Phytoseiulus* for distribution to other plants that become infested later. The damage suffered by these predator-producing plants would cost the grower a negligible sum.

The difficulties of controlling spider mites chemically because of the widespread resistance to acaricides suggests that biological control will be increasingly used. Proper exploitation depends upon the establishment of facilities for commercial mass production of the predator.

Chemical control

In view of the world-wide importance of tetranychid mites as pests of agricultural and horticultural crops much effort has been expended to develop chemicals for their control. The facility with which strains, resistant to one or more acaricides, can be selected by repeated applications, has increased the problem of control. The frequent applications required to give adequate control under glass, where the life cycle is short, provide favourable conditions for the selection of these resistant strains. Furthermore, the narrow margin between effective acaricidal and phytotoxic concentrations, as well as the difficulty of obtaining effective spray coverage due to the natural, or induced, configuration of the plants, results in a proportion of the mites receiving only sub-lethal concentrations.

The development of resistance is most pronounced in cucumber houses where these factors act severely. The selection of mites resistant to azobenzene occurred in several nurseries in 1948, following the use of this compound for about 4 years. Later, after about the same period, the use of parathion, which supplanted azobenzene, led to resistance to this material also. Subsequently, mites have become resistant to nearly all the acaricides which can safely be used on this crop. The problem is almost as acute in glasshouses in which successive crops of chrysanthemums are grown 'all the year round'. Mites in tomato and carnation houses, however, have not developed resistance so readily for they remain susceptible to fenson and chlorfenson after their use for up to 13 successive years. On the other hand, resistance to organophosphorus and/or specific acaricides has developed on some carnation and rose nurseries.

The complexity of the patterns of resistance in different strains of both *T. urticae* and *T. cinnabarinus*, even within individual glasshouses, complicates the choice of effective compounds. Growers must learn from experience which chemicals, when efficiently applied, fail to give satisfactory control. Other chemically-unrelated products must then be tried. Resistance to one material within a chemically-related group of compounds does not necessarily confer resistance to other compounds within that group. For example, experience with diazinon and azinphosmethyl used against mites resistant to parathion, has shown that, although these compounds were initially effective, strains resistant to them were rapidly selected.

The ease with which acaricide-resistant strains are produced has increased the need to exploit cultural and hygienic preventive measures. Since infestations of annual crops such as cucumbers, tomatoes and beans derive almost entirely from mites which left infested plants the previous autumn and hibernated within the glasshouse, prevention of winter diapause is essential. Older plants are frequently allowed to remain almost neglected within unheated glasshouses in the early autumn and any mites remaining are gradually deprived of adequate nutrients. Such conditions induce hibernation and also the formation of 'ropes' of migrating individuals which provide a ready means of introduction into other glasshouses. Care should therefore be taken to avoid dispersal of red spider mite in this way on plants, clothing or equipment.

Although it may sometimes appear uneconomic to apply acaricides to plants which have almost ceased cropping, it is usually essential to do so in September. To ensure that no immature stages develop under such diapause-inducing conditions the plants, and any weeds, should be promptly removed, or desiccated, when cropping ceases, or at the latest by mid-September. Desiccation can be hastened by severing the stems of tomatoes, cucumbers and beans, and then scorching the foliage either by spraying with a 1 in 50 dilution of 38–40% formaldehyde (a respirator and protective clothing are necessary) or by fumigating with formaldehyde (p. 376). Burning sulphur at the rate of $\frac{1}{4}$ lb/1000 ft^3 also hastens desiccation but may corrode metal work and cause deterioration of some paints. The structure and supporting wires in houses in which sulphur has been burnt must be thoroughly washed down to remove any phytotoxic, soluble, metal salts. However, neither of these chemicals will kill mites already hibernating in cracks and crevices.

Plants should be propagated in houses free from both active and hibernating mites, for very few acaricides can be applied with safety to seedlings and tender young plants. The latter are particularly susceptible to injury when raised at relatively high temperatures in poor light conditions.

Mites are normally susceptible to many organophosphorus compounds, both systemic and non-systemic, as well as to the specific caricides (p. 23). As non-systemic organophosphorus acaricides have little effect on the eggs frequent applications are necessary. Quinomethionate, binapacryl and dinobuton can be used on tolerant plants and, as yet, their limited use has not selected resistant strains. Resistance to petroleum emulsions has not occurred, in spite of its use on some nurseries for 40 years, presumably because its acaricidal action is physical. New compounds are frequently developed for mite control but many are too phytotoxic for use on most glasshouse plants. Naled, an organophosphorus compound which can be vaporized from heating pipes, has proved satisfactory against red spider mite on some cucumber and tomato nurseries but more information is required for its safe and effective use on other plants.

Control of spider mites on specific crops

Cucumber

It is anticipated that the predatory mite (p. 196) will eventually become widely used to control red spider on cucumbers and other crops but meanwhile chemical control is necessary. To avoid the use of chemicals on young cucumbers which are very susceptible to injury a clean start is most desirable.

Seedlings and young plants may be sprayed with parathion (preferably using a 20% concentrate) or, if 70 °F (21 °C) can be maintained for at least 4 hr, smoked with parathion (0·25 g/1000 ft^3) or sulfotep (1 g/1000 ft^3). These measures will only kill active mites and consequently H.V. sprays of demeton-S-methyl (0·02%), dimethoate (0·03%), oxydemeton-methyl (0·02%) or schradan (0·1%) should be applied when at least four true leaves are formed on the plants, provided that the stipulated intervals between application and cutting are observed (pp. 374–380). Except with dimethoate, this necessitates application within a few days of planting out. The use of schradan is restricted, by official recommendation, to main crop cucumbers, to which it can be applied only once at least 6 weeks before cutting, and at a rate not exceeding 100 gal/acre at 0·1%. If applied to pot-plants the area is that which will be occupied on planting out.

If the mites are resistant to these organophosphorus compounds, H.V. sprays of dicofol (0·01%) may be effective, but may cause injury to young plants, particularly when these are abnormally tender due to poor light or other adverse conditions. The spray should, if possible, be restricted to the infested leaves, preferably when the temperature is about 65 °F (18 °C), but not during sunshine or when the plants require watering. A wetting agent should be added to sprays prepared from emulsifiable concentrates of dicofol. A glasshouse-grade petroleum emulsion (diluted × 100) should be used if the mites are resistant to both dicofol and organophosphorus acaricides, and should be applied as a fine spray to completely cover the leaf surfaces. The rapid growth of young cucumbers often enables them to withstand moderate infestations without appreciable detriment to fruit yields and the grower must decide whether the degree of infestation at this stage justifies the risk of spray damage.

Before mid-May, initial infestations on established plants can often be checked by spraying visibly infested leaves with dicofol (0·01%), petroleum emulsion (diluted × 100) or tetradifon (0·01%), avoiding, as much as possible, the wetting of tender young leaves on the leading and lateral shoots. If a heavy infestation necessitates over-all treatment, either dicofol or petroleum sprays, or, where mites are not resistant to organophosphates, smokes of parathion or sulfotep may be used without appreciably 'hardening' the plants. Dimethoate (0·03%) may also be used up to 1 week before cutting cucumbers.

From May onwards, established plants may be sprayed with dicofol, tetradifon or petroleum emulsion at the above concentrations, or with quinomethionate or dinobuton at 0·012%. The two latter can be used to advantage if the plants are attacked by cucumber powdery mildew or if the mites are resistant to both dicofol and tetradifon. Applications should be repeated at intervals of 10–14 days until the attack is adequately controlled, and, except with petroleum emulsions, a wetting agent should be added to ensure thorough coverage. An interval of 10 days should elapse between applying petroleum emulsion sprays and vaporizing sulphur to control powdery mildew or applying quinomethionate or dinobuton.

To avoid the large labour requirement for H.V. spraying, aerosols of dicofol (2 g/1000 ft^3), dicofol with diazinon (1·8 g and 1·5 g/1000 ft^3), or tetradifon aerosols (2 g/1000 ft^3) may be used where the mites are not resistant to these compounds. To obtain satisfactory control, even of susceptible mites, several applications of aerosols or smokes at 7- to 10-day intervals are necessary. As dicofol and, more particularly, tetradifon, control red spider mites by their action on eggs and larvae, the immediate control may be poor. They are, however, persistent, particularly when applied as sprays, although incomplete spray coverage of heavily fruiting cucumbers often necessitates routine applications at 10- to 14-day intervals. L.V. spraying is unsatisfactory. Diazinon, alone or in combination with dicofol, gives immediate control of active stages of non-resistant mites and is of value both where heavy infestations occur, and where thrips or aphids are also present. Azinphos-methyl aerosols have also been used but, as mites resistant to diazinon are almost invariably resistant to them also, the use of this more poisonous compound is not recommended. Smoke generators containing either tetradifon or combinations of azobenzene and parathion, or azobenzene and sulfotep are also available. Although expensive, they are easier to apply.

To obtain satisfactory control, even of susceptible mites, several applications of aerosols or smokes should be made at 7- to 10-day intervals. Trimming and 'tying in' should precede the application of acaricides.

The acaricides suggested for cucumbers are also suitable for use on melons and marrows.

Tomato

Tomatoes are less liable to damage by acaricides than cucumbers. Aerosol applications of persistent specific acaricides, repeated if necessary at 14- to 21-day intervals, normally give adequate control though seedlings may be damaged. Mites on young plants in pots or in beds before mid-May, can be controlled by spraying with dicofol (0·01%) or tetradifon (0·01%) or, provided the necessary interval is allowed before picking, with demeton-S-methyl (0·02%), dimethoate (0·03%) or oxydemeton-methyl (0·02%). Sprays of diazinon (0·02%) can be used if aphids or whitefly are also present, and picking will not commence before 2 weeks.

After the middle of May, or earlier if the plant growth is fairly hard, aerosols of dicofol, tetradifon, chlorfenson or fenson can be used at 2 g, 1 g, 5 g and 5 g/1000 ft^3 respectively. One or two applications, with an interval of 14–21 days, give good control of most strains in tomato houses. To kill adult mites rapidly when very heavy infestations have been allowed to develop in mid-summer, the addition of diazinon at 1·5 g/1000 ft^3 is of benefit. Tomatoes may be sprayed with petroleum emulsions (diluted × 100) but, as thorough wetting of the plants is required, they are less effective than the substances previously mentioned. Oedema of tomato plants may result from their use before May.

Mites can also be controlled with smokes of tetradifon, azobenzene combined with parathion, or azobenzene alone. Two or three applications at 7- to 10-day intervals will be required in the case of azobenzene; it may be necessary to make a second application of either of the others after 14–21 days.

Chrysanthemum

The problem of spraying densely planted beds of chrysanthemums effectively, the occurrence of strains resistant to many acaricides, and marked differences in the tolerance of cultivars to pesticides, make it difficult to give general recommendations for the control of mites on this crop.

Where the mites are not resistant to organophosphorus pesticides the systemic compounds demeton-S-methyl, oxydemeton-methyl and schradan overcome, to some extent, the problem of achieving the effective coverage required with most non-systemic compounds. Even at the normal concentrations of 0·02%, 0·02% and 0·08% respectively, some varieties are prone to damage by H.V. sprays of these compounds. Soil drenches of the first two at 0·02%, applied at 2 gal/yd^2, are less phytotoxic and can be safely used on most cultivars. Soil drenches are, however, considerably more expensive than sprays in spite of their rather more persistent protection. Recent chemical and biological investigations (Worthing, 1968) indicate that, with soil applications, demeton-S-methyl is more effectively transported to buds than oxydemeton-methyl and also has a longer life in the plant. These insecticides should be applied 3 or 4 weeks after planting with a second application 4–6 weeks later.

Where mites are resistant to the above compounds, sprays of dicofol (0·01%) or tetradifon (0·01%) may be used. Aerosols of dicofol (2 g/1000 ft^3) or tetradifon (1 g/1000 ft^3) may be used before the density of the foliage prevents their diffusion among the plants. Diazinon aerosols (1·8 g/1000 ft^3), used against susceptible mites on young, established plants are also effective against leaf-miner and some aphids. Several applications at 10- to 14-day intervals may be required.

Quinomethionate sprays may be preferable where the mites are resistant both to organophosphorus and specific acaricides, or where the plants have powdery mildew. Proprietary combinations of petroleum emulsion with both a fungicide and DDT are of particular merit where mildew or rust diseases, caterpillars or thrips are present. Complete coverage is essential to obtain efficient results with petroleum sprays which should be used at the dilution advised by the particular manufacturer. Their enhancement of the appearance of the foliage is considered beneficial by some growers.

If neglect results in mite migration to open blooms, H.V. sprays of dicofol, schradan, diazinon or petroleum emulsion; aerosols of dicofol, diazinon, or parathion smokes should be applied. To minimize the risk of infection by flower diseases, sprays should be applied only when conditions promote rapid drying. Petroleum sprays should not be used in the late autumn and winter. There is, as yet, insufficient information on the tolerance of the flowers of different cultivars to H.V. quinomethionate sprays, though damage to several varieties with white flowers has been reported. Low volume applications are unsafe.

Certain new organophosphorus and carbamate systemic insecticides tested against *Myzus persicae* (p. 119) have shown promise against red spider mite.

Carnations

Carnations tolerate most acaricides but as aphids are usually a concurrent pest, systemic organophosphorus insecticides are preferred. H.V. sprays of demeton-S-methyl (0·02%), dimethoate (0·03%), formothion (0·045%) or oxydemeton-methyl (0·02%), repeated, if necessary, at intervals of 4–6 weeks give good control in the absence of resistance. Demeton-S-methyl, however, imparts an unpleasant smell to blooms cut within 4 days of application. Sufficient wetting agent must be included in the sprays to wet the foliage.

Soil drenches of these compounds at two-thirds the above concentrations give excellent control of non-resistant strains when applied at 2 gal/yd^2 to first-year plants and 3 gal/yd^2 to older plants, or 3–4 fl. oz/6 in. pot. Their use is not advised during the late autumn and winter when the transpiration rate may be insufficient to accumulate lethal concentrations in the plant tissues.

Where simultaneous control of aphids is not required, and atomizing equipment is available, aerosols of chlorfenson (6 g/1000 ft^3), dicofol (2 g/1000 ft^3), fenson (6 g/1000 ft^3) or tetradifon (1 g/1000 ft^3) provide good control with ease of application. Smokes of tetradifon (1 g/1000 ft^3), azobenzene plus parathion, or azobenzene alone may be used. Carnations are very tolerant to azobenzene smokes which can be used at 8 g/1000 ft^3 (i.e. twice the normal rate) on established plants.

Roses

Roses tolerate aerosols of dicofol and these provide the simplest control of mites which are not resistant to this chemical. Aerosols of diazinon are effective against some strains. Tetradifon may be used safely and effectively on most varieties although on 'White Butterfly' and 'Cinderella' it may cause some defoliation.

Sprays of dicofol (0·01%), dimethoate (0·03%), formothion (0·05%), oxydemeton-methyl (0·02%) also give good control in the absence of resistance to the respective compounds. Demeton-S-methyl sprays (0·02%) may be used provided buds are not cut within 4 days of spraying. H.V. sprays of binapacryl (0·02–0·03%), prepared from an emulsifiable concentrate formulation are recommended in Holland. Both binapacryl and quinomethionate are effective against strains of red spider mite which are resistant to organophosphorus compounds and dicofol though they are phytotoxic. Spraying resting bushes with a preparation of endosulfan (Thiodan) is also recommended, but its use is not accepted in Britain. Quinomethionate could also be used at this time as mites on roses do not hibernate.

A new compound, known as chlorphenamidine, gave the best results against red spider mite on roses in a trial in 1965 in Guernsey.

Miscellaneous plants

The control measures for red spider mite on other plants grown under glass must be dealt with collectively. Some plants which are known to be susceptible to damage by acaricides are listed in the Appendix (p. 382).

The best acaricides, provided the plants are tolerant, are dicofol (0·01%) and tetradifon (0·01%) or the systemic compounds demeton-S-methyl (0·02%), formothion (0·05%), dimethoate (0·03%) and oxydemeton-methyl (0·02%). The last named is probably the least phytotoxic systemic compound, though many ornamentals are susceptible to damage and it may be necessary to employ non-systemic organophosphorus compounds such as diazinon (0·02%), parathion (0·01%) or sulfotep smokes on some subjects. These will require several applications, at 7- to 14-day intervals, to eradicate the pest.

REFERENCES

BOUDREAUX, H. B. (1963). Biological aspects of some phytophagous mites. *A. Rev. Ent.*, **8**, 137–54.

BOUDREAUX, H. B. & DOSSE, G. (1963). Concerning the names of some common spider mites. *Recent Adv. Acarol.*, **1**, 350–64.

DOSSE, G. (1952). Greenhouse Spider Mite, *Tetranychus urticae* f. *dianthica* and its control. *Höfchenbr. Bayer PflSchutz-Nachr.*, **5**, 239–67.

HELLE, W. & BUND, VAN DE, C. F. (1962). Crossbreeding experiments with some species of the *Tetranychus urticae* Group. *Entomologia exp. appl.*, **5**, 159–65.

HUSSEY, N. W. (1967). Provisional programme for the use of the predatory mite *Phytoseiulus riegeli* to control red spider mite on cucumbers. *Rep. Glasshouse Res. Inst.*, **1966**, 140–3.

HUSSEY, N. W. & PARR, W. J. (1963a). The effect of Glasshouse Red Spider mite (*Tetranychus urticae* Koch) on the yield of cucumbers. *J. hort. Sci.*, **38**, 255–63.

—— (1963b). Dispersal of the Glasshouse Red Spider mite. *Entomologia exp. appl.*, **6**, 207–14.

HUSSEY, N. W., PARR, W. J., & GOULD, H. J. (1965). Observations on the control of *Tetranychus urticae* Koch on cucumbers by the predatory mite *Phytoseiulus riegeli* Dosse. *Entomologia exp. appl.*, **8**, 271–81.

KINGHAM, H. G. & GOULD, H. J. (1963). Observations on Glasshouse Red Spider Mite on cucumbers in the Lea Valley. *Expl. Hort.*, **10**, 15.

PARR, W. J. & HUSSEY, N. W. (1966). Diapause in the Glasshouse Red Spider (*Tetranychus urticae* Koch) a synthesis of present knowledge. *Hort. Res.*, **6**, 1–21.

SPEYER, E. R. (1928). The Red Spider Mite. *J. hort. Sci.*, **7**, 161–71.

WORTHING, C. R. (1968). Evaluation of insecticides for control of *Myzus persicae* on year-round chrysanthemums. *J. hort. Sci.*, **43**, 197–214.

Bryobia kissophila van Eyndhoven (the ivy mite). Experts disagree about the exact identity of the *Bryobia* sp. which occasionally causes damage to cultivated ivy and whose closest relatives are more commonly known as pests of outdoor tree and bush fruits. However, at least one species, *B. kissophila*, is now regarded as restricted to *Hedera*.

Species of *Bryobia* (Fig. 6.9) are distinguished from other *Tetranychidae* by the hairy, 6-segmented legs, the first pair of which are extremely long. The flattened body is covered with clavate hairs and has an obvious peripheral 'step' on its dorsal side. The eggs are perfectly smooth, dark-red and 0·2 mm in diameter; they are slightly sticky and hence often covered with fine dust particles. Newly hatched larvae are orange-red but, as feeding continues, they gradually become greener. Both the protonymphal and deutonymphal stages are brownish at first but later, through assimilation of chlorophyll, become greenish. In the dormant stages the first pair of legs are folded together underneath the body and appear greatly shortened. In general, *Bryobia* has

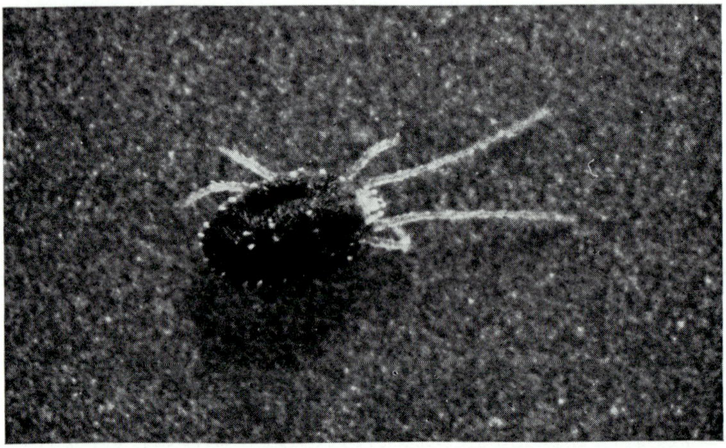

FIG. 6.9. *Bryobia* mite, adult (×35). (By kind permission of the Plant Pathology Laboratory, Ministry of Agriculture, Fisheries and Food.)

a striking preference for the upper side of the leaves and, unlike other *Tetranychidae*, does not produce a silken web while feeding. Infested leaves become mottled with greyish-yellow spots and, in severe cases, plant growth may be stunted.

No males are known in *B. kissophila* and reproduction is parthenogenetic. Closely related species attack many alpine plants including *Androsacea*, gentians, *Primula* and *Saxifraga* spp. They frequently appear in large numbers in the autumn when they may migrate from weed hosts and invade sheds, glasshouses and other buildings.

Control

Bryobia mites are easily controlled by most chemicals which are effective against non-resistant strains of spider mites; malathion is also effective. There is little difficulty, therefore, in selecting a non-phytotoxic product to control other pests which may also be present. Persistent compounds such as dicofol and tetradifon require fewer applications than non-persistent compounds, and systemic compounds can be used on plants where the leaves are difficult to spray on the undersides. Sometimes the convenience of aerosols compensates for the greater number of applications needed. If plants tolerate the following chemicals they can be sprayed with demeton-S-methyl at 0·02%, dimethoate at 0·03%, formothion at 0·045% or oxydemeton-methyl at 0·02%, or watered with these compounds at 0·012%, 0·02%, 0·04% and 0·012% respectively. About half a pint is required for a 6 in. pot and *pro rata* for plants in pots of other sizes. A single application of any of these four insecticides may suffice, particularly if the insecticide is applied as a soil drench. For spray application, on waxy-leaved plants which are difficult to wet, demeton-S-methyl is advised in preference to oxydemeton-methyl.

Some plants, susceptible to attack by *Bryobia* mites, are sensitive to systemic organophosphorus compounds and, on these, dicofol sprays at 0·01% or tetradifon at 0·01% will kill the eggs and immature stages of the mites and, by their persistence, provide satisfactory control after one or two applications. Sprays containing diazinon at 0·02%, malathion at 0·05% or parathion at 0·01% will kill active stages, though two or more applications may be necessary to destroy newly-hatched mites as eggs are not killed by these chemicals. Parathion should only be used when the co-presence of another pest requires its use or when the host plant is damaged by the other insecticides. If two or three applications are given at 10- to 14-day intervals petroleum emulsions (glasshouse grade), at a dilution of 1 in 100, are also effective and can be used safely on many plants. Several applications of parathion or tetradifon smokes, or of malathion, diazinon, or dicofol aerosols will probably be effective and may be most suitable where there is difficulty in spraying thoroughly at high volume. No information is yet available regarding the value of sprays containing quinomethionate, or dinitrophenolic compounds, or of the fumigants dichlorvos and naled; they are, in general, more phytotoxic than the other chemicals previously mentioned.

Bryobia may also be a household pest invading buildings, in the early spring, from new and well-fertilized grassland. A 3-ft grass-free barrier around the building should be sprayed with 0·2% dicofol to reduce the nuisance.

REFERENCES

VAN EYNDHOVEN, G. L. (1955). *Bryobia* from *Hedera*, apple and pear (Acar, Tetran.). *Ent. Ber. Amst.*, **15**, 340–7.

GABEL, M. (1959). Beitrage zur Kentniss der Gattung *Bryobia*. *Z. angew. Zool.*, **19**, 191–247.

KREMER, F. W. (1956). Studies on the biology, epidemiology and control of *Bryobia praetiosa* Koch. *Höfchenbr.Bayer PflSchutz-Nachr.*, **9**, 190–252.

LANGENSCHEIDT, M. (1966). *Bryobia* als Eindringling in Neubeuten und Hinweise auf Bekämpfungsmöglichkeiten. *Anz. Schadlingsk.*, **39**, 149–51.

MATHYS, G. (1957). Contribution à la connaissance de la systématique et de la biologie du genre *Bryobia* en Suisse romande. *Bull. Soc. ent. Suisse*, **30**, 189–284.

Family TARSONEMIDAE

The family Tarsonemidae are minute mites (less than 250 μ long) with shining, oval and segmented bodies. Two pairs of smooth setae and a pair of clavate pseudostigmatic organs arise from the dorsal surface near the second pair of legs. The hind legs are characteristic, being shorter and thinner than the other legs and composed of only three segments (four on other legs). They are terminated by two setae, one of which is extremely long and the other relatively short. In the males, which are even smaller than the females, the fourth pair of legs is enlarged, the femur making a scissor-like joint with the next segment so forming a pair of 'claspers' for holding the female deutonymphs. The Tarsonemidae are a widely distributed group of mites, many of which feed on fungal hyphae. Some species are, however, primarily phytophagous and cause serious injury to plants of economic importance.

The nature of the differences between species of Tarsonemid mites makes identification a matter for the specialist. Thus, the arrangement of the apodemes, or thickened rods, which serve as muscle attachments, is one of the principal characters used in specific separation but these structures are only visible in preparations viewed under the high-power microscope.

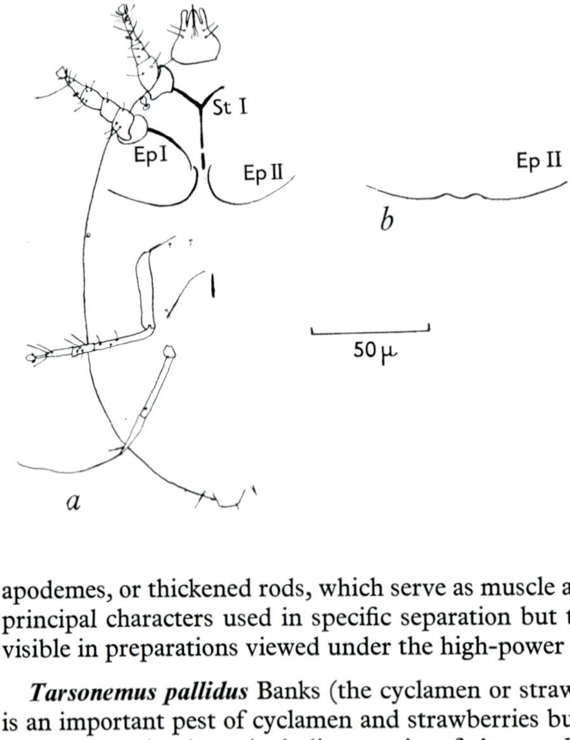

FIG. 6.10. (*a*) *Tarsonemus pallidus* Banks (female); (*b*) *T. confusus* Ewing—epimeron II.

Tarsonemus pallidus Banks (the cyclamen or strawberry mite) (Fig. 6.10*a*) is an important pest of cyclamen and strawberries but has also been recorded from many other hosts including species of *Amaranthus*, *Antirrhinum*, *Azalea*, *Begonia*, *Chrysanthemum*, *Crassula*, *Fuchsia*, *Gerbera*, *Gloxinia*, *Impatiens*, *Pelargonium*, *Petunia*, *Saintpaulia* and *Verbena*.

The pale, amber-coloured females congregate within unopened buds and between the adpressed halves of the dorsal surfaces of folded leaflets. The white, opaque, smooth, oval and relatively large eggs ($125 \times 75\ \mu$) are laid within this shelter, both near the base of the midrib at its junction with the petioles, and in the marginal indentations. Each female may lay as many as three eggs per day though the total fecundity has not been accurately determined. The species is largely parthenogenetic, males comprising only about 5% of the population.

The incubation period of the eggs at 68 °F (20 °C) is 4 days, increasing to 13 days at 59 °F (15 °C). The embryo, which is visible through the egg-shell for some days before hatching, emerges as a six-legged, almost transparent, pearly-white larva. This larva has the front two pairs of legs situated close to the head and the third pair on the posterior half of the body. Each larva measures 200×80 μ and, at the posterior end, has a prominent triangular protuberance.

FIG. 6.11. Damage to cyclamen by *Tarsonemus pallidus* Banks. (Courtesy of Glasshouse Crops Research Institute.)

After 4–7 days the larval cuticle is loosened from the body, but still covers it like a 'pupal' case. The legs are withdrawn within this loose skin, the two front pairs are directed forwards, closely adpressed to the head, while the third pair lie straight behind. This 'immobile' pupa does not develop into an adult for at least 4 days so that the life cycle usually requires 10–14 days. The males often pick up and carry female 'pupae', with the anterior end to the rear, until the female matures and copulation is possible. Mature females measure 230×120 μ, and the males 170×95 μ.

Under glass, numbers of *T. pallidus* increase rapidly in the shelter of the outer scales of both the leaf and flower buds. It is not unusual to find hundreds of eggs within an area no larger than the diameter of a pencil. To the naked eye such concentrations of eggs resemble a fine coating of dust. As the leaflets expand the decrease in humidity drives the mobile stages back down to the crown to seek other young buds. The foliage, expanding from infested buds, is usually curled, twisted, brittle, and covered with small scabs. Infested flower buds may wither and die (Fig. 6.11), though where few mites are present, the flowers may merely be discoloured or flecked with dark spots. Damage is often more severe than the number of mites would suggest and it has been postulated that toxins are excreted during feeding which induce distortion of plant growth.

Tarsonemus confusus Ewing (Fig. 6.10*b*), known in America from tomato leaves, has been found in European glasshouses on species of *Azalea*, *Cissus*, *Cyclamen*, *Gloxinia*, *Pilea*, and *Saintpaulia* (Karl, 1965).

At about 68 °F (20 °C) the eggs hatch after 4 days and the larval life, both active and resting, is completed within 4 days. Oviposition begins 2 days after maturation of the female which, in common with related species, produces about 1 egg per day. The damage caused is similar to that created by *T. pallidus*.

Control

The best control of cyclamen mite is obtained by spraying with endrin at 0·025% or endosulfan at 0·05%—the use of these materials is now permitted on ornamental plants under glass. Sodium selenate, which has been recommended in the U.S.A. for soil application to cyclamen and other ornamentals, is not permitted in Britain.

The most satisfactory alternative is dicofol; two or three applications, at 28-day intervals, of H.V. spray containing 0·03% dicofol giving satisfactory control. Emulsifiable concentrate formulations are most effective. Additional wetting agent may be necessary to thoroughly wet leaves and buds. To prevent detrimental malformation of leaves and injury to flower buds treatment should commence at the first sign of damage. If control measures are delayed it is advisable to remove flower buds as mites sheltered within them are not killed, and the flowers will probably be distorted; furthermore, cyclamen flowers may be damaged by sprays of dicofol at the above concentration. Where there is difficulty in spraying the undersides of the leaves, e.g. *Saintpaulia*, it may be necessary to dip the plants in 0·03% dicofol. In the U.S.A. the use of schradan drenches is advised on *Peperomia* and *Saintpaulia*, and the application of 0·05% schradan at the rate of $\frac{1}{4}$ pint/4-in. pot at 3- to 4-week intervals is worth a trial. Soil drenches should not be applied when the soil is dry or when transpiration is rapid, i.e. under hot, dry conditions. Other commonly used systemic compounds are of little value.

Immersion of cyclamen, rochea (*Crassula*) and *Saintpaulia* plants for 7 min in water at a constant temperature of 15 °C was recommended for small-scale treatment; although effective, it is liable to injure the plants. Although the hot-water treatment of strawberry runners is effective in controlling *S. pallidus*, it is unsuitable on plants to be forced because it retards growth. The use of endrin or endosulfan, on strawberry plants being forced in glasshouses, is permissible, provided applications are not made after the opening of the first flowers.

Tarsonemus confusus will almost certainly be controlled by the measures which are advised for cyclamen mite.

REFERENCES

FRENCH, N., VERNON, J. D. R. & WOODVILLE, H. C. (1968). Cyclamen Mite, *Steneotarsonemus pallidus*, on chrysanthemum. *Pl. Path.* **12**, 143.

GARMAN, P. (1917). *Tarsonemus pallidus*, a pest of geraniums, *Maryland Agr. exp. Stn Bull.*, **208**, 327–42.

KARL, E. (1965). Studies on the morphology and ecology of *Tarsonemidae* of garden plants, I. *Tarsonemus pallidus* Banks. *Biol. Zentbl.*, **84**, 47–80.

MOZNETTE, G. F. (1917). The cyclamen mite. *J. agric. Res.*, **10**, 373–90.

VRIE, M. VAN DE (1957). Observations on the biology and control of the tarsonemid mite on strawberries. *Meded. TuinbVoorl-Dienst.*, **20**, 433–9.

WIESMANN, R. (1941). Untersuchungen über die Biologie und Bekämpfung der Erdbeermilben *Tarsonemus pallidus* Banks. *Landw. Jahrb. Schweiz.*, **55**, 259–329.

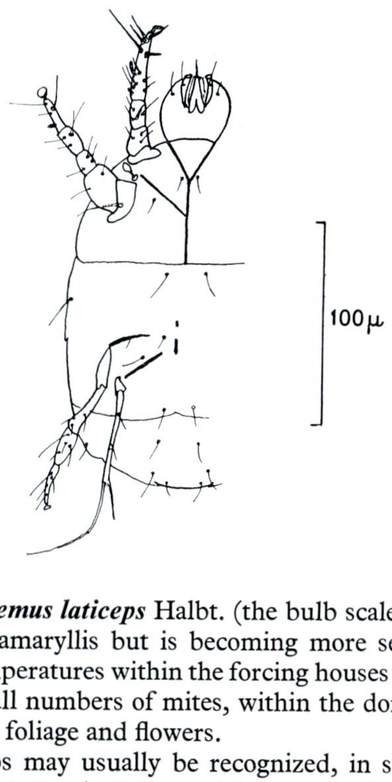

FIG. 6.12. *Steneotarsonemus laticeps* Halbert (female).

Steneotarsonemus laticeps Halbt. (the bulb scale mite, Fig. 6.12) is a well-known pest of amaryllis but is becoming more serious on forced narcissus bulbs. High temperatures within the forcing houses encourage rapid reproduction so that small numbers of mites, within the dormant bulbs, may increase and damage the foliage and flowers.

Infested bulbs may usually be recognized, in storage, by their abnormal dryness. The outer scales adhere tightly to the infested bulb which is often small and somewhat soft. Diagnosis can be confirmed by cutting transversely through the bulb about 6 mm below the apex. Small particles of brown, discoloured tissue are exposed, especially at the angular points in the scales. If the latter are parted the discoloured tissue shows as brown streaks down the inner face of the scales from apex to base. Mites of all stages are found on the discoloured area; as soon as they are exposed the active stages seek fresh crevices in which to hide and only the eggs, cast skins and dead mites remain visible.

Superficial feeding by the mites removes the surface wax from the leaves so that the foliage becomes bright green. Yellow flecks and streaks, resembling the symptoms of 'yellow stripe' virus, commonly occur on affected foliage.

FIG. 6.13. 'Saw-tooth' damage to narcissus by bulb scale mite. (J. Hesling.)

Severely affected bulbs lose vigour and the foliage senesces prematurely. The concentration of mites in the neck region contaminates the flower buds as they emerge. In extreme cases the flower bud is killed but, in less severe cases, although the bud is killed the flower stem continues to extend bearing the blind bud. Even where the mites are present in small numbers the bud growth is impaired and weak or malformed flowers develop. Outdoors, due to the low temperature, the mites do not increase so rapidly when the bulb is making soft and rapid growth and, up to flowering, growth of narcissus is normal. A few yellow streaks and blotches sometimes appear on the basal portions of the leaves and flower stems while many affected leaves develop a 'saw-tooth'

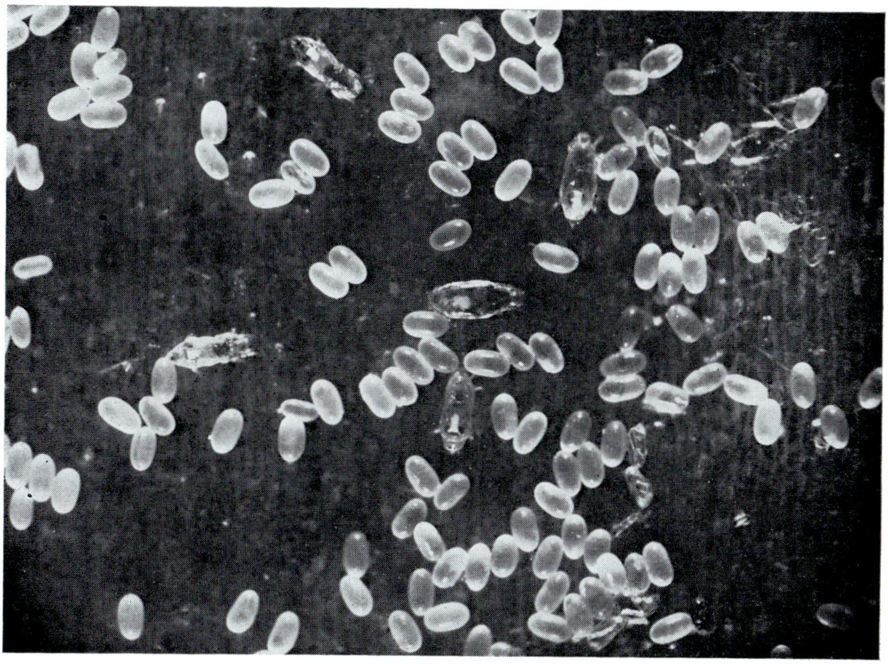

FIG. 6.14. Adults and eggs of bulb scale mite in neck region of bulb (× 40). (A Shell photograph.)

disfiguration (Fig. 6.13). In late April and early May higher temperatures encourage rapid increase by the mites which spread freely over the leaves. Where feeding induces premature wilting, so depleting the storage of food reserves within the bulb, the latter is reduced in size and vigour.

Each female mite lays, singly, up to 30 eggs, which, at room temperature, hatch in about 11 days. The only available data suggest that females live for at least a month. The larvae are active for about 16 days after hatching; the 'pupal' stage occupying a further 3 days. At bulb-forcing temperatures these developmental periods are considerably reduced; indeed, the total life cycle may then be completed within a fortnight.

The rapidly increasing mite population congregates in the neck of the forced bulb where individuals become visible to the naked eye at the bases of the young leaves (Fig. 6.14). At first, the mites remain within the shelter of the ensheathed leaves, but, as the population rises, they become generally distributed over the leaves. Some eggs are laid several inches above the ground though these invariably desiccate and fail to hatch.

As the foliage dies back the mites wander along the dying leaves to adjacent bulbs. By mid-June almost all the mites have withdrawn within the bulb scales to survive the dormant season deep in the bulb tissues. Some mites apparently remain on the outside of the bulbs, or even in the soil, for clean bulbs grown in soil from which infested, dormant bulbs have been removed may become infested, during forcing, by mites from the infested soil.

Although, within a single bulb, the mite population fluctuates throughout the year, eggs and immature forms are present at all times. At lifting, in late July, the scale mites are concentrated in the neck region. As the bulb is then fully turgid and solid the mites cannot penetrate between the tightly adpressed scales. During storage the bulbs lose water and tend to shrink so that air-spaces develop between the fleshy scales. The mites then spread downwards along these crevices so producing the yellowish strips of damaged tissue already referred to. As callus grows over the injury these areas of superficially damaged tissue later become brown. By October and November the population declines and the individual mites, though continually working towards the basal region of the bulb, become sluggish. As soon as root action recommences the bulbs regain full turgor and many mites are then crushed between the swelling scales, indeed, in healthy and symmetrical bulbs, the only survivors are in the neck region. In misshapen and damaged bulbs, which do not swell so uniformly, death of mites by crushing is less extensive.

S. laticeps is also a pest of amaryllis on which it causes red streaks and spots on the leaves as these emerge from the neck region of the bulb. These reddish lesions elongate as the leaves and flower stems grow and ultimately a scar several inches in length may result (Fig. 6.15). In severe cases the flowers shrivel or the flower stem remains short. Storage of amaryllis bulbs, at low temperature and humidity, in a well-lit storage room reduces the risk of serious infestation.

Control

The best way of avoiding loss due to bulb scale mite is to ensure that it is not present in bulbs intended for forcing. Suitable bulbs should be selected from stocks which, following heat treatment, have been grown for only a year or two in clean ground. If this mite or other pests are suspected, a number of bulbs should be cut through and carefully examined. Bulbs showing the described symptoms should be selected for further microscopical examination.

Few chemicals are effective against this pest: indeed, it is suggested that some may aggravate the trouble by killing predatory mites and symphilids. The only established treatment eradicating bulb scale mite in narcissus bulbs is hot-water treatment for 4 hr at 110 °F (43·3 °C) or 3 hr at 112 °F (44·4 °C). Fumigation with methyl bromide, by exposing the bulbs in a gas-tight chamber to the vapour from 3 lb of methyl bromide for $4\frac{1}{2}$ hr at 60 °F (16 °C) has been used in some countries, but this procedure may damage the bulbs.

A degree of control, sufficient to reduce injury to flowers, can be obtained by immersing the bulbs, when fully dormant, for 1 hr in water maintained at 110 °F (43·3 °C). The necessary equipment is described in the section on the control of bulb and stem eelworm (p. 292). Although this treatment is not

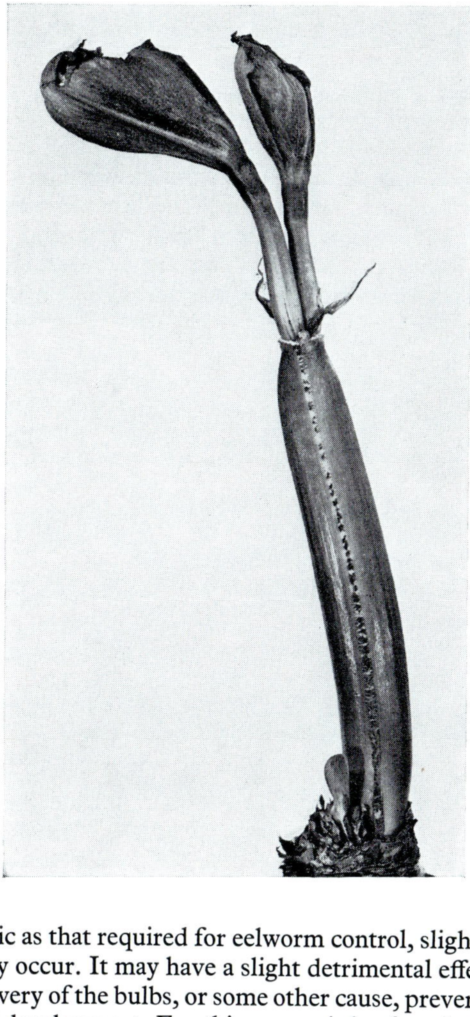

FIG. 6.15. Damage by bulb scale mite to amaryllis. (A Shell photograph.)

so drastic as that required for eelworm control, slight effects on flower production may occur. It may have a slight detrimental effect on flower production if late delivery of the bulbs, or some other cause, prevents treatment at the correct stage of development. For this reason it is often desirable to force bulbs from clean stocks hot-water treated a year or two previously. Hot water may also interfere with the conditioning of bulbs to promote early flowering.

The use of thionazin, as a dip for the control of bulb scale mite in narcissus, has recently been approved under the Agricultural Chemicals Approval

Scheme. Immersion for $2\frac{1}{2}$ hr in 0·23% thionazin (prepared from an emulsifiable concentrate) is required and the dip should be stirred constantly. Bulbs should be treated before root development is visible and planted within a week of treatment. The use of thionazin as a drench treatment on boxed bulbs is not recommended.

Thionazin is a Part II substance under the provisions of the Agriculture (Poisonous Substances) Regulations and is used at a most poisonous concentration in bulb dips; consequently the precautions imposed by legislation and those given additionally by the manufacturers must be strictly observed. Rubber gloves should be worn when handling treated bulbs.

Where bulb scale mite has not been controlled before 'boxing' the bulbs, chemical treatments may prevent malformation and destruction of flowers. The best results are obtained by drenching the boxed bulbs with 0·1% endrin or 0·1% endosulfan a few days after bringing them into the heated glasshouse. Due to its greater persistence endrin appears to be slightly superior. These drenches, if watered on with a fine rose, at the rate of about 2 pints per 24 normal sized bulbs ($\frac{1}{2}$ gal/yd^2), will reduce the damage to flowers and foliage. A wetting agent, additional to that in the emulsifiable concentrates of these insecticides, is required; the amount necessary being that recommended by the manufacturer for H.V. sprays.

However, when applied in this manner, neither chemical will eradicate bulb scale mite so that infested bulbs should be destroyed or heat-treated before being planted out.

Endrin, endosulfan and thionazin are not available to private gardeners. Bulb scale mite damage can be reduced by exposing the bulbs to frost, after they have produced several inches of growth under glass. A worthwhile measure of control can also be obtained by spraying the foliage and flowers, at intervals of 7–10 days, with a petroleum emulsion diluted to 1 in 100. This treatment, which can also be given to amaryllis, detracts from the appearance of the plants by destroying the waxy 'bloom' on the foliage and flower stem.

Information is not available on the suitability of thionazin for the control of bulb scale mite on *Amaryllidaceae*. Bulbs of many species can, however, be hot-water-treated without damage. A period of $1\frac{1}{2}$ hr at 110 °F (43·3 °C) is sufficient for all but large bulbs which may require up to 2 hr. Gradual cooling after removal from the water bath is advocated, for if bulbs are cooled rapidly, retarded growth may result.

REFERENCES

FOX-WILSON, G. (1939). Contributions from the Wisley Laboratory. 87. *Amaryllis* Pests. *Jl R. hort. Soc.*, **64**, 318–26.

HODSON, W. E. H. (1934). The bionomics of the bulb scale mite *Tarsonemus approximatus* Banks var. *narcissi* Ewing. *Bull. ent. Res.*, **25**, 177–85.

WINFIELD, A. L. (1964). Chemical control of bulb scale mite on forced narcissus. *Expl. Hort.*, **11**, 69–77.

Hemitarsonemus latus (Banks) Ewing (the broad mite, Fig. 6.16) is becoming a more important pest of glasshouse ornamentals. While it has caused serious damage to begonia, *Cissus* sp. and *Fatshedera* sp. it has also been recorded on

Acalypha wilkesiana, *Browallia* sp., *Capsicum* sp., chrysanthemum, *Clethra* sp., cyclamen, dahlia, egg-plant, *Ficus pumpila*, fuchsia, gerbera, *Hedera helix*, *H. colchica*, *H. canariensis*, *Hibiscus rosa-senensis*, *Impatiens sultani*, *Matthiola incana*, and *Rhoicissus* sp. Recently Fjelddalen has found this mite on tomato.

During their average life of 10 days females produce about 5 eggs daily. The large eggs (115 $\mu \times 70$ μ), which are flattened and smooth on the ventral surface and have 5–6 rows of round, white tubercles on the dorsal surface, adhere firmly to the leaf surface. They hatch within 2 or 3 days and, in glass-house conditions, pass through the nymphal stages within a further 4 days. The species can be parthenogenetic, the unfertilized females producing only male offspring. In most populations females outnumber males by 4:1. The mites live gregariously under the leaves. On soft plants, including *Impatiens*, they swarm on buds and petioles but, on harder growth (e.g. *Clethra*), attack is confined to the buds and young leaves.

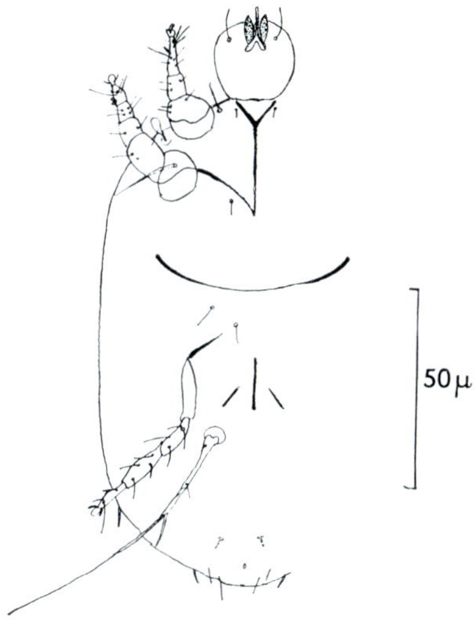

FIG. 6.16. *Hemitarsonemus latus* Ewing (female).

As in other Tarsonemid species, the active males carry the female 'pupae' before copulation and this habit accounts for the dispersal of the mite to suitable young growth; they may also be carried by insects. Infestations tend to develop quite suddenly in the rising temperatures of late spring and early summer and then slowly decline from late summer.

Most feeding takes place on young, tender and newly expanded foliage, chiefly on the undersides of leaves. Damage symptoms include bare new growth, puckered, down-curved and shiny, brittle leaves, generally stunted growth, abscission of the buds causing severe bud-drop—especially in *Impatiens*, and deformation of the blooms of gerbera and chrysanthemum.

REFERENCES

FJELDDALEN, J. (1953). Skuddtoppmiden. *Soertrykk av Gartneryket*, **30**, 527–30.
— (1960). Nye vertplenter for Skuddtoppmidd. *Soertrykk av Gartneryket*, **8**, 120–3.

FOX-WILSON, G. (1950). The Broad Mite. *Jl R. hort. Soc.*, **75**, 69–74.
GADD, C. H. (1946). Observations on the yellow Tea Mite *Hemitarsonemus latus* (Banks) Ewing. *Bull. ent. Res.*, **37**, 157–62.
KARL, E. (1965). Untersuchungen zur Morphologie und Ökologie von Tarsonemiden gärtnerischer Kulturpflanzen. II. *Biol. Zentbl.*, **84**, 331–57.

Hemitarsonemus tepidariorum Warburton (the fern mite, Fig. 6.17) has been known, for the past 60 years or so, to cause damage to the greenhouse fern *Asplenium bulbiferum* but its distribution is apparently restricted to England.

The female lays one or two eggs (113 μ × 76 μ) daily, depositing them singly, or in small groups, hidden within distorted pinnae, on the upper surface of

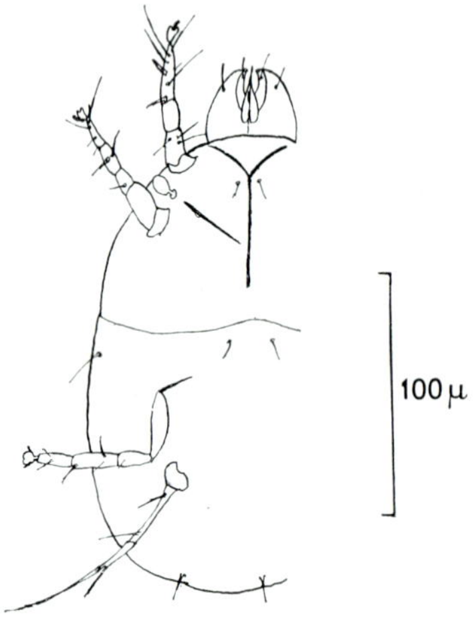

FIG. 6.17. *Hemitarsonemus tepidariorum* Warburton (female).

the leaf stalks especially near the tips of the fronds or between the fleshy scales at the apex of the rhizome. They hatch within 2–4 days into six-legged larvae which remain active for 6–14 days before entering a quiescent 'pupa' stage for a further 3–4 days. The adult females measure 229 μ × 149 μ and the males 155 μ × 95 μ.

Adults may be found throughout the year but during the winter the males are rather rare. From late spring to early summer both sexes become very numerous and it is at this time that most damage occurs.

H. tepidariorum sucks the sap from superficial cells of the youngest growth and so causes minute depressions which subsequently turn brown and give the leaf a speckled appearance. In severe infestations the leaf may turn completely brown and die. Undamaged areas around local feeding patches continue to grow normally causing the pinnae to be distorted, swollen or even gall-like. This check in assimilation causes a dwarfed and unhealthy plant.

The mites are readily spread from nursery to nursery on infested stock. On the greenhouse bench they disperse by walking considerable distances, especially on damp surfaces.

Control

As these mites readily fall from infested plants and are able to travel over the soil, or benches, to other plants new introductions should be carefully examined for damage before being taken into a glasshouse containing other potential hosts.

The insecticides effective against cyclamen mite (see p. 208) are equally effective against broad mites, as also is hot-water treatment. Two or three applications of dicofol (0·03%), at intervals of 2–3 weeks, will eliminate the pest. Control can also be obtained by spraying several times, at 5- to 7-day intervals, with parathion (0·01%).

DMC ('Dimite') and the proprietary product 'Aramite' are used in America to control broad mite, but dicofol is normally advised on account of its lower hazard to workers. Sprays must be thoroughly applied to wet the undersides of the leaves, additional wetter being used if necessary. The staging and glasshouse floor should also be thoroughly sprayed at 7-day intervals to kill fallen mites.

Repeated applications of diazinon aerosols (0·18 g/1000 ft^3) or parathion smokes, may be used where convenience is important and the host plants are tolerant.

The difficulty of spraying the undersides of leaves of certain plants, such as *Saintpaulia*, can be overcome by using soil drenches of the systemic schradan 0·05–0·1%. This has also given good control of fern mite on potted *Asplenium*, when applied in amounts sufficient to wet the contents of the pots. *Aspleniums* should not be propagated from bulbils taken from infested plants. If necessary, bulbils can be freed from mites by immersion for 20 min in water maintained at 110 °F (43 °C).

Amateurs may find that routine light dusting, at weekly intervals, with flowers of sulphur is the most convenient means of keeping broad mites in check.

REFERENCE

CAMERON, W. P. L. (1925). The Fern Mite (*Tarsonemus tepidariorum* Warburton). *Ann. appl. Biol.*, **12**, 93–112.

Identification of glasshouse Tarsonemidae

The texts of Schaarschmidt (1959) and Beer (1959), should be consulted for extensive keys and morphological details of all known species. However, the following outline key will serve, with the host plant damage, to differentiate the species discussed.

1.(2) Mouthparts in both sexes retracted into head (gnathostoma)...... *Steneotarsonemus laticeps* (Fig. 6.12)

Mouthparts in both sexes visible anteriorly.................. 2

2.(5) Claws on tarsus of third leg normal...........................

3.(4) Sternal apodeme I (central rod) broken. Epimeron II as (Fig. 6.10a) *Tarsonemus pallidus*

4.(3) Sternal apodeme I unbroken. Epimeron II (Fig. 6.10b)........ *T. confusus*

5.(2) Tarsus of third leg in both sexes with the claw hidden by the empodium 6

6.(7) Sternal apodeme I joined to epimeron I .... *Hemitarsonemus latus* (Fig. 6.16)

7.(6) Sternal apodeme I not joined to epimeron I .... *Hemitarsonemus tepidariorum* (Fig. 6.17)

REFERENCES

BEER, R. E. (1959). A revision of the *Tarsonemidae* of the Western Hemisphere (Order, Acarina). *Univ. Kansas Science Bull.*, **36**, Pt. II.

BEER, R. E. & NUCIFORA, A. (1965). Revisione dei generi della famigla Tarsonemidae. *Boll. zool. agric. Bachic.*, **2**, 19–44.

SCHAARSCHMIDT, L. (1959). Systematik und Ökologie der Tarsonemiden. In H. J. Stammer. *Beitrage zur Systematik und Ökologie Mitteleuropaischer Acarina* Band **1**, Teil 2. 713–823. Leipzig.

Family PHYTOPTIPALPIDAE

Brevipalpus spp. (the false spider mites) are less important than the Tetranychidae but recently several species have damaged cultivated plants.

Pritchard & Baker (1951) were unable to confirm the identity of many of the species described in the last century, including *B. obovatus* Donnadien, to which name most English occurrences have been credited.

Brevipalpus spp. are minute, red or yellowish, mites (Fig. 6.18) with the body only 250 μ long (about 303 μ if the rostrum and genitalia are included). The palpi are 4-segmented, the body flat and egg-shaped, and the cuticle has a reticulate pattern. The legs are usually wrinkled.

The mites crowd under the leaves producing a brown, scurfy discolouration extending along either side of the midrib to the base of the leaf. The damage often induces premature senescence. On *Hedera* the leaves tend to be cupped while the tips of the lobes become necrotic. Following heavy attack, affected plants are completely defoliated and buds may fail to develop or, at most, produce only stunted, yellowish growth. Fine, spotted-bronze or rusty-brown areas may occur along the veins or over entire leaves on other hosts.

After a preoviposition period of 8 days at 68 °F (20 °C) each female lays about 60 eggs. Reproduction is largely parthenogenetic, males occurring only in small numbers. The bright-red, elliptical eggs (96 $\mu \times 67 \mu$) are laid with the long axis perpendicular to the leaf, usually in a crevice or along the groove by the midrib, often in dense clusters of several hundred. At 65–70 °F (18–21 °C) the eggs hatch in about 18 days and the nymphal stages are completed in 26 days. Unlike most true spider mites, *Brevipalpus* does not produce a silken web.

Brevipalpus spp. have been recorded as pests of *Ageratum* spp., azalea, fuchsia, campanula, *Cissus* sp., coleus, *Elettaria cardomonum*, gardenia, ornamental *Hedera*, orchids, and *Peperomia* sp. One species *B. russulus* Boisduval causes the cacti *Echinocactus* sp. and *Mammillaria* spp. to become uniformly reddish-grey. Other species are commonly found on *Phalaenopsis* orchids where pits are formed on the upper leaf surfaces.

Control

Few tests have been made to assess the relative merits of recently introduced acaricides such as dicofol, tetradifon, tetrasol and quinomethionate. It is evident, however, that false spider mites are, in general, resistant to some of the organophosphorus compounds which are effective against many other mites. Diazinon appears exceptional, as excellent control of *Brevipalpus inornatus* and *B. australis* is obtained with H.V. sprays at 0·05%, and this compound may also be effective against other species. Some host plants are however susceptible to injury by diazinon.

The specific acaricides chlorbenzilate (0·04%), dicofol (0·05%) and DMC (0·04%) have also proved effective against these and other species occurring on orchids in the U.S.A. Empson (1961) obtained good control of *B. obovatus* on ornamental ivy under glass by spraying with chlorbenzilate at 0·02%. There is some evidence that control can be obtained by giving two or three applications of 0·01% dicofol at intervals of 7–10 days, although a single application of three times this concentration is insufficient.

Repeated applications of petroleum sprays (1 in 100 dilution of a glasshouse grade petroleum emulsion) will also control false spider mites but may be injurious to some plant species liable to infestation (see p. 382).

REFERENCES

BAKER, E. W. (1949). The genus *Brevipalpus* (*Acarina: Pseudoleptidae*). *Am. Midl. Nat.*, **42**, 350–402.

DOSSE, G. (1955). Beitrage zur biologie, verbreitung und Bekämpfung der Milbe, *Brevipalpus obovatus* Donn. *Z. angew. Zool.*, **37**, 437–46.

— (1957). Die ersten Feinde von *Brevipalpus inornatus* (Banks) (*Acarina: Phytoptipalpidae*) in europaischen Gewachshausen. *Pflanzenschutzberichte*, **18**, 13–17.

EMPSON, D. W. (1961). *Brevipalpus obovatus* Donn. in Staffordshire. *Pl. Path.*, **10**, 126.

MCGREGOR, E. A. (1916). The privet mite in the South. *J. econ. Ent.*, **9**, 556–60.

MANGLITZ, G. R. & CORY, E. N. (1953). Biology and control of *Brevipalpus australis*. *J. econ. Ent.*, **46**, 116–19.

MORISHITA, F. S. (1954). Biology and control of *Brevipalpus inornatus* (Banks). *J. econ. Ent.*, **47**, 449–56.

PRITCHARD, A. E. & BAKER, E. W. (1951). The False Spider Mites of California (Acarina: Phytoptipalpidae) *Univ. Calif. Publs Ent.*, **9**, 1–93.

Family ERIOPHYIDAE

This group of minute, almost wormlike, mites (Fig. 6.19) are either free-living on plants, or gall-makers.

The long body is annulate and bears, in all developmental stages, only two pairs of anteriorly placed legs. Eriophyids are so small that there is no tracheal system, air merely diffusing internally through the integument. Eyes, heart and circulatory system are also lacking, but a few sensory setae are present. The legs lack tarsal claws but do have characteristic rayed or 'featherclaws'. The genitalia are situated just behind the second pair of legs. The minute mouthparts are short and stylet-like for piercing.

Because there has been no comprehensive study on this group the nomenclature of many records is probably confused. Since the mites are so small

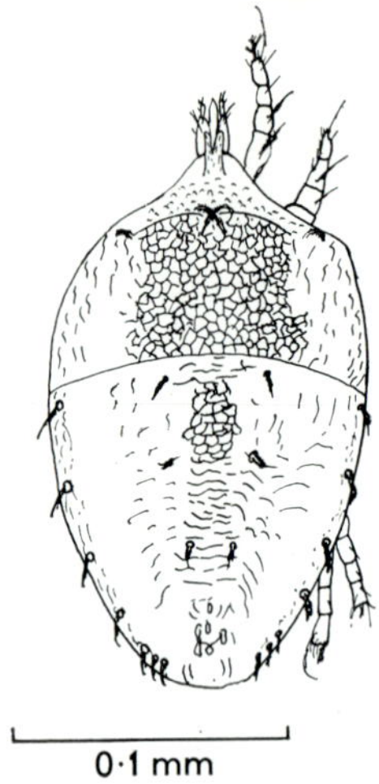

FIG. 6.18. *Brevipalpus* sp. (female).

(200 μ long) close observation will undoubtedly extend the list of potential glasshouse pests within this group.

The Eriophyids exhibit considerable host specificity. If uncontrolled they may blast buds or deform, or russet, both leaves and fruit. The chrysanthemum is affected by two species *Epitrimerus alinae* Liro, and a species of *Paraphytoptes* (near *chrysanthemi* Kieffer). *E. alinae* was discovered attacking chrysanthemum cultivars Fina, Golden Seal and Cordillion in Cheshire in 1954, but it is believed to be generally distributed in Britain. The mites occur in large numbers amongst the hairs on the flower stems, the highest populations being found 3–6 in. below the florets. The feeding causes russetting of the stem, particularly at the bases of the upper petioles, followed by drooping and premature fall of the leaves.

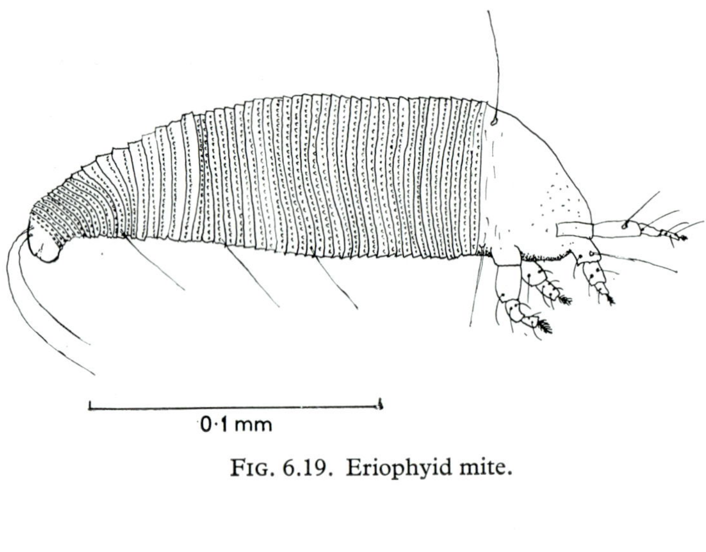

FIG. 6.19. Eriophyid mite.

The *Paraphytoptes* species, which has been referred to as the 'chrysanthemum semi-bud mite', was first recognized in Scotland in 1950. The minute yellow mites live freely within the buds and on the leaves. Infested cuttings produce stunted plants, often with pale and backwardly curling, apical leaves which tend to become brittle and readily broken off. Buds of infested plants are often undersized (Fig. 6.20).

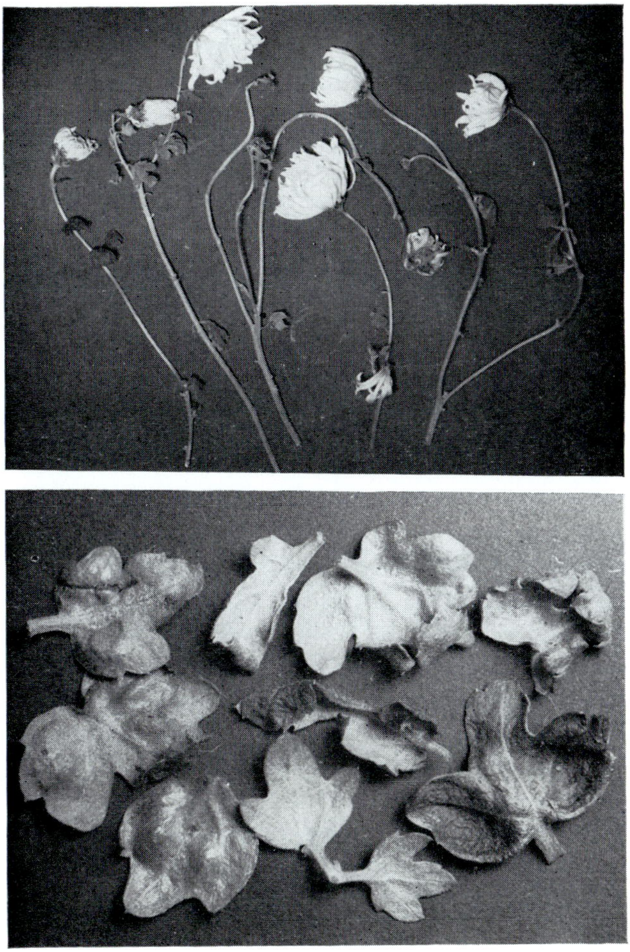

FIG. 6.20. Eriophyid damage to chrysanthemums. (Reproduced from *Entomologist's Monthly Magazine*, **100** (1964) P. W. Miles, NAAS. Trawscoed.)

In many parts of the world the tomato is affected by two species of Eriophyid mites *Aceria lycopersici* and *Aculus lycopersici*. One of these has been recorded outdoors in the Channel Islands, and brief details of both species are included here.

Aceria lycopersici (Wolffenstein), known as the tomato erineum mite, causes white and densely hairy patches on the stems and petioles; *Aculus* (*Vasates*) *lycopersici* (Massee), the tomato russett mite, causes marked russeting of the stems and bronzes the leaves. They may be distinguished morphologically by the abdominal rings which, in *Aceria*, are differentiated into broad, smooth half-rings dorsally and narrow tuberculate bands ventrally; in *Aculus* there is no such dorsoventral differentiation of the microtuberculate rings. Another species *Aceria paradianthi* Keifer, the carnation bud mite, has been recorded causing a 'grassiness' of carnation similar to that induced by boron deficiency.

Free-living eriophyids are favoured by low relative humidity and high temperatures. They lay minute eggs (45–60 μ) near leaf hairs, in surface striations, or alongside leaf veins. At 81 °F (27 °C) and a relative humidity of 30% each female produces 40–50 eggs at a rate of 2–3 per day which hatch in 3 days. Individual mites may live as long as 80 days. Under glass, reproduction is continuous, each generation being completed within 12–15 days.

Control. See p. 224.

REFERENCES

HALL, C. C. (1967). A look at eriophyid life cycles (Acarina: *Eriophyidae*). *Ann. ent. Soc. Amer.*, **60**, 91–4.
KEIFER, H. H. (1959). Eriophyid Studies 27. *Occ. Pap. Bur. Ent. Calif.*, **1**, 18 pp.
LAMB, K. P. (1953). A revision of the gall-mites (*Eriophyidae*) occurring on Tomato (*Lycopersicum esculentum* Mill), with a key to the *Eriophyidae* recorded from solanaceous plants. *Bull. ent. Res.*, **44**, 343–50.
MILES, P. M. (1964). Chrysanthemum semi-bud mite in Wales. *Ent. mon. Mag.*, **100**, 97.
MORRISON, G. D. (1950). A mite (*Phytoptus* ?) on chrysanthemum plants. *Trans. High. Agric. Soc. Scot.*, **42**, 15–19.
RICE, R. E., & STRONG, F. E. (1962). Bionomics of the Tomato Russett Mite, *Vasates lycopersicae* (Massee). *Ann. ent. Soc. Amer.*, **55**, 431–5.
VERNON, J. D. R. (1957). An eriophyid mite causing damage to chrysanthemums. *Pl. Path.*, **6**, 111.

Family PYEMOTIDAE

Siteroptes graminum Reuter (Fig. 6.21). The *Pyemotidae* are a family of minute mites usually only about 75 μ in length. They are, for the most part, fungal feeders and it is by feeding on the fungus *Sporotrichum poae* that *S.*

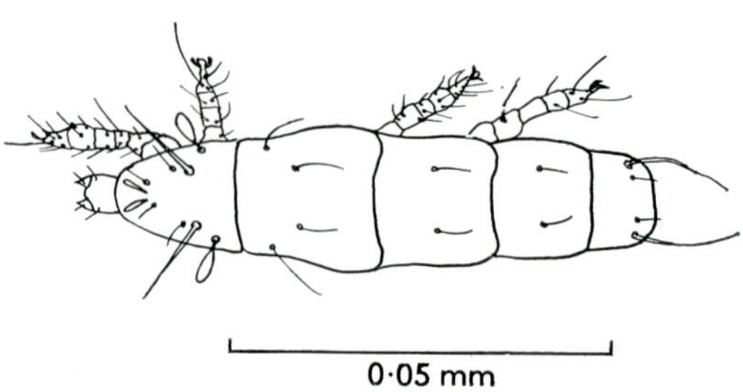

FIG. 6.21. *Siteroptes graminum* (female).

graminum becomes a vector of carnation bud rot. The spores of *Sporotrichum* are spread by the adult mites to new buds. Buds of any size may be invaded by the complex and, despite the rapid growth of the internal rot, the growth appears quite normal for at least 16 days following infection. By the time the calyx appears abnormal, all the petals and sepals have rotted.

The mite has a remarkable life cycle. After attachment to suitable food, the body (hysterosoma) swells with fluid and becomes a sac up to 500 times the original body volume. Within this sac eggs develop and mature to adult mites. Copulation takes place within the mother's body, and the males, usually

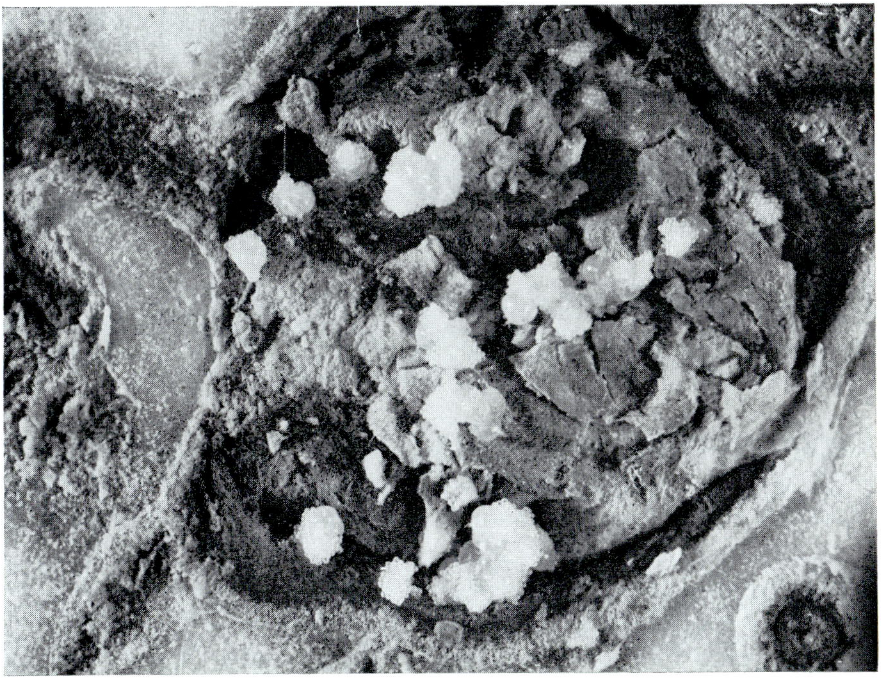

FIG. 6.22. Eggs of *Pediculopsis* mite on gladiolus corms (×10). (Reproduced from *Plant Pathology* **1** (1952) by permission of HMSO.)

constituting about 5% of the normal brood, often leave the swollen hysterosoma before the mass birth of their sisters. The female's swollen body then collapses and she usually dies. Virgin females are facultatively parthenogenetic.

Another closely related species of *Pediculopsis* associated with *Botrytis* rot, (*Botrytis gladiolorum* Timmermens), has been found on *Acidanthera* and *Gladiolus* corms (Fig. 6.22). In this case, however, large numbers of eggs are produced which hatch, within 2 days, into minute, active larvae. This Pyemotid is unable to live on healthy bulbs and in fact feeds entirely on the hyphae of *Botrytis*. While they undoubtedly carry *Botrytis* spores their precise role as vectors of the disease is unknown.

Control of Eriophyidae and Pyemotidae

These mites so rarely cause economic damage to glasshouse plants that little reliable information on their control exists. Vernon found that *Epitrimerus alinae* on chrysanthemums was controlled by a single spray of demeton-methyl at 0·04%. Attacks of an Eriophyid mite causing bronzing and stunting of tomato plants in Guernsey was successfully countered by two applications of 0·01% parathion at an interval of 10 days. On some hosts, when temporary unsightly deposits can be tolerated, wettable sulphur at 0·4% may be satisfactory. *Aculus lycopersici* can be controlled by repeated applications of sulphur dust. Dicofol (0·02%) should be tried where the plants are intolerant of both organophosphorus compounds and sulphur. Malathion, parathion, dicofol, binapacryl, quinomethionate, wettable sulphur and lime sulphur are used to control related species occurring on outdoor hosts. On the other hand, several specific acaricides such as chlorfenson, chlorbenzilate, and 'Aramite' appear to be ineffective against these mites. Organophosphorus compounds are, therefore, likely to prove most satisfactory for the control of Eriophyid mites on glasshouse plants. Where considerations of human safety and plant tolerance permit, sprays of systemic compounds such as demeton-S-methyl (0·02%), dimethoate (0·03%), formothion (0·05%) or oxydemeton-methyl at (0·02%) or, alternatively, soil drenches of these compounds at two-thirds of these concentrations, are likely to prove effective. Sprays of non-systemic compounds, such as diazinon (0·02%), malathion (0·1%) or parathion (0·01%) should be satisfactory if two or three applications are made at 10- to 14-day intervals.

Siteroptes graminum is unlikely to tolerate the relatively low humidity which should be maintained in carnation houses where, of course, sources of *Fusarium poae* should be absent. Chadwick & Robinson controlled this mite, which was associated with the fungus on cereals grown experimentally in glasshouses, by spraying with 0·05% endrin at weekly intervals. The systemic compound demeton had little effect. Mites within buds, or otherwise protected, were not killed by several sprayings at 10-day intervals.

As alternatives to endrin, dicofol and quinomethionate are suggested. If bud rot occurs on carnations, diseased flowers should be removed as soon as they are detected and weekly applications of these compounds made until diseased flowers no longer occur. The humidity in the glasshouse should be kept as low as possible.

The investigations by Harrison indicate that the *Pediculopsis* sp. occurring on gladiolus and *Acidanthera* corms was able to establish only on fungus-infested corms and, in consequence, their presence suggests that the corms are of poor quality and unsuitable for growing in glasshouses.

REFERENCES

CHADWICK, W. J. & ROBINSON, A. C. (1958). A rot of smutted inflorescences of cereals by *Fusarium poae* in association with the mite *Siteroptes graminum*. *Phytopathology*, **48**, 232–4.

COOPER, K. W. (1937). Reproductive behaviour and haploid parthenogenesis in the grass mite, *Pediculopsis graminum* (Reut.) (*Acarina, Tarsonemidae*). *Proc. natn. Acad. Sci.*, **23**, 41–44.

COOPER, K. W. (1940). Relations of *Pediculopsis graminum* & *Fusarium poae*. A central bud rot of carnation. *Phytopathology*, **30**, 853–9.
HARRISON, I. R. (1952). *Pediculopsis* sp.—A mite found on *Acidanthera* and Gladiolus corms. *Pl. Path.*, **1**, 119–20.

Family TYROGLYPHIDAE

Rhizoglyphus echinopus Fumouze and Robin (the bulb mite) is a smooth, colourless and glistening mite about 0·5–0·7 mm in length with reddish-brown appendages. *R. echinopus* is found on hyacinth and narcissus and all kinds of

FIG. 6.23. *Rhizoglyphus echinopus* on narcissus bulb (×20). (J. Hesling.)

vegetables and ornamental bulbs (Fig. 6.23). The mites are rarely found as isolated individuals, large populations develop whenever conditions are suitable. Eggs may be found at all seasons of the year, each female laying at least 50. The period of development is greatly influenced both by the available food and temperature; at about 68 °F (20 °C) the life cycle is completed in 17–27 days though, at 77 °F (25 °C), it may be as short as 9 days.

The six-legged larvae possess a pair of coxal rods, or 'brustiele', between the first and second pairs of legs. These rods are 3·5 times longer than their greatest diameter and have a small, spherical, apical dome. After a period of activity the larva becomes inert and, from this resting form, emerges the *protonymph* which has four pairs of legs but no coxal rods. From the resting protonymph emerges the *tritonymph*, a form closely resembling the adult but with only a vestigial genital opening. Between the proto- and trito-nymphal stages a hypopal, or deutonymphal, form may occur which serves both to distribute the species and to assist survival in adverse conditions. In severely rotted bulbs many thousands of hypopae may be present. The hypopus is brownish, oval in shape and dorso-ventrally flattened with the lower surface concave. The mouth is lacking but, on the posterior portion of the ventral surface, is a plate from which suckers are projected to attach the organism to vector insects such as narcissus flies, sciarids, and scatopsids (all of which occur frequently in decaying bulbs).

While *R. echinopus* is not a primary pest of bulbs it is often responsible for serious losses, as the slightest injury or bruising of the bulb will enable mites to breed. Once the mites are established, they rapidly extend the damage and reduce the whole bulb to a rotten pulp.

During storage the mites readily wander from bulb to bulb and may act as vectors of various pathogenic fungi.

While normally associated with narcissus bulbs this mite has been found on *Eucharis* sp., lilies, orchids, gladioli, hyacinth and tulip bulbs, as well as dahlia tubers and freesia corms. In America the closely related *R. hyacinthi* has been found on iris bulbs.

Control

Infestations of bulb mite are almost invariably an indication that the bulbs have either been invaded by other pests (e.g. bulb flies, bulb scale mite or eelworms), have suffered mechanical damage, or have been stored under unsatisfactory conditions. This mite is frequently associated with *Fusarium* spp. in gladioli and freesia corms, and with fungal or bacterial rots of dahlia tubers. When narcissus, hyacinth, lily or tulip bulbs are intended for forcing, a representative sample should be examined and, if infestation is general, the bulbs should be rejected. If the infestation is restricted to only a small proportion of the bulbs any which are soft should be rejected. Infested freesia and gladioli corms are usually soft or even mushy and obviously unsuitable for planting.

Bulb mites may be controlled by hot-water treatment; immersion for 1 hr in water maintained at 110 °F is sufficient for all bulbs except, possibly, large lily bulbs. This treatment is unsuitable for tulips required for forcing, and should only be given to narcissi at the correct developmental stage. (See stem and bulb eelworm, p. 293.)

It is probable that 0·23% thionazin, when used as a pre-planting dip to control eelworm or bulb scale mite, will control bulb mite, but definite information is lacking.

REFERENCES

EYNDHOVEN VAN, G. L. (1963). The *Rhizoglyphus echinopus* of Fumouze and Robin. *Mitt. schweiz. ent. Ges.*, **36**, 48–9.

HODSON, W. E. H. (1928). The Bionomics of the Bulb Mite, *Rhizoglyphus echinopus* Fumouze and Robin. *Bull. ent. Res.*, **19**, 187–200.

HUGHES, A. M. (1961). The Mites of Stored Food. *Minist. Ag. Fish. Food. Tech. Bull.*, **9**, 287 pp. H.M.S.O.
OBREEN, H. C. (1967). Contribution to the biology and taxonomy of *Rhizoglyphus echinopus. Meded. Rijksfac. Landbouwet. Gent.*, **32**, 602–6.

Tyrophagus similis Volgin (***Tyroglyphus dimidiatus*** (Hermann)) (the 'French fly'). For many years growers have known as 'French fly' the mites causing characteristic damage to young cucumber plants. In actual fact they are Tyroglyphid mites, and are introduced into the houses in straw or manure. In the warm environment the mites become active and wander upon the newly set-out plants. They usually feed within the growing point where they eat minute holes (up to 4 mm in diameter) in the youngest leaves. As the latter expand the damage enlarges and quite large holes may develop. In severe infestations the buds may be so severely damaged as to cause 'blindness'. Infestations usually terminate naturally within a few weeks (normally before the end of March, after one or two generations have been completed on the cucumbers) as if the mites no longer find plant tissues palatable. These mites occasionally infest French beans, causing similar damage.

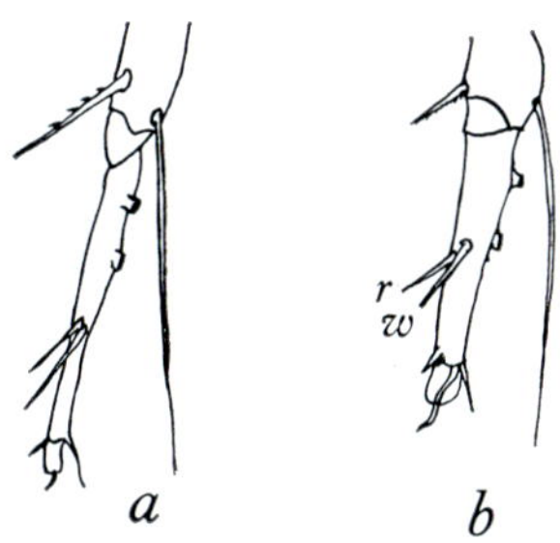

FIG. 6.24. Separation of *Tyrophagus* sp. by tarsus IV of male: (*a*) *T. longior*; (*b*) *T. similis*.

The scientific name of the species involved has undergone many changes in recent years so that early records are uncertain. Recent cases have been identified as *T. similis* although there are many early records of *T. longior* (Gerv.) as the species involved. In America another species *T. neiswanderi* has caused similar damage. This is particularly confusing because, in a recent case, where damage was caused to the young leaves of a mature cucumber crop in mid-summer, the mite was identified as *T. longior* a species more commonly associated with stored grain and straw, though there are records of damage to betroot, cyclamen and tomato seed. In the outbreak on mature cucumbers the mites did not actually eat holes through the leaves but skeletonized the dorsal surface. It is possible, therefore, that there is a specific difference in the feeding behaviour of these mites on cucumber. The most convenient character to differentiate *longior* and *similis* concerns tarsus IV of the male. In the latter species, the distal sucker is at the same level as setae *r* and *w* but in *longior* is nearer the base of the segment (Fig. 6.24). *T. similis* is usually associated with plants but it has not been confirmed as a truly phytophagous species. At 72 °F (22 °C) the entire life cycle is completed within 24 days though some individuals may mature in as little as 17 days. The adult females, which live for as long as 3 weeks, produce from 40–60 eggs which are covered with a distinct sculpturing. These eggs hatch within 6 days and the three active larval stages take 3, 4 and 5 days respectively.

Control

Although mild infestations of French fly have little effect on the growth of either cucumbers or French beans, growers are advised to adopt control measures at the first sign of damage, particularly when cucumbers are growing in bales of straw. The most convenient method of control is by 'smoking' with parathion, one or two applications usually sufficing on crops of French beans, and cucumbers when grown in beds prepared from loam and dung.

Mites may continue to emerge from straw bales, or beds containing a high proportion of unrotted straw in the dung, for several weeks. Repeated smokings may therefore be needed to prevent serious damage to cucumbers. Under these circumstances, better control is obtained by spraying both the plants, and the surfaces of the straw bales, with 0·01% parathion, bearing in mind that at least 4 weeks must elapse between the application of parathion sprays and cutting fruits for market. The heads of the plants should be thoroughly drenched and the surfaces of the bales well wetted. To avoid scorching tender new growth, spraying should not be done in bright sunlight. When using parathion the safety precautions required by regulations must be observed.

Persistent damage to the young leaves of established cucumbers by *Tyrophagus* sp. is rare. Where damage is sufficiently severe, spraying with dicofol at 0·01% will probably provide the most suitable control. The regular use of diazinon, dicofol, tetradifon or petroleum emulsions for the control of red spider should prevent attacks by French fly.

REFERENCES

JARY, S. G. & STAPLEY, J. H. (1937). Investigations on the insect and allied pests of cultivated mushrooms IX. *Tyroglyphus dimidiatus* (Hern.) (*longior* Gerv.). *Jl. S-east agric. Coll. Wye*, **40**, 119–29.

JOHNSTON, D. E. & BRUCE, W. A. (1965). *Tyrophagus neiswanderi*, a new acarid mite of agricultural importance. *Ohio Ag. Res. Center Res. Bull.*, **977**, 17 pp.

ROBERTSON, P. L. (1959). A revision of the genus *Tyrophagus*, with a discussion on its taxonomic position in the *Acarina*. *Aust. J. Zool.*, **7**, 146–181.

Chapter 7

Slugs and Snails

PHYLUM MOLLUSCA

Class GASTEROPODA

Family LIMACIDAE: Slugs

Slugs (Fig. 7.1) have soft, unsegmented bodies, and exude a slimy mucus from epidermal glands as well as from a large slime gland opening below the mouth. They have a well-developed head bearing two pairs of elongate, hollow, retractile tentacles. The eyes are carried at the apex of the upper,

FIG. 7.1. Keeled slug (×2). (Crown Copyright by permission of Controller of HMSO.)

larger pair. The mouth, partially concealed on the ventral surface of the head, has a sharp mandible and a tongue, or *radula*, armed with rows of teeth which, by sawing together, bite food with a clipping action. Slugs can contract and elongate their bodies to a remarkable degree. When fully extended the head protrudes far beyond the mantle (an internal shield corresponding to the shell in snails) but, when contracted to almost a quarter of its extended length, the head and tentacles are concealed below the mantle.

When temperatures and, above all, humidity, are suitable, slugs are active throughout the year, but both dry and cold weather restrict their nocturnal activity. During inclement weather they migrate below plant debris or deep into the soil. Normal activity on the soil surface occurs on still, humid nights, but during the day they shelter in moist, dark places below stones and boxes or amongst rank vegetation, especially along the north side of buildings.

FIG. 7.2. Slug damage to leaves and bulb of tulip. (Crown Copyright by permission of Controller of HMSO.)

Slugs are usually introduced into glasshouses in soil, compost or on the bottoms of boxes or pots which have been in contact with the ground for some time. When the immediate surroundings of the house are covered with rough vegetation they may also crawl through bottom ventilators. They are omnivorous feeders preferring plant food but occasionally accepting insects, toadstools, manure or decaying vegetation. Only surface tissues of plants are rasped by young slugs, whereas older individuals eat irregular holes. Seedlings may be completely eaten and damage to blooms of flowering plants can be severe. Karlin & Naegele (1960) found that, while some species tend to be climbers, attacking the flowers and leaves of tall plants, others browse surface vegetation, and a few feed almost entirely on roots. Severe damage is often caused to the bulbs and roots of tulip, iris and lily. Where bulbs are covered with straw prior to forcing, new foliage may be holed or even grazed off (Fig. 7.2).

Three common outdoor species and two alien species, *Limax nyctelius* Bourguignot and *Lehmannia poirieri* (Mabille), have been recorded from glasshouses in Britain (Quick, 1960). Eleven species have been found in glasshouses in New York State (Karlin & Naegele, 1960) while eight species were recorded by Plate & Fromming (1953) in Germany.

Agriolimax reticulatus (Müller) (the grey field slug) is harmful to cereals and clovers and is common throughout Britain. It is $1\frac{1}{2}$ in. long, varying from pale cream to almost black with the longitudinal grooves darker than the overall mottled colouration. Individuals tend to become darker in winter. When disturbed, it produces copious quantities of white slime.

As in other slugs individuals are hermaphroditic and normally each partner is fertilized at a single copulation. Mating occurs throughout the year. Ten to thirty days after mating the spherical eggs, each surrounded by a transparent, gelatinous envelope, are laid in batches of 10–30 in soil crevices 1 in. below the surface. Eggs may also be found beneath stones or boxes laying on damp ground. In glasshouses hatching takes place within 2 weeks. The young slugs are similar to the parents and at first tend to be colonial, feeding in groups. Growth is slow and they do not reach full size for 6 months, but in the high temperature within greenhouses slugs may lay fertile eggs when only 6 weeks old. Adults live more than a year. *A. reticulatus* is predominantly a surface feeder attacking germinating seedlings and occasionally orchids, cacti and even flowers of aster and dahlia.

Arion hortensis Fer. (the garden slug), when full grown, is slightly over 1 in. long. It is dark brown to black with a sprinkling of lighter yellowish-brown dots. The sole is paler yellow and it produces yellowish mucus as it glides along. Hatching from the yellow, sticky eggs occurs, outdoors, in early spring. The young mature during September and eggs are laid during the autumn and winter. Individuals probably live no longer than a year. *A. hortensis* is a largely subterranean species attacking roots and bulbs.

Milax budapestensis (Hazay) (the keeled slug) is 2 in. long when fully extended. The back is nearly black with an orange keel running from the mantle to the end of the body. This predominantly burrowing species mates beneath stones and logs in the autumn and winter. The yellowish leathery eggs develop to maturity by the following autumn. *M. budapestensis* is an important pest of bulbs.

Family HELICIDAE: Snails

Snails are closely related to slugs but they differ in possessing a conspicuous shell within which the body can be retracted (Fig. 7.3). While no data is available for Britain, 14 species are known to occur in American glasshouses (Karlin & Naegele, 1960) and 23 in houses in Germany (Plate & Fromming,

1953). However, common outdoor species undoubtedly invade our greenhouses from time to time, especially where these contain perennial pot-plants which are cleaned only occasionally.

Snails cause similar damage to slugs. As with the common garden snail, *Helix aspersa* Mull., most live as adults for several years. The latter species, which is often found in small garden greenhouses, is up to 2 in. long with a grey-brown shell consisting of five distinct whorls. Eggs are laid on the soil surface and then covered with particles.

FIG. 7.3. The garden snail (× 1½). (By kind permission of the Plant Pathology Laboratory, Ministry of Agriculture, Fisheries and Food.)

Control of slugs and snails

Most infestations in glasshouses and frames result from slugs entering from outside, either by crawling through cracks in the structure or being brought in with pots and boxes which have been standing outside. A border free from grass and weeds should therefore surround all glasshouses, frames or mushroom houses, and standing-out grounds should be grass- and weed-free so as not to harbour slugs or snails. Pots and boxes should be stored away from weeds and debris.

Where slugs or snails are present in vacant glasshouses, or are causing damage they can be controlled by poisonous baits or by spraying the soil surface and plants with a suspension of metaldehyde.

Baits can be prepared by mixing ½–1-lb metaldehyde powder or 1-lb Paris green (copper aceto-arsenite) and ½-lb dried blood with 28 lb of bran, which should be sufficiently moist to make the chemical adhere to the bran but not wet enough to make the bran flakes adhere to one another. Addition of 1 part 30% carbaryl to 40 parts of bait is said to improve the catch. Proprietary pelletized baits containing metaldehyde are easier to apply and equally effective, although more expensive. These baits usually contain fungicidal substances to increase their persistence by preventing decay. For amateurs, baits consisting of 1-oz metaldehyde powder mixed with 3 lb of bran or other feeding stuff such as damp, spent tea-leaves can be used. These baits should be evenly broadcast over the soil surface, at about 56 lb/acre (2 oz/100 ft^2), when the surface soil is warm and damp so that no watering will be required for several days. For best results, the atmospheric humidity should be low after the application of a metaldehyde bait, as the slugs are destroyed by desiccation. Death follows very copious slime production as a consequence of irritation by metaldehyde. When the temperature is low and atmospheric humidity high, better results are obtained with Paris green bait than metaldehyde. Pellets of a new carbamate, methiocarb, may also be distributed at 1 oz per 15 yd^2.

Where slugs are attacking lettuces or other plants giving cover during the daytime, metaldehyde suspensions are often more effective than baits. They should be applied at the equivalent of 2-oz metaldehyde (5 fl. oz 20% suspension) per 100 yd^2 in sufficient water to thoroughly wet the soil and plants. With seedlings and tender young plants application should be restricted as much as possible to the soil. Amateurs may trap slugs and snails beneath cut potatoes, rhubarb or cabbage leaves, boards or tiles, placed on the ground.

REFERENCES

FROMMING, E. (1954). *Biologie der mitteleuropaischen Landgastropoden.* Duncker & Humblot, Berlin. 404 pp.

HYMAN, L. H. (1967). *The Invertebrates.* Vol. 6. *Mollusca.* McGraw-Hill, New York. 792 pp.

KARLIN, E. J. & NAEGELE, J. P. (1960). Biology of the Mollusca of Greenhouses in New York State. *Cornell Univ. ag. exp. Stat. Memoir,* **372**, 35 pp.

PLATE, H. P. & FROMMING, E. (1953). *Die tierischen Schadlinge unserer Gewachshauspflanzen.* Duncker & Humblot, Berlin. 298 pp.

QUICK, H. E. (1960). British Slugs. *Bull. Br. Mus. nat. Hist.,* **6**, 103–226.

WEBLEY, D. (1965). Aspects of trapping slugs with metaldehyde & bran. *Ann. appl. Biol.,* **56**, 37–45.

Chapter 8
Nematodes or Eelworms

Nematodes are worm-like but differ in structure, organization and behaviour from other worms so that they are placed in a distinct group of the Animal Kingdom, and, collectively, they have the same 'rank'—Class*—as such well known groups as the Insects and Mammals. Many nematodes are animal parasites, in which role they are best known, e.g. roundworms, threadworms and whipworms are nematodes parasitic in man and domestic animals. The name *eelworm* is used, loosely, for small nematodes which live in habitats ranging from the bodies of insects to the sea-bed. Such habitats have one essential factor in common—moisture, with biological activity. Wherever there is such moisture, eelworms may be expected; they occur in trees, house gutters, vinegar, beer mats—in fact they are almost ubiquitous.

Eelworms have very diverse feeding habits. Those that feed within or upon plants are pests in horticulture, and are usually found either within plants or in the soil around roots. If the word 'parasite' is used in a loose sense, this group may conveniently be called 'plant-parasitic eelworms'.

Eelworm structure and biology (Figs. 8.1 and 11.2)

Eelworms are minute (commonly less than 1 mm long) and can only be seen clearly through a microscope. A few soil-inhabiting forms reach a length of 8 mm but even these are virtually invisible in soil because they are translucent and thin. Typically, eelworms are worm-like or eel-like with elongate bodies.

Eelworm cuticle is tough and impermeable; it surrounds the nematode and lines the lumen of the mouth, oesophagus, excretory pore and rectum. The cuticle is transversely annulated and surrounds a hypodermis and bands of muscle cells. The latter form several elongate sheets, one cell thick, along the eelworm's length. The muscle bands are dorso-ventral and dorso-lateral in position; between them the hypodermis is deeper and so forms dorsal, ventral and two lateral longitudinal bands known as the dorsal, ventral and lateral 'chords'. These contain nerves and, in the lateral chords only, an excretory canal. The cuticle immediately above the lateral chord is often not clearly annulated; instead, it is frequently infolded along its length to produce several longitudinal lines or 'incisures', whose number and spacing are

* Inevitably, not all taxonomists agree with this. However, all must agree that a high-ranking taxon is deserved and 'class' seems a very convenient choice.

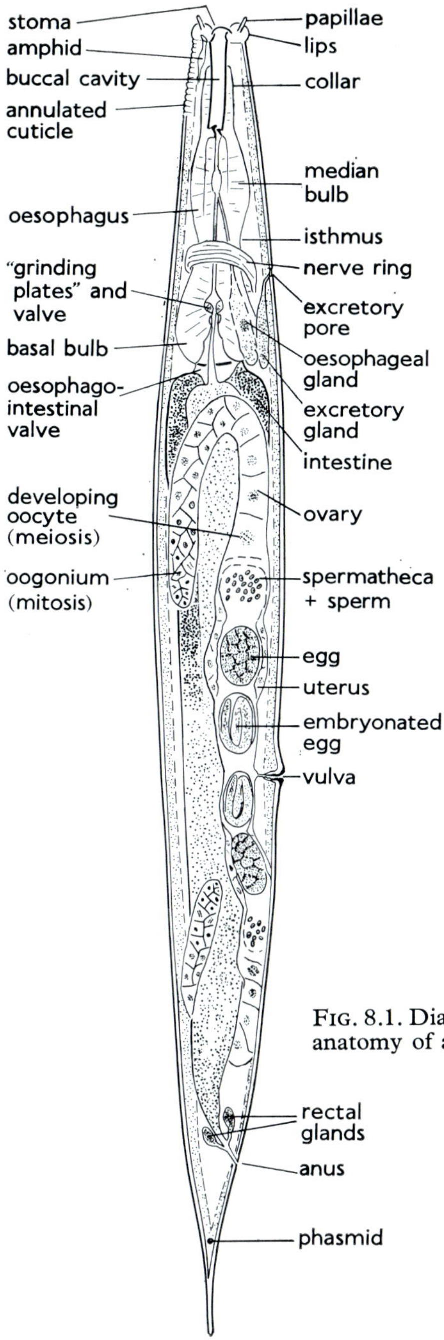

FIG. 8.1. Diagram showing the anatomy of a female eelworm.

important in eelworm identification. The cuticle has few major openings: these are the mouth, anus, vulva and excretory pore.

The anterior end of an eelworm, or 'head' (there is no true 'head' in nematodes) bears a terminal mouth or stoma leading into a buccal cavity which is modified according to feeding habits. The buccal cavity leads into the lumen of the 'oesophagus' (properly, this is a pharynx) which is a muscular tube with valves and, sometimes, 'grinding plates'. Muscles and valves act together to suck in the food and force it into the intestine; if present, grinding plates break up the food particles before their digestion in the intestine. Ducts from glands (usually three) lying within, or very close to, the oesophagus empty into its lumen and discharge substances which assist feeding. The eelworm intestine is a simple tube which leads from the oesophagus, via a rectum, to the anus. Because the muscles of the oesophagus force food into the intestine, its contents are under some pressure and their uncontrolled escape is prevented by the valves and muscles of the oesophagus and by the muscles associated with the anus.

Eelworm reproductive organs, or gonads, lie in the 'body-cavity' between the gut and the outer tube formed by the cuticle, hypodermis and muscle bands. Male and female gonads consist merely of a single, or pair of, tubes. In the female, one end of each tube opens to the outside at the vulva, the other, blind, end contains the ovary which produces eggs. Developing eggs pass down the tube or 'oviduct' and, after fertilization and formation of an egg shell, enter the distal part of the tube (the uterus), from which they are released through the vulva at egg-laying. Eelworm eggs are usually elongate-oval in shape; each contains one embryo which in some species is ready to hatch when the egg is laid.

In the male, the tubular gonad contains, proximally, a spermatogonium producing cells which divide to form sperm. The male gonad empties with the anus into a common opening, the cloaca. Associated with the cloaca are intromittent organs, the spicules, which at pairing can be exserted into the vulva to assist the transfer of sperm. In some male eelworms, accessory pieces, the gubernaculum and telamon, probably serve to guide the spicules, and sometimes the cloacal opening is surrounded by semi-rigid wing-like projections of the cuticle which form a bursa to envelop the vulval opening during pairing.

A collar of nerve cells and nerve tissues, the nerve-ring, usually encircles part of the posterior third of the oesophagus, behind the median muscular bulb when this is present. Nerves run from the nerve ring to the head, on which are sensory papillae and amphids—paired lateral openings believed to be chemoreceptors. The visible eelworm excretory system is a simple tube opening ventrally, usually in the oesophageal region, at an excretory pore. The position of the pore is important in identification. The rest of the excretory system, like the nervous system above, is extremely difficult to observe.

Eelworm life cycle

Most eelworm pests have simple life cycles. After pairing, the female lays eggs where she is feeding, and the larvae feed like their parents. Larval eelworms moult four times, the adult being the 'fifth stage'. Modifications of this basic life cycle in pest eelworms are given later with species descriptions; however, the most important life history adaptations are:

1. Some eelworms reproduce parthenogenetically (see p. 107). The chance introduction of a single larva might lead to crop damage after eelworm multiplication, e.g. *Meloidogyne*, root-knot eelworm.

2. Under adverse conditions some eelworms persist in a state of suspended animation (cryptobiosis) until favourable conditions return (*Ditylenchus* spp., *Aphelenchoides* spp.).

3. In *Heterodera* (p. 243) larval eelworms are stimulated by host-root exudates to hatch *en masse* from encysted eggs (see p. 247). In the absence of a host crop some larvae may remain, encysted in soil, for many years.

It follows that eelworms in a cryptobiotic state may be transferred unwittingly in apparently innocuous material such as straw and peat. Further, it is frequently impractical to 'starve out' some species, and crop rotations may need to be very long.

Eelworm feeding

Many nematodes have simple intestines, but the forms of the oesophagus, buccal cavity, and mouth often differ remarkably between species and the differences reflect feeding habits. Eelworms which ingest organic matter and bacteria (microphagous forms) have tubular mouths and a muscular oesophagus and suck in their food almost continuously as they scavenge through the suspension of bacteria and debris which is both their home and food. These eelworms are of little importance as plant pests; however, they are often associated with diseased plants because they thrive in the rotting plant material for which some other pathogenic organism is responsible. Microphagous eelworms are very common in mushroom beds, where they may indirectly affect cropping (see p. 354). Plant-parasitic eelworms feed on the contents of plant cells, and to enter plant tissues and/or to obtain their food, they must be able to pierce the cell walls. For this, they carry a mouth-spear or stylet. When a stylet-bearing eelworm feeds, its lips are invariably adpressed to the plant cell, and are so positioned that the spear thrust is at right angles to the cell wall. To counter this thrust the eelworm must be supported by adpressing itself to plant cells, soil particles, or some similar rigid background. When the eelworm has penetrated the cell, it sucks the contents through the hollow spear. Some eelworms do this almost at once; the cell is soon drained, and the nematode moves on. In this way the continuous feeding by many worms causes the destruction of thousands of plant cells. The ensuing plant symptoms are necrosis or death of the tissues; however, if the eelworms attack the growth zone (meristem), stunting, distortion, or 'blindness' may occur. Basically, this form of damage is from merely mechanical injury to the plant cells. Some eelworms, having pierced the cell, inject salivary fluid into it via the stylet, and, in the course of from several minutes to as long as a day (in some species), the salivary fluid digests the contents. Later, the digested material is sucked into the nematode, which then withdraws its stylet before proceeding to another cell. This type of feeding—extra-corporeal digestion with secreted enzyme activity—often results in a rapid and extensive plant reaction and abnormal growth, such as enlarged cells and galls, or distorted stems, foliage and roots. This form of damage is perhaps due more to the complex chemical activity and plant reaction than to simple mechanical injury, although the latter must occur when the cells are pierced. The true nature of eelworm feeding and of the ensuing plant damage is far from understood; some of the changes brought about in attacked plants suggest that substances are produced that are akin (in symptoms) to some herbicides used in weed control. However (unlike some weedkillers), eelworms are not selective and may produce symptoms of a similar nature in both broad-leaved and narrow-leaved (monocotyledonous) plants.

Some eelworms are vectors of disease organisms. *Xiphinema*, *Longidorus* and *Trichodorus* species are perhaps more important as vectors of soil-borne plant-virus diseases than as pathogens in their own right; their pathogenicity is, nevertheless, well proven and it is now believed that some may be more important as pathogens than hitherto believed. Some eelworms carry bacteria; *Pelodera lambdiensis* probably spreads 'bacterial blotch' of mushrooms, although it is probably harmless to mushroom production itself. Disease complexes involving eelworms, fungi, bacteria and viruses are probably common, and worthy of greater study than hitherto (Weischer, 1968).

Eelworms in the soil

All plant-parasitic eelworms, including even those with a life cycle that is normally completed in plant shoots and leaves, may be found in the soil. Some eelworms, e.g. *Heterodera*, remain dormant in soil awaiting another host crop after their host plant has died. Other eelworms, e.g. *Aphelenchoides* spp., migrate in the soil from dying or senescing hosts to a new plant home. Eelworm movement in soil is greatly influenced by soil type; the physical properties and conditions of a soil are paramount in this influence, and have been deeply studied by Wallace (1963).

Most soil consists of 'units' of many different sizes, ranging from single mineral particles, as in pure, fine sand, through 'soil-crumbs' to 'clods' which are, mainly, closely packed aggregations of mineral particles, colloidal substances and organic matter. In natural soils there are pore spaces between particles and both between and within crumbs and clods; the pores are of different sizes and in different number according to the structure of the soil. In a waterlogged soil the pore spaces are filled with water (except for entrapped air bubbles) and under these conditions microbial activity often uses up the dissolved oxygen in the soil water so that stagnation follows. As soil water drains, however, its place in the pore spaces is taken by air, and then the residual soil moisture contains much dissolved oxygen. The large pore spaces drain first, while the smallest pore spaces in a soil of very fine compacted particles need high capillary-suction forces to empty them. During drainage the soil water remains as a film on the soil units and particles, the film getting thinner as more water drains away.

Nematodes in soil live in the pore spaces and move to plants in the water films. Eelworms require dissolved oxygen and may die in anaerobic, waterlogged soil. In a very dry soil, there is insufficient water for movement so that migration is hindered, and extreme dryness may be lethal.

In all soils direct nematode travel is probably most hampered by the size of the soil pores, for, where the pore necks are too small, the eelworms cannot pass through them, and possible paths through the soil become long and tortuous. Eelworms are usually unable to move soil particles aside.

An understanding of the eelworm's soil environment is important for control by liquid and gaseous nematicides. These substances usually move through the pore spaces and finally reach the nematode via the water film in which it lives. Excessive soil water hinders eelworm control. Thus, it may dilute the nematicide, retain it near its injection point, or hinder soil-atmospheric diffusion of gas by filling the soil pores. Soil pore spaces in compacted, fine soils, and within some soil crumbs or clods, are slow to drain, and in them eelworms may survive chemical treatment for they can only be reached by a nematicidal substance diffusing in the soil water; sufficient diffusion may not occur before the fumigant disperses. For these reasons, soil to be fumigated should be moist and deeply cultivated to a seed-bed tilth. Peachey & Chapman (1966) give useful information on the physicochemical aspects of soil fumigation.

REFERENCES

DECKER, H. (1963). *Pflanzenparasitäre Nematoden und ihre Bekämpfung*. V.E.B. Deutscher Landwirtschaftverlag, Berlin. 376 pp.

JONES, F. G. W. & JONES, MARGARET G. (1964). *Pests of field crops*. Edward Arnold, London. 406 pp.

PEACHEY, J. E. (1962). The chemical control of plant-parasitic nematodes by soil fumigation. *N.A.A.S. Quarterly Rev.*, **24**, 34–45.

PEACHEY, J. E. & CHAPMAN, M. E. (1966). Chemical control of plant nematodes. *Technical Communication No. 36 Commonwealth Agric. Bureaux, Farnham Royal.* 119 pp.

SASSER, J. N. & JENKINS, W. R. (Eds.) (1960). *Nematology*. Chapel Hill, Univ. N. Carolina Press. xvi + 480 pp.

SOUTHEY, J. F. (Ed.) (1965). *Tech. Bull. Minist. Agric. Fish. Food.*, **7**, 2nd ed., 282 pp. H.M.S.O., London.

THORNE, G. (1961). *Principles of nematology*. McGraw Hill, New York. xiv + 553 pp.

WALLACE, H. R. (1963). *The biology of plant parasitic nematodes*. Edward Arnold, London. viii + 280 pp.

WEISCHER, B. (1968). Wechselwirkungen zwischen nematoden und anderen schaderregern an nutzpflanzen. *C.r. VIII symp. int. nemat. Antibes Sept. 1965* (published 1968. E. J. Brill, Leiden).

ROOT-KNOT NEMATODES AND CYST-NEMATODES

Genus *Meloidogyne* Goeldi 1892 (root-knot eelworm; 'club')

Root-knot eelworms are important pests of glasshouse crops. In tropical and subtropical countries they are pests on a field scale; in cooler climates damage is largely restricted to crops in greenhouses where the higher temperatures and continuous cropping greatly favour their increase. Species from warmer climates often occur in greenhouses because of their importation in soil or in plant material.

Infested plants are generally unthrifty and tend to wilt. The eelworms attack the roots which usually produce galls, sometimes of massive size (Fig. 8.2). The extent of the swelling is not necessarily correlated with the number of eelworms or the degree of damage to the plant, but varies both with the species of *Meloidogyne* and plant attacked; some plants gall little despite many eelworms in their roots. The absence of galls may not mean absence of eelworm; equally, because plant root-galls may be produced in other ways, their presence

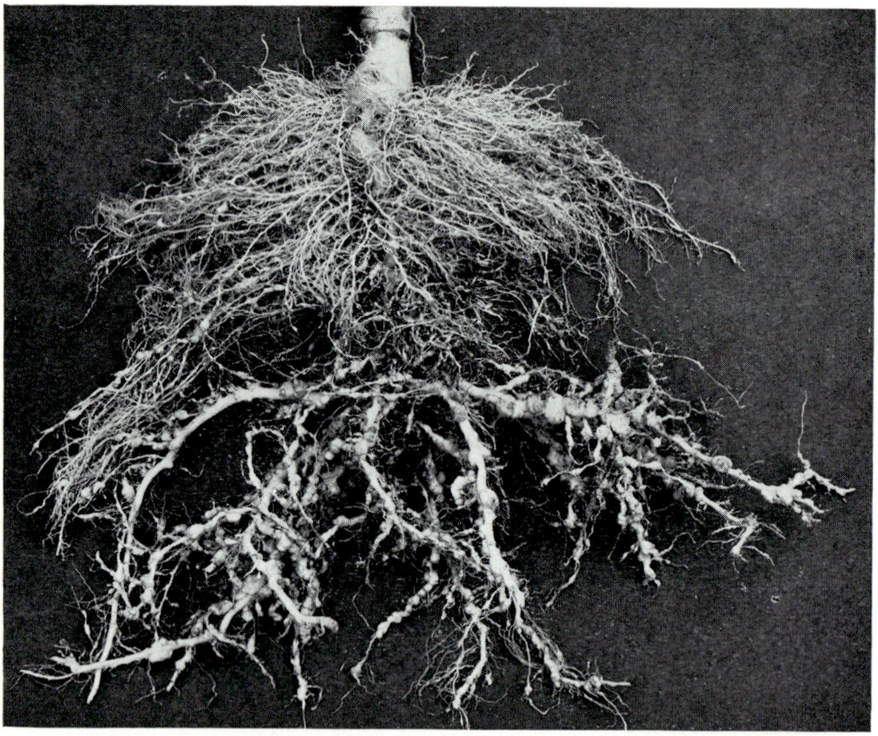

FIG. 8.2. Tobacco root grown in root-knot nematode infested soil treated with EDB to a depth of 9 in. This photograph demonstrates clearly both the effect of treatment and the nematode symptoms, and shows how the nematode persists at depth. In sandy soils treatment to a depth of 15 in. may be necessary. (Photo: R. A. C. Daulton.)

may *not* be a sure diagnosis of root-knot eelworm. It is desirable to examine eelworm-suspect roots by dissection and/or staining to confirm the diagnosis—staining is the surer method if the attack is recent and larval worms are preponderant. This examination may need to be taken farther, for species of *Meloidogyne* must be correctly identified if some kind of crop rotation is needed, for example, when a chemical control measure cannot be applied. Expert opinion is needed for identification, and also to avoid confusion with *Nacobbus*—another sedentary saccate eelworm which can cause root-galls—

and with *Heterodera rostochiensis* (p. 245), the potato-root eelworm, which sometimes causes minor galling on roots of tomato, and is especially harmful to its seedlings. Root-galls may also be caused by *Xiphinema* and *Hemicycliophora*—nematodes which live in soil and 'browse' on plant roots.

The minute (350–450 μ long) second-stage larvae invade roots from soil. Invasion and migration in the root occupy several hours, the larva forcing its way between, or through, the plant cells and coming to rest with its head close to the root stele. During this time the eelworm damages plant cells, these are affected by its salivary secretions and react by enlarging. Soon, some of them adjacent to the nematode head respond to the continuous stimulus to form special 'giant cells' (p. 242), which seemingly act as 'nectaries' and nourish the developing eelworm. Adulthood is reached in from roughly 3 to 8 weeks—the time taken being dependent on temperature, plant nutrition and eelworm species. The male is worm-like and, roughly, $\frac{3}{4}$ mm long, but the female swells markedly just before maturity, to become an oval, or spheroidal, sac about $\frac{1}{2}$ mm in diameter with a narrow neck. She contains little but reproductive organs, and large glands which secrete a mass of jelly—the 'gelatinous matrix'—into which hundreds of eggs are laid. Sometimes the female eelworms are completely embedded in the galled root tissue, but many protrude sufficiently so that they can be reached and fertilized by the male which, when mature, forsakes his feeding site to find them. Male root-knot eelworms do not feed when they leave the root, and, when spent, perish in the soil. The female, however, persists at the feeding site and may produce eggs for as long as 10 weeks. Root-knot eelworm eggs are usually unembryonated when laid, but soon develop into a larva which moults before hatching, so beginning the life cycle again. Where there is a root-knot eelworm infestation, a root fragment or a soil crumb may harbour dozens of larvae or eggs. This fact cannot be overstressed when considering the practicability of control, and the limitation of eelworm spread. Even a single egg or larva may have disease potential, for some species of *Meloidogyne* are parthenogenetic. Root-knot eelworm larvae move only a few inches per month in soil, their dispersion is predominantly by carriage in, or on, plant material, in soil clinging to footwear, implements and machinery, and in irrigation water. So-called 'virgin land' may not be free of root-knot eelworm, because some species are indigenous, even in temperate parts of the world. There is good evidence that newly hatched larvae may be attracted to a root, but even random movement would ensure that some larvae met and invaded their host in a greenhouse border soil with dense root growth.

Adults and eggs of root-knot eelworm, unlike those of the cyst-nematodes (*Heteroda* spp.), are unable to resist drying. In moist conditions, at low temperature (10 °C; 50 °F), the eggs of *Meloidogyne incognita acrita* may survive in soil for over a year; with high temperatures survival is shorter.

Attacked plants are generally unthrifty, wilting readily under only slight lack of water. The leaves may be discoloured. Clearly, such symptoms are non-specific and the root systems of suspect plants must be examined. *Their removal from soil should be done carefully, avoiding spillage of soil or root material, for if the infestation is confirmed any spill will mean that the pest has been spread.*

Attacked roots frequently are stunted, abnormally swollen and galled. Often there are relatively few fibrous roots; small rootlets may be killed if the attack is heavy, or the plant very sensitive. *Meloidogyne*-induced galls are distinguishable from some other root-galls (e.g. bacterial nodules) because they tend to be 'woody' and are clearly of root 'material'. Minor galls may occur if the eelworms have merely tried, unsuccessfully, to enter the root. The 'giant cells' referred to earlier (p. 241) are modified cells from within, or very close to, the root stele, and, because of their form and location, may be considered to be the reason for poor plant growth. Each giant-cell complex (syncitium) occupies much of the stele and must interfere with the conduction of water to the leaves and shoots. Nevertheless, with favourable growing conditions and no shortage of water, some plants may tolerate attack—especially if part of their root system has avoided it because of uneven eelworm distribution in the soil. Thus, root-knot eelworm may multiply unnoticed, and yet devastate a following crop.

There are many records, especially from agriculture, of *Meloidogyne* acting in association with fungal and bacterial pathogens to produce a disease complex. In horticulture, *Fusarium* wilt of tomatoes is increased by root-knot nematodes which speed up the appearance of disease symptoms even in varieties chosen as wilt-resistant. Carnations are more severely wilted by *Fusarium oxysporum f. dianthi* if root-knot eelworm is present. Root-wounding by root-knot eelworms increases the wilting of carnations infected with the bacterium *Pseudomonas caryophylli*, and the wilting of tomato infected with *P. solanacearum*. It is very probable that disease complexes involving eelworms are more common than is realized.

There are at least seventeen species of *Meloidogyne*, some having biotypes with different food-plant preferences, so that identification requires a specialized knowledge, and the reader is referred to Franklin (1965a), and to papers by Franklin (1965b), Golden & Birchfield (1965), Kirjanova & Ivanova (1965) and Santos (1967), in which are described more recently discovered species of potential importance to horticulture.

The lists of host plants, and of plants resistant to these eelworms, are extensive, and detailed information is given in Goodey *et al.* (1965) and Franklin & Hooper (1959). Davidson & Townshend (1967) and Townshend & Davidson (1962) list weed hosts of *Meloidogyne incognita* and *M. hapla* respectively. Greenhouse plants which commonly suffer are begonia, carnation, chrysanthemum, cucumber, lettuce and tomato.

Plant breeders have produced tomato varieties and 'tomato' root-stocks which are resistant to some *Meloidogyne* species. Eelworm-resistant rootstocks (which must be grafted with a scion from a commercial variety) are available with inbuilt resistance to other (fungal) diseases also. Some 'enemy plants' which produce nematicidal substances (*Crotalaria, Tagetes*) have been grown in infested soil in an attempt to control eelworm, but they offer little prospect of successful use under glass where economics commonly dictate that the land be kept in continuous profitable use.

REFERENCES

DAVIDSON, T. R. & TOWNSHEND, J. L. (1967). Some weed hosts of the southern root-knot nematode *Meloidogyne incognita*. *Nematologica*, **13**, 452–8.

ELMILIGY, I. A. (1968). Three new species of the genus *Meloidogyne* Goeldi 1887 (Nematoda: Heteroderidae). *Nematologica*, **14**, 577–590.

FOX, J. A. (1967). Biological studies on the blueberry root-knot nematode (*Meloidogyne carolinensis*). *Diss. Absts.*, **28**, 1311.

FRANKLIN, M. T. (1965a). *Meloidogyne*—Root knot eelworms. *Tech. Bull. Minist. Agric. Fish. Food.*, **7**, 2nd ed., 59–88. H.M.S.O., London.

— (1965b). A root-knot nematode, *Meloidogyne naasi* n.sp. on field crops in England and Wales. *Nematologica*, **11**, 79–86.

FRANKLIN, M. T. & HOOPER, D. J. (1959). Plants recorded as resistant to root-knot nematodes (*Meloidogyne* spp.). *Tech. Commun. Bur. Helminth., St. Albans.* No. 31, 33 pp.

GOLDEN, A. M. & BIRCHFIELD, W. (1965). *Meloidogyne graminicola* (*Heteroderidae*) a new species of root-knot nematode from grass. *Proc. helminth. Soc. Wash.*, **32**, (2), 228–31.

GOODEY, J. B., FRANKLIN, M. T. & HOOPER, D. J. (1965). The nematode parasites of plants catalogued under their hosts. Farnham Royal, England. Commonw. Agric. Bur., 140 pp.

KIRJANOVA, E. S. & IVANOVA, T. S. (1965). [Eelworm fauna of *Pelargonium roseum* L. in the Tadzhik S.S.R.] *Izv. Akad. Nauk. tadzhik S.S.R., Otd. biol. Nauk, No. 1 (18)* pp. 24–31. (*Meloidogyne tadshikistanica* n.sp.).

SANTOS, M. SUSANA N. DE A. (1967). *Meloidogyne ardenensis* n.sp. (Nematoda: Heteroderidae), a new British species of root-knot nematode. *Nematologica*, **13**, 593–8.

TOWNSHEND, J. L. & DAVIDSON, T. R. (1962). Some weed hosts of the northern root-knot nematode, *Meloidogyne hapla* Chitwood, 1949, in Ontario. *Can. J. Bot.*, **40**, 543–8.

WHITEHEAD, A. G. (1968). Taxonomy of *Meloidogyne* Goeldi, 1887 (Nematoda: Heteroderidae) with descriptions of four new species. *Trans. Zool Soc. Lond.*, **32**, 263–401.

Heterodera rostochiensis Woll. 1923 (potato-root or potato-cyst eelworm)

This eelworm is also known as the potato cyst-nematode and 'Golden Nematode'. The devastation that it can cause is well known to potato-growers, but it is not sufficiently realized that it thrives on, and severely damages, tomatoes. Because of the long history of potato-growing in most temperate countries, this eelworm has become widespread and there can be no guarantee that land is free from it, especially in potato-growing areas. Surveys in the United Kingdom showed that, in some areas, up to 90% of tomato-greenhouse acreage was infested—sometimes with disastrous results. It may be true to say that in some areas this eelworm is a worse pest of tomato than the closely related root-knot eelworm, *Meloidogyne* (p. 239).

Host plants of potato-root eelworm are found in the *Solanaceae*. Besides potato and tomato, potato-root eelworms can live on *Solanum dulcamara*, a hedgerow plant, and, in isolated cases, on *S. nigrum* (black nightshade). The host status of the latter plant is far from clear, for both the plant and the eelworm so vary in different geographical localities that, here and there, *S. nigrum* is able to support the eelworm. *S. nigrum* is sometimes found in greenhouses and its possible role in carrying-over eelworm should not be ignored.

S. miniatum and *S. sarachoides* are also hosts. Potato-root eelworm has its origin on *Solanums* in South America. It has been dispersed throughout the world with seed potatoes, and any ornamental Solanums should be regarded as potentially capable of transmitting eelworm on their roots, and of increasing it; seed potatoes, brought into greenhouses for 'chitting', may also introduce adhering eelworm cysts which drop off on to the soil. Potato-root eelworm is not only a hazard to tomato production. Its presence as a contaminant of plants for export may result in their being denied entry to an importing country. In these circumstances, it is extremely difficult, or impossible, to disinfest the offending plants without harming them. Nurseries

FIG. 8.3. *Heterodera trifolii* cysts on red clover roots (×10). (J. Hesling.)

growing stock for export should exercise great care to avoid eelworm infestation.

On tomato, potato-root eelworm produces symptoms of unthriftiness, dwarfing, and a tendency to wilt—symptoms like those produced by root-knot eelworm. Numbers of eelworm in soil which would be considered almost harmless to potato will severely stunt young tomato plants and seedlings, and bring growth to a standstill. The leaves of such plants have a most noticeable purplish hue—in this way resembling plants grown at too-low a temperature.

Transplanting into clean soil will often allow their quick recovery but, of course, the ensuing crop will be retarded (and the eelworm transferred to the clean soil). The nature of the invasion and feeding of potato-root eelworm larvae in a tomato root closely resembles that of root-knot eelworm (p. 240).

Unlike root-knot eelworm, potato-root eelworm does not provoke massive root-galling. However, minor galling is common, especially on the roots of tomato seedlings which may be so swollen and stunted that the damage is mistaken for that caused by *Meloidogyne*; sometimes an abnormally fibrous root system develops. Where the eelworm attack is severe, roots may be killed outright; when in a necrotic condition they are prone to invasion by fungi whose presence may mask the identity of the true pathogen.

As with *Meloidogyne*, plant symptoms are non-specific. The identity of the pest can be established by staining and/or dissecting plant roots (see p. 247 for method of distinguishing *Meloidogyne* and *Heterodera*). When the female *Heterodera* are mature, however, they are just visible to the naked eye as minute (1 mm) white, yellow, or brown spheres (eelworm cysts) on the plant roots (see below for life-history details), and their presence is an adequate diagnosis of potato-root eelworm. Sugar-beet nematode (*H. schachtii*) has occasionally been reported from tomato (Golden & Shafer, 1959). Its cysts on roots are lemon-shaped and closely resemble those of *H. trifolii* (p. 249) (Fig. 8.3).

LIFE CYCLE (Fig. 8.4)

Larvae of potato-root eelworm enter roots from soil, they prefer young rootlets which are often penetrated just behind the root-tip; natural wounds, at emerging side-roots, often provide entry points. Each larva migrates a short distance within the root to a suitable feeding site and there stops. Around or near its head, which is usually close to the root stele, 'giant cells' appear and nourish the larva through three moults (the first moult occurs within the egg) to maturity. Male and female eelworms develop differently. At first, each larva swells to a swollen-sausage shape, but later the female becomes oval to spherical, while the male develops as a coiled worm in an elongate sac formed from an earlier larval cuticle. When mature, the male emerges from this sac and moves off into the soil to find and fertilize a female. The female swells so that her body bursts the root epidermis and protrudes from it; in this condition she is fertilized. The female remains attached to the root by her head and neck. Within her swollen body, which is now known as an eelworm 'cyst', are a degenerate but functional gut, two large ovaries which produce hundreds of eggs, and glandular tissues. At first the eelworm cyst is pearly-white in colour, but later changes to yellow, and finally to golden-brown. At this stage the female dies but the embryonated eggs within her remain viable, sometimes for many years. The cyst hardens as it changes to a brown colour and so forms a tough protective shell for the eggs. When the roots die, or are disturbed, the cysts drop off into the soil. The gut, glands and cytoplasm in the cyst decay; often the head breaks off, so leaving an aperture through which the hatched larvae may subsequently emerge; another exit is always provided when

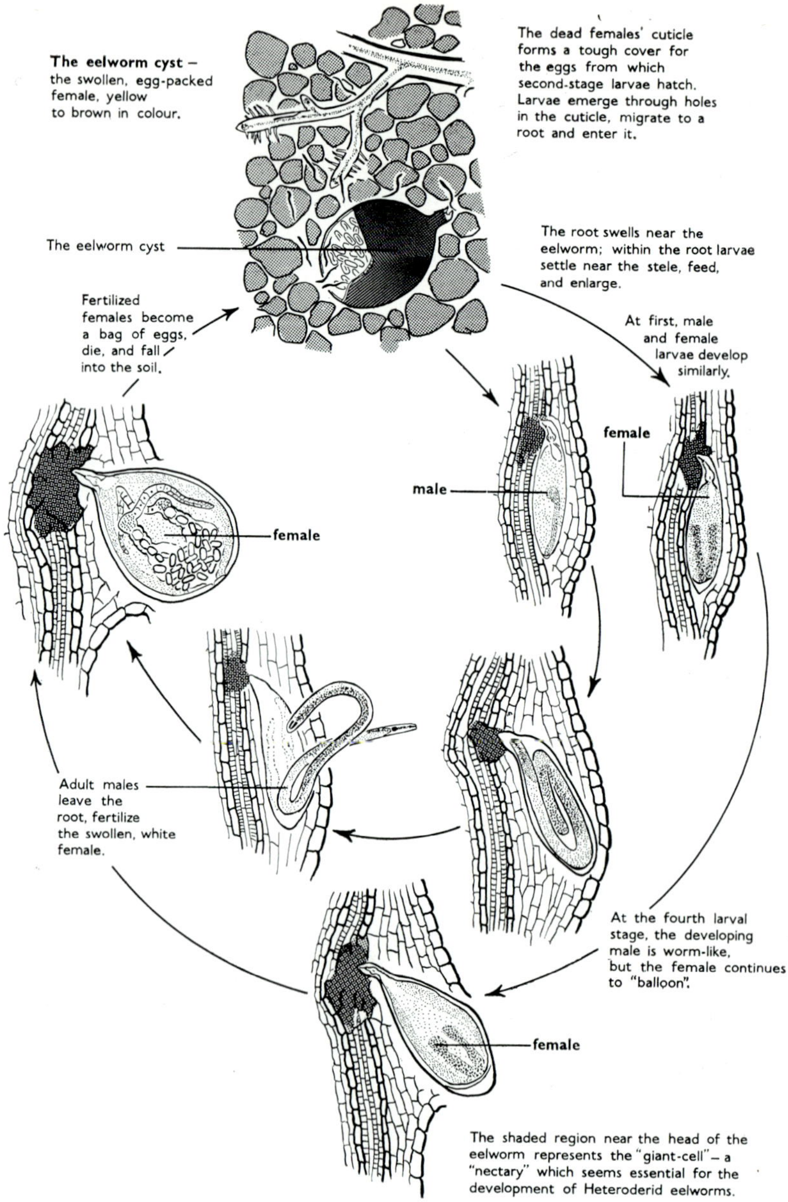

FIG. 8.4. Life cycle of potato-root eelworm, *Heterodera rostochiensis*.

a thin circumvulval patch in the cyst wall disintegrates to leave a 'fenestra'—a character important in identification.

The life cycle, from invasion to cyst formation, occupies about 5 weeks, but varies with temperature. Outdoors, there is usually only one generation annually for relatively few eggs hatch in their first year but, under glass, eggs hatch readily and several generations are possible. Potato-root eelworm exhibits a response to host roots which is a remarkable adaptation for survival. When host plants are grown in infested soil their roots produce a substance known as 'root diffusate', which stimulates the larvae to hatch and emerge from the cysts. Without host-plant stimulation emergence is relatively slight, and some larvae may remain unhatched for 10 years or longer. Thus, most of the eelworms wait for a host plant to grow nearby in the soil before they emerge.

The chemical nature of root diffusate is not fully understood; it has been suggested that a laboratory-produced diffusate might be used to induce eelworm hatching in the absence of a host plant so that the emerged larvae would ultimately starve in the soil where they are unlikely to survive for more than 6 months. However, root diffusate loses its hatching activity in about 4 days in organic soils; furthermore, there are great practical difficulties in incorporating the substance in soil on a field scale. Some larvae may not respond, and would persist to provide an ideal nucleus for a subsequent infestation.

Potato-root eelworm increases at a variable rate that is greatly dependent on the initial infestation level. Highest increases occur from low infestations. In these conditions the larvae do not compete for food in the roots and many mature to form large cysts containing up to 850 eggs. Where soil infestations are high and the larval attack is heavy, roots are killed or disorganized so that only a small percentage of the larvae mature to cysts. In overcrowded conditions relatively more larvae develop into males, for in this species sex seems to be influenced by the conditions that the larva finds within the root. Unlike some *Meloidogyne* species, potato-root eelworm is not parthenogenetic.

The mode of spread of potato-root eelworm is essentially the same as that of *Meloidogyne* (see p. 241) but with the important difference that potato-root eelworm eggs, within cysts, are able to withstand desiccation. For example, cysts can be carried in potato sacks and attached to seed potatoes. They may be blown about in dry windy weather and when dry, or partly empty, they float well and can then be carried in flood or irrigation water. Sewage sludge, often sold as fertilizer, may contain viable cysts. Potato-root eelworm is easily identified by the white, yellow, or golden-brown spheroidal cysts on roots.* If roots are examined by dissection or staining before cyst formation, there may be difficulty in distinguishing the developing larvae from those of *Meloidogyne* (root-knot eelworm). The latter invokes more root-galling on tomato than potato-root eelworm, and its early developmental stages are recognizable by possession of a spike-like 'tail'. Different isolates of potato-root eelworm have different host preferences in the *Solanaceae* and some, known as 'pathotypes' are able to thrive on potatoes bred as 'eelworm-resistant'. The significance of pathotypes in relation to tomato growing is unknown; most commonly grown tomato cultivars seem susceptible to all eelworm isolates.

* Tomato is a host of 2 recently described *Heterodera* species with spheroidal cysts: *H. virginiae* (Miller & Gray, 1968) and *H. mexicana* (Vela, 1968). Both may have been introduced into other countries but are most unlikely to be widespread.

Tomato rootstocks K, KVF and KN and KNVR are resistant to some fungal root diseases; the last two are also resistant to root-knot eelworm *Meloidogyne incognita*. None of these rootstocks is resistant to potato-root eelworm, and the eelworm will increase on them. However, the rootstocks tolerate higher eelworm populations in the soil than commercial tomato varieties and good yields have been obtained from commercial varieties grown on rootstocks planted in infested soil. If soil sterilization is inconvenient, or disadvantageous, and the eelworm population is below 135 eggs/g of soil, the use of rootstocks may permit a successful crop to be grown (Graham, 1966); such use of rootstocks merely postpones, and cannot replace, soil sterilization to control eelworm. Dunn & Hughes (1964) showed that the combined effect on tomato growth of *Heterodera rostochiensis* and the fungus *Rhizoctonia solani* was much greater than that of the eelworm alone.

REFERENCES

DUNN, E. (1967). The eelworm problem in glasshouse tomatoes. *Scottish Agriculture*, **46**, 87–9.

DUNN, E. & HUGHES, W. A. (1964). Interrelationship of the potato-root eelworm *Heterodera rostochiensis* Woll., *Rhizoctonia solani* Kuhn and *Colletotrichum atramentarium* (B & Br.) Tonb. on the growth of the tomato plant. *Nature, Lond.*, **201**, 413.

GOLDEN, A. M. & SHAFER, T. (1959). Susceptibility of tomato (*Lycopersicon esculentum*) to the sugar-beet nematode (*Heterodera schachtii*) *Pl. Dis. Reptr*, **43**, 1196–7.

GRAHAM, C. W. (1966). Potato root eelworm and tomato rootstocks. *Plant Path.*, **15**, 76–85.

GUILE, C. T. (1967). On cyst colour changes, bionomics and distribution of potato cyst-eelworm (*Heterodera rostochiensis* Woll.) pathotypes in the East Midlands. *Ann. appl. Biol.*, **60**, 411–9.

MCKENNA, L. A. & WINSLOW, R. D. (1963). The effects of various infestation techniques on the development of the potato root eelworm, with observations on the length of the life cycle in a heated glasshouse. *Res. exp. Rec. Minist. Agric. Nth. Ire.*, **12**, 97–101.

MILLER, L. I. & GRAY, BETTY J. (1968). Horsenettle cyst nematode, *Heterodera virginiae* n. sp., a parasite of solanaceous plants. *Nematologica*, **14**, 535–43.

SOUTHEY, J. F. (1965). Potato root eelworm. *Tech. Bull. Minist. Agric. Fish. Food.*, **7**, 2nd ed. 171–8. London, H.M.S.O.

VELA, A. C. (1968). Taxonomy, life cycle, and host-range of *Heterodera mexicana* n. sp. (Nematoda: Heteroderidae). *Diss. Abstr.* B. **28**, No. 4375.

Heterodera cacti Filipjev and Schuurmans Stekjoven, 1941. (cactus cyst-nematode; cactus eelworm)

This species is probably indigenous in Central and South America which is the home of many cactus species; its occurrence on roots of cacti grown in glasshouses in temperate zones is doubtless due solely to its importation.

The life history of *H. cacti* resembles that of other bisexual *Heterodera* species, e.g. *H. rostochiensis*, p. 245. White eelworm cysts, which later turn brown, are found on the roots; they are rounded lemon-shaped. *H. cacti* cysts resist drying. In glasshouses, cactus eelworm spread is most likely to occur when larvae are splashed from pot to pot during watering, or when rooted, infested 'offsets' are moved into their own pots. By inference from the biology of *H. rostochiensis*, which resists drying, cactus eelworm eggs may be expected to remain viable in the cyst in soil for several years.

Symptoms of attack on cacti are unthriftiness and unusually slow growth. In some infested *Epiphyllum* species the leaves are abnormally reddish in colour, though this discolouration is not a specific symptom of eelworm but more a general indication of an ineffective root system. *H. cacti* attacks, and may destroy, the roots. The plant produces fresh roots which are attacked in turn. The root damage is the reason for the poor plant growth.

Cactus cyst-nematode is the only *Heterodera* described from *Cactaceae* and the presence of cysts on the roots seems sufficient to identify the pathogen. For details of the morphology and for precise identification Adam (1932), Southey (1957) and Hesling (1965) may be consulted. *H. cacti* has been recorded from roots of *Euphorbia* species, succulent plants whose natural home is Southern Africa (Kirjanova & Terenteva, 1961).

Infestations of *H. cacti* in soil can be controlled by one of the methods outlined on p. 258 *et seq*. However, infestations of potted cacti may, in some cases, be best dealt with by removing the plant from the soil, cutting off all roots and washing off all soil, followed by re-rooting in a sterilized compost in a clean pot. The infested compost should be sterilized or disposed of safely.

REFERENCES

ADAM, W. (1932). Note sur *Heterodera schachtii* Schm. parasite des cactus. *Bull. Mus. Hist. nat. Belg.* **8**, 1–10.

HESLING, J. J. (1965). *Heterodera:* morphology and identification. *Tech. Bull. Minist. Agric. Fish. Food*, **7**, 2nd Ed., 103–30.

KIRJANOVA, E. S. & TERENTEVA, T. G. (1961). [*Heterodera cacti* (Nematoda: Heteroderidae) in hot houses in Leningrad.] In: Skryabin, K. I. & Turligina, E. S. [Eds.] 'Voprosi fitogelmintologii'. *Moscow: Akademiya Nauk* SSSR. pp. 74–83. [In Russian.]

SOUTHEY, J. F. (1957). Observations on *Heterodera cacti* Filipjev et Sch. Stekhoven and *Meloidogyne* spp. on imported cactus plants with a list of new host records. *Nematologica*, **2**, 1–6.

Heterodera trifolii Goffart, 1932; (Oostenbrink, 1949) (clover cyst-nematode)

This species is occasionally reported as a pest of carnations. It is widespread in the world, occurring where its natural hosts, clovers (red, white, crimson and alsike) and vetches, grow; runner bean, dock and chickweed are also hosts. *H. trifolii* may therefore be readily introduced into glasshouses in unsterile loam.

Larvae invade the roots of the host plant from soil; within the plant their development (which takes 4–6 weeks depending on temperature and other factors) resembles that of the potato-root eelworm (p. 245). Males are unknown, however, and the species reproduces parthenogenetically; also, the cyst is lemon-shaped, not spheroidal (see Fig. 8.3). It is at first white, and changes colour through yellow to brown. When the cyst is on the root and in the white or yellow phase, there is often an extrusion of gelatinous material from the protruding body, this is the 'egg sac' into which some eggs are 'laid'. Many of the eggs, whether in the egg sac or within the cyst, hatch soon after the contained larva is developed. In this way, at favourable temperatures, several generations can be completed in one year and on some hosts very large populations are soon achieved. In the absence of a host plant, however, *H. trifolii* populations probably decline rapidly because larvae freely hatch without awaiting stimulation from a host plant, and so, unlike *Heterodera rostochiensis*, *H. trifolii* is unlikely to survive in fallow soil for many years. Cysts from soil are frequently empty. Eggs within cysts cannot withstand prolonged desiccation, thus, spread as dry viable cysts—which is so important with potato-root eelworm—is unlikely; also, carnation culture offers little chance for eelworm spread on plant materials and containers.

Infested carnation plants are dwarfed, growth is slow, the foliage is yellowed, and some plants wither completely. Attacked roots are stunted, and produce a profusion of stunted side-roots. Cysts in the white, and yellow, colour phase are easily seen on the roots; these, and the brown cysts, should be removed for examination and identification. Correct identification requires specialist knowledge and reference should be made to Hesling (1965) and Tobar Jimenéz (1964).

REFERENCES

HESLING, J. J. (1965). *Tech. Bull. Agric. Fish. Food*, **7**, 2nd ed., pp. 103–30. H.M.S.O., London.

RITTER, M. (1960). Les nematodes nuisibles de l'oeillet. *C.R. Journées Florales*. (Nice-Antibes) May 1960. ed. Services Agricoles des Alpes-Maritimes. Nice 1–13 pp.

TOBAR JIMÉNEZ, A. (1964). Especies del género *Heterodera* A. Schmidt 1871 (Heteroderidae: nematoda) de la provincia de Granada. II. La *Heterodera schachtii* A. Schmidt, 1871 y sus diferencias morfológicas con la *H. trifolii* Goffart, 1932. *Revta. ibér. Parasit.*, **24**, 105–9.

Heterodera fici Kirjanova, 1954 (fig cyst-nematode)

Heterodera fici is a little-studied, but seemingly widespread, species which attacks *Ficus elastica* (rubber plant) and *F. carica*, the edible fig. *H. fici* is probably indigenous in those parts of the world where edible figs are grown as an outdoor crop and where *F. elastica* thrives out of doors—i.e. in Mediterranean countries and in California, U.S.A. It is most likely to be imported with plants and soil from these areas.

The life history of *H. fici* resembles that of *H. rostochiensis*. The mature females form rounded, lemon-shaped cysts on the roots; cysts are often inconspicuous, being deeply embedded in the root surface (Kirjanova, 1954).

Little is known of the biology of *H. fici* and of the damage to its hosts. When its numbers are high the roots must suffer damage, and unthrifty, slow growth results (Scognamiglio, 1959).

Ficus elastica is commonly propagated for sale as a pot-plant by striking 'leaf cuttings'; if these are struck and grown on in clean soil *H. fici* is unlikely to be a problem.

REFERENCES

KIRJANOVA, E. S. (1954). [Progress and prospects for the development of phytonematology in the U.S.S.R.] In *Trudy probl. temat. Soveshch.*, **3**, 32–4. (In Russian.)

SCOGNAMIGLIO, A. (1959). Specie di nematode nuova per L'Italia l'*Heterodera fici*', Kir. 1954. *Progresso agric.*, **5**, (6), 1–7.
TOBAR JIMENEZ, A. (1963). Contribución al conocimiento de la '*Heterodera fici*' Kirjanova, 1954 (Heteroderidae: nematoda) y su diferenciación morfológica de la '*H. humuli*' Filipjev, 1934. *Revta ibér. Parasit.*, **23**, 341–5.

MIGRATORY ROOT-PARASITIC EELWORMS

In addition to the cyst-nematodes and root-knot nematodes, which invade plant roots and exist as *sedentary* parasites within them, there is a large group of soil-inhabiting eelworms which 'browse' on superficial and deeper root cells, or which enter and leave root tissue in the course of their feeding or life history. This group includes many species from several genera, and in agriculture—and perhaps especially in warm climates—some cause serious plant disease and crop loss (Mountain, 1961). In protected cultivation, however, the economic importance of migratory root-parasitic eelworms has not been assessed, perhaps because the host crops they damage are not widely grown, or because eelworm attack is infrequently reported, for its accurate and unequivocal diagnosis is sometimes difficult, e.g. because of complications by other soil pathogens. Nevertheless, migratory root-parasitic eelworms are a potential hazard most likely to become a reality on perennial greenhouse plants in continuous growth, for under such conditions eelworm increase can occur over a number of years and control treatments cannot be applied because of the risk of damaging the plants. Southey (1967) has recently reviewed this group of eelworms.

Some examples of migratory root-parasitic eelworms are listed below (Table 8.1), followed by information on those genera and species that have been recorded as, or seem most likely to be, glasshouse pests.

TABLE 8.1. Migratory root-parasitic eelworms: principal genera of potential importance as pests of protected crops

Protected crops subject to attack	Common name	Eelworm genus	Symptoms
Roses, carnations	Ring nematodes	*Criconemoides*	Stubby roots, root necrosis and proliferation
	Spiral nematodes	*Helicotylenchus*	
Tomato	Sheath nematodes	*Hemicycliophora*	Root-tip gall
	Lance nematodes	*Hoplolaimus*	
Strawberry	Needle nematodes	*Longidorus*	Root-tip galls, few feeder roots. Vector of soil borne NEPO virus
Carnation	Pin nematodes	*Pratylenchus*	Stunted and yellow top growth, stunted roots with abnormal proliferation

TABLE 8.1 (*continued*)

Protected crops subject to attack	Common name	Eelworm genus	Symptoms
See text, below	Root-lesion, meadow nematodes	*Pratylenchus*	Root lesions, stunting, rot, and absence of feeder roots
Many ornamentals and citrus	Burrowing nematode	*Radopholus*	Root lesions and discolouration
Amaryllis	Spiral nematodes	*Rotylenchus*	Root lesions and discolouration
Amaryllis	Spiral nematodes	*Scutellonema*	Root lesions and discolouration
	Stubby-root nematodes	*Trichodorus* (Fig. 8.5)	Stubby roots, discoloured roots, arrested root growth, proliferation. Vector of soil borne NETU virus
Azalea	Stunt-nematode, stylet nematode	*Tylenchorhynchus*	Poor root growth, stunted plants
Citrus spp.	Citrus nematode	*Tylenchulus**	Abnormally dark roots, side-roots short and swollen. Soil clings to infested roots. Trees lack vigour, become unproductive
Roses, strawberry	Dagger nematodes	*Xiphinema*	Root kinking, especially near tip, 'curly tip'. Vector of soil-borne NEPO virus

* The developing stages and adults are sedentary root parasites, the newly hatched larvae are migratory. *Tylenchulus semipenetrans*, the citrus nematode, is a serious and frequent pest where citrus fruits are grown and, if introduced, could become a problem of citrus grown under glass.

Genus *Pratylenchus* Filipjev, 1936 (root-lesion nematodes; meadow nematodes; lesion nematodes). Root-lesion nematodes are world-wide in distribution and some species seem more common in certain climatic zones. Thus *P. neglectus*, *P. scribneri* and *P. vulnus* are associated with warm temperate and sub-tropical areas, while *P. penetrans* and *P. pratensis* are common in cool, temperate climates. It has been suggested that high numbers of certain species are correlated with soil type, but the evidence is conflicting and unreliable because of the varying suitability of the plant hosts grown in the soils and because some species show seasonal fluctuations in numbers. There is also a suggestion that there may be 'races' of *Pratylenchus* species; also, early records of host ranges are unreliable because of the possible misidentification of the eelworm involved. In general, it can be said that the genus *Pratylenchus* is polyphagous, and probably the species also. Some plants support greater eel-

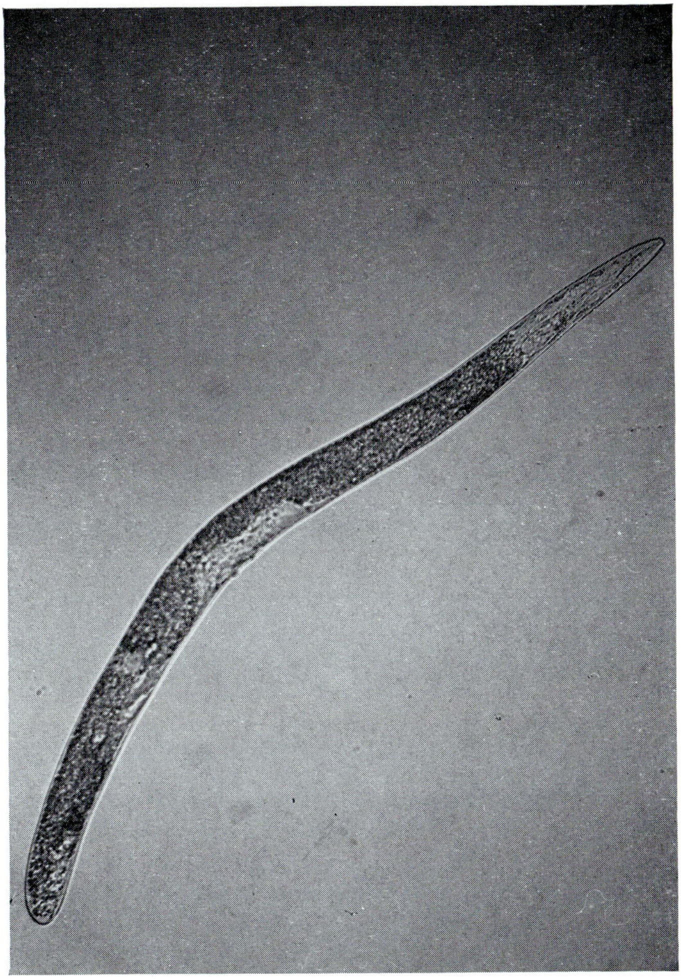

FIG. 8.5. *Trichodorus* sp. female. (Spreads diseases caused by NETU viruses.) (J. Hesling.)

worm increases than others, but little is known of the effect of root environment and secondary pathogens on the eelworm reproduction. These facts emphasize the great difficulty, if not impossibility, of advising eelworm control by crop rotation. *Pratylenchus* species are widespread in sandy soils.

In the main, *Pratylenchus* species feed on plant roots, but plant tubers stolons, etc., and in some cases, stem bases, may be attacked. In glasshouses, damage by indigenous, or imported, *Pratylenchus* is most likely on perennial crops or on long-lived subjects that are potted-on as they grow. It is a wise precaution to use sterilized soil for propagation of perennial subjects. Where soil sterilization for annual crops is a regular routine, *Pratylenchus* seems less likely to become a problem.

On roots, *Pratylenchus* species cause longitudinal discoloured (usually brownish or blackish) lesions, often of irregular shape. Sometimes the necrosis extends as a girdle around the root cortex, which may slough off, leaving the root stele exposed. Very often, secondary invading fungi colonize the lesion and may mask the correct cause of disease. New roots are often formed above the nematode-damaged area; in severe cases feeder roots may be absent.

Different species of *Pratylenchus* seem to differ markedly in their pathogenic effect. With some species, e.g. *P. pratensis*, plant roots may support high numbers of eelworm with little sign of injury; in other cases, e.g. *P. penetrans*, it is claimed that a few nematodes in soil (e.g. 1 in 50 g) can cause marked damage.

Pratylenchus usually enter roots from soil but may feed entirely ecto-parasitically on small roots. Adult and larval stages may invade; there seems to be no special infective stage. Young and old roots may be attacked and the nematodes live and breed within the root where the entire life cycle is completed. When the root rots, the nematodes migrate from it to the soil and into other roots. There is no evidence that any stage, including eggs, is resistant to drying, although the nematodes become 'quiescent' in soil in periods of drought and are reactivated when normal soil moisture returns. Little is known of the spread of root-lesion nematodes in soil; most spread seems likely to be passive, as when the eelworms are transported as contaminants of implements and machines, in draining water, or within roots of transplants.

Pratylenchus species are small, most are less than 750 μ in length. They are recognizable by the rounded tail (of the female), the stout, conspicuous mouth-spear with its large basal knobs, and the position of the vulva which is usually at about three-quarters of the length from the head. There is one ovary. The male is usually slightly smaller than the female and has a near-pointed tail with a bursa surrounding its tip. In some species males are unknown, in others, males are rare. The identification of the two dozen or so species is difficult; much early confusion was cleared by Loof's (1960) publication and this, and that of Seinhorst (1968), should be consulted when species identification is attempted.

Pratylenchus species are sometimes difficult to extract quantitatively from soil and decaying plant roots. It is sometimes necessary to 'incubate' soil and roots in shallow, well-aerated water for several days to ensure that the nematodes are extracted (de Maeseneer & Van Den Brande, 1960).

For a fuller, interesting and stimulating account of the root-lesion nematodes the reader is referred to Thorne (1961). Further details are given below of those species which have been recorded from glasshouse crops, and which may be important as pests. Root-lesion nematodes in soil can be controlled by the methods described on pp. 259–269.

Pratylenchus vulnus Allen and Jensen 1951 (walnut root-lesion nematode). Sher & Munnecke (1958) showed that rose cuttings in soil infested with *P. vulnus* (and other nematodes) grew poorly in comparison with those in sterilized soil. *P. vulnus* was shown to be responsible for the poor growth, which consisted of stunted plants with leaf chlorosis and small root systems devoid of feeder roots; it may be important as a pest of glasshouse roses. *P. vulnus* has also been recorded from peach, fig, dwarf (bush) bean and lucerne. It reproduces rapidly at 25 °C (Lownsbery *et al.*, 1967).

Pratylenchus penetrans (Cobb, 1917) Filipjev and Schuurmans Stekhoven, 1941. Cobb's root-lesion nematode is a widespread species often associated with poor growth in many crops. Examples of plants damaged by *P. penetrans* and which may be grown or propagated under glass are begonia, clematis, chrysanthemum, lily, narcissus and strawberry. Experimental studies suggest that only adults and fourth-stage larvae penetrate roots (Kable & Mai, 1968).

Pratylenchus scribneri Steiner, 1943 (Scribner's root-lesion nematode). Christie & Birchfield (1958) claim that this nematode is a very serious pest of amaryllis. This conclusion was based on studies made on bulbs planted in Florida (U.S.A.) fields, and could perhaps apply to amaryllis grown under glass in infested soil, or from infested bulbs (for the nematodes are able to penetrate the basal plate of the bulb and, to a slight extent, the bulb tissue). The possibility of their spread within bulbs should not be overlooked.

The eelworms penetrate the root epidermis and establish 'nests' in the cortical parenchyma. These nests are noticeable as small, red spots on the roots. Ultimately, as the nematodes increase, the spots enlarge into definite lesions which are dark red to brownish-red in the centre, with a brighter red margin. The root system is ultimately destroyed and the foliage growth suffers accordingly. In some cases the bulbs decrease in size because of the sloughing-off of the outer scales. It should be noted that the red lesions on roots of amaryllis are not specifically due to *P. scribneri* but a common response to root injury of any sort.

P. scribneri is also known as a pest of *Cymbidium* orchids. Here, symptoms are poor growth, yellowing of outer leaves and reduction of flowering. The roots have necrotic lesions, or rot, and the pseudobulbs may show black, necrotic lesions (Sher, 1959).

Pratylenchus convallariae Seinhorst, 1959. This species has been recorded from the continents of Europe and North America and is universally reported as a serious pest of lily of the valley, *Convallaria majalis* L. Damage to this plant was shown as long ago as 1919, when the eelworm was known as *P. pratensis*, but the separate identity of *P. convallariae* was not demonstrated until 40 years later (Seinhorst, 1959). Cayrol & Ritter (1962) gave a comprehensive account of its biology and pathogenicity. Where lily of the valley is grown intensively, the injury by *P. convallariae* may ultimately eliminate an economic flower crop. Lily of the valley is often forced under glass and the possibility of losses due to *P. convallariae* must not be overlooked, for the nematode is easily spread on planting-stock 'pips'.

The eelworm enters the fleshy roots, and after damaging the root cortex will ultimately pass through the endodermis into the conducting tissues of the root stele. Secondary disease organisms ultimately invade the damaged root, and complete rot ensues.

Above ground, evidence of eelworm attack is shown in the summer following planting by the presence of unthrifty, yellowed plants scattered in the crop. These are centres from which the disease spreads, and at the beginning of winter there are patches of failing plants surrounding the initial foci of infection, which by now are severely damaged. During the following year the damage increases and poor plants are widespread. When the plants are lifted the necrotic condition of the roots is clearly apparent. Lifted 'pips' from such infested stocks produce a poor crop of inferior flowers.

P. convallariae lives and breeds within the roots of its host plant. The duration of life cycle varies with temperature and other factors, and is about 2 months. Eelworm numbers in roots reach a peak as the attack progresses and then fall off as the secondary rots set in; at this stage the number of microphagous nematodes (*Rhabditis* spp.) in the roots may become large.

P. convallariae is spread in infested roots of lily of the valley 'pips'. It is difficult to control without damaging the plant. Planting material must be carefully selected as 'healthy', and may be subjected to 60 min immersion in water at 43 °C (110 °F).

P. convallariae is morphologically very similar to *P. penetrans*; for accurate identification the papers by Seinhorst (1959, 1968) should be consulted.

REFERENCES

CAYROL, J. C. & RITTER, M. (1962). Recherches sur la maladie vermiculaire du muguet provoquée par *Pratylenchus convallariae* (Seinhorst 1959). *Annls Épiphyt.*, **13**, 301–28.

CHRISTIE, J. R. & BIRCHFIELD, W. (1958). Scribner's lesion nematode, a destructive parasite of amaryllis. *Pl. Dis. Rptr*, **42**, 873–5.

D'HERDE, J., DE MAESENEER, J., & VAN DEN BRANDE, J. (1962). Fatigue du sol dans la culture du bégonia. *Revue Agric. Brux.*, **15**, 1471–81.

D'HERDE, J. & VAN DEN BRANDE, J. (1963). Wortelrot bij chrysanten. *Meded. LandbHoogesch. OpzoekStns Gent*, **28**, 630–7.

KABLE, P. F. & MAI, W. F. (1968). Influence of soil moisture on *Pratylenchus penetrans*. *Nematologica*, **14**, 101–22.

LOOF, P. A. A. (1960). Taxonomic Studies on the genus *Pratylenchus* (Nematoda). *Tijdschr. Plziekt.*, **66**, 29–90.

LOWNSBERY, B. F., HUANG, C. S., & JOHNSON, RUTH. N. (1967). Tissue culture and maintenance of the root-lesion nematode, *Pratylenchus vulnus*. *Nematologica*, **13**, 390–4.

DE MAESENEER, J. & VAN DEN BRANDE, J. (1960). Nieuwe inzichten in het onderzoek van Grondmonsters op de aanwezigheid van *Pratylenchus* sp. *Meded. Landb-Hoogesch. OpzoekStns Gent*, **25**, 1040–6.

MOUNTAIN, W. B. (1961). Studies on the pathogenicity of *Pratylenchus*. *Recent Advances in Botany*. University of Toronto Press. pp. 414–7.

NONG, L. & WEBER, G. F. (1965). Pathological effects of *Pratylenchus scribneri* and *Scutellonema brachyurum* on Amaryllis. *Phytopathology*, **55**, 228–30.

OOSTENBRINK, M. (1956). Over de invloed van verschillende gewassen op de veermeerdering van en de schade door *Pratylenchus pratensis* en *Pratylenchus penetrans* (*Vermes, Nematoda*), met vermelding van een afwijkend moeheidsverschijnsel bij houtige gewassen. *Tijdschr. PlZiekt.*, **62**, 183–203.

SEINHORST, J. W. (1959). Two new species of *Pratylenchus*. *Nematologica*, **4**, 83–6.

— (1968). Three new *Pratylenchus* species with a discussion of the structure of the cephalic framework and of the spermatheca in this genus. *Nematologica*, **14**, 497–510.

SHER, S. A. (1959). Nematodes on ornamentals. *Calif. Agric.*, **13**, 21.
SHER, S. A. & MUNNECKE, D. E. (1958). Nematodes and bacteria on rose. *Calif. Agric.*, **12**, 9–10.
SOUTHEY, J. F. (1959). Some records of root-lesion eelworms *Pratylenchus* spp. in England. *Pl. Path.*, **8**, 130–2.
— (1967). Migratory root eelworms. *N.A.A.S. Quarterly Rev.*, **77**, 30–8.
THORNE, G. (1961). *Principles of Nematology*. McGraw Hill, New York and Maidenhead. pp. 203–24.

Xiphinema Cobb, 1913 (dagger nematodes). *Xiphinema* species are widespread in the world but seem to thrive better and be more common in warm-temperate to tropical regions. About 44 species, and 10 species *inquirendae* are known; in most cases little is known of their pathogenicity or biology. Interest in *Xiphinema* was remarkably stimulated when certain species were found to be vectors of soil-borne viruses (1958 *et seq.*), and many species are newly discovered.

Some species, e.g. *X. vuittenezi*, seem to multiply only slowly, others, e.g. *X. index* may increase manyfold in 1 year. Soil populations of other species, e.g. *X. americanum*, show seasonal fluctuations. *X. americanum* on roses multiplies about twentyfold in 1 year; under glass higher increase rates are possible. From their slow rate of development and increase, it is inferred that some *Xiphinema* may live for 2 years or longer in favourable soils.

Many *Xiphinema* species are associated with roots of trees, shrubs and perennial plants. Some species have a wide host range and may browse on many different plants; thus, control by crop rotation seems unlikely. Most *Xiphinema* are found in the top 3–9 in. of the soil, but some have been located at depths of 3 ft, and their presence at greater depths where roots have penetrated should not be unexpected. The nematodes are, therefore, more than usually difficult to eradicate by usual soil sterilization methods which are effective in only the top 15 in. or so of soil. However, in many cases, the nematode's increase rate seems insufficiently rapid to allow it to regain pest status except after several years following treatment, so normal soil sterilization procedures seem adequate. Studies of the extent of *Xiphinema* spread in soil suggest that it moves little more than 1 ft/year of its own accord; it is unlikely that any stage survives drying.

Perennial glasshouse crops, such as roses, are most likely to become damaged by *Xiphinema*, for the nematode may have many years in which to increase in the undisturbed soil.

Xiphinema spp. feed on plant roots; where soil is heavily infested there may be marked stunting and stubbiness of roots with minor galling, especially near root tips. The galling may be confused with that due to root-knot eelworm, *Meloidogyne* (p. 241). On roses, cv. 'Better Times', *X. americanum*-induced galling consists, characteristically, of the enlargement of the root tip and a curling of the end of the root; there may be some necrosis and shrivelling on the 'inside' of the curve. This response has been termed 'curly-tip' (Schindler, 1957). In general, *Xiphinema* species seem to cause local necrosis, discolouration, and lesions on roots; when in this condition roots are subject to secondary attack by fungi and bacteria. Above-ground symptoms, on roses, are poor growth and chlorosis. *X. diversicaudatum* is known to transmit strawberry latent ringspot virus which is associated with leaf malformation, chlorosis, and dwarfing in glasshouse roses (Harrison, 1967).

Xiphinema are relatively large eelworms (a length of 5 mm for adults is common) and slow moving. They may remain stationary for many hours while feeding; in this time the mouth-spear is thrust into the root tissue while the nematode lies in the soil by the root.

All *Xiphinema* have a long mouth-spear with, at its base, a spear extension of almost equal length and with broad basal flanges. Larval *Xiphinema* hatch from eggs laid in soil and may be recognized by the absence of reproductive organs and the possession of two mouth-spears; rarely, adults have two spears (Dalmasso, 1967). One of the larval spears is in current use, the other is a 'spare' and is positioned behind the 'current' spear which it replaces when the nematode moults, for, during moulting, the 'current' spear is cast with the cuticle. *Xiphinema* may be confused with *Longidorus* (needle nematodes)—another large soil-inhabiting genus also with a long mouth-spear. In *Longidorus*, however, there are no basal flanges to the spear extension, and the spear guide-ring is close to the spear tip; in *Xiphinema* the guide-ring is near to the spear base.

Recent intensive taxonomic studies on the genus *Xiphinema*, the discovery of new species and the synonymization, or rejection, of many species described before 1958, make the identification of species difficult; papers to be consulted are by Luc & Tarjan (1963); Luc & Dalmasso (1963); Sturhan (1963); Cohn & Martelli (1964); Coomans (1964); Luc *et al.* (1964); Heyns (1965); Martelli *et al.* (1966); Brown (1967).

D'Herde & Van Den Brande (1964) and Flegg (1967) describe useful methods of extracting these large nematodes from soil.

REFERENCES

ADDOH, P. G. (1964). Ph.D. thesis. Biological study of *Xiphinema americanum*. *Diss. Abstr.*, (1963), **26**, 1853.

BROWN, R. H. (1967). *Xiphinema monohysterum* n.sp. (Nematoda: Dorylaimidae) from Southern New South Wales. *Nematologica*, **13**, 633–7.

COHN, E. & MARTELLI, G. P. (1964). Studies on *Xiphinema ingens* Luc & Dalmasso, 1964 and the male of *Longidorus brevicaudatus* (Schuurmans Stekhoven 1951) Thorne, 1961 (Nematoda: Dorylaimidae). *Nematologica*, **10**, 192–6.

COOMANS, A. (1964). *Xiphinema basilgoodeyi* n.sp. with observations on its larval stages (Nematoda: Dorylaimina). *Nematologica*, **10**, 581–93.

DALMASSO, A. (1967). Cas de femelles de *Xiphinema* possédant deux odontostyles. *Nematologica*, **13**, 474–6.

DAVIS, R. A. & JENKINS, W. R. (1960). Nematodes associated with roses, and the root injury caused by *Meloidogyne hapla* Chitwood, 1959, *Xiphinema diversicaudatum* (Micoletzky, 1927) Thorne, 1939, and *Helicotylenchus nannus* Steiner, 1945. *Univ. Maryland Agric. Exp. Stn Bull. A*. **106**. 16 pp.

D'HERDE, J. & VAN DEN BRANDE, J. (1964). Distribution of *Xiphinema* and *Longidorus* spp. in strawberry fields in Belgium and a method for their quantitative extraction. *Nematologica*, **10**, 454–8.

FLEGG, J. J. M. (1966). Once-yearly reproduction in *Xiphinema vuittenezi*. *Nature, Lond.*, **212**, 741.
— (1967). Extraction of *Xiphinema* and *Longidorus* species from soil by a modification of Cobb's decanting and sieving technique. *Ann. appl. Biol.*, **60**, 429–37.
— (1968a). The occurrence and depth distribution of *Xiphinema* and *Longidorus* species in Southeastern England. *Nematologica*, **14**, 189–96.
— (1968b). Life cycle studies of some *Xiphinema* and *Longidorus* species in Southeastern England. *Nematologica*, **14**, 197–210.
FLORES, H. & CHAPMAN, R. A. (1965). Development of *Xiphinema americanum* associated with tobacco or sod in the field and with roses or grapes in the greenhouse. *Phytopathology*, **55**, 1058.
GRIFFIN, G. D. & DARLING, H. M. (1964). An ecological study of *Xiphinema americanum* Cobb in an ornamental spruce nursery. *Nematologica*, **10**, 471.
HARRISON, B. D. (1967). The transmission of strawberry latent ringspot virus by *Xiphinema diversicaudatum* (Nematoda). *Ann. appl. Biol.*, **60**, 405–9.
HARRISON, B. D., PEACHEY, J. E. & WINSLOW, R. D. (1963). The use of nematicides to control the spread of arabis mosaic virus by *Xiphinema diversicaudatum* (Micol.). *Ann. appl. Biol.*, **52**, 243.
HEYNS, J. (1965). Four new species of the genus *Xiphinema* (Nematoda: Dorylaimoidea) from South Africa. *Nematologica*, **11**, 87–99.
LUC, M. & DALMASSO, A. (1963). Trois nouveaux *Xiphinema* associés à la vigne (Nematoda: Dorylaimidae). *Nematologica*, **9**, 531–41.
LUC, M., LIMA, M. B., WEISCHER, B. & FLEGG, J. J. M. (1964). *Xiphinema vuittenezi* n.sp. (Nematoda: Dorylaimidae). *Nematologica*, **10**, 151–63.
LUC, M. & TARJAN, A. C. (1963). Note systématique sur le genre *Xiphinema* Cobb, 1913. *Nematologica*, **9**, 111–5.
MARTELLI, G. P., COHN, E. & DALMASSO, A. (1966). A redescription of *Xiphinema italiae* Meyl. 1953 and its relationship to *Xiphinema arenarium* Luc et Dalmasso, 1963 and *Xiphinema conurum* Siddiqi, 1964. *Nematologica*, **12**, 183–94.
PEACHEY, J. E., GREEN, C. D. & GREET, D. N. (1965). The effects of D.D. and Nemagon on eight-year-old glasshouse roses infested with dagger nematode. *Expl. Hort.* **13**, 49–50.
SCHINDLER, A. F. (1957). Parasitism and pathogenicity of *Xiphinema diversicaudatum*, an ectoparastic nematode. *Nematologica*, **2**, 25–31.
STURHAN, D. (1963). Beitrag zur systematik der Gattung *Xiphinema* Cobb, 1913. *Nematologica*, **9**, 205–14.
YASSIN, A. M. (1968). Transmission of viruses by *Longidorus elongatus*. *Nematologica*, **14**, 419–28.

Control of root-infesting eelworms

It is vitally important to prevent the introduction of eelworms into glasshouse soils because their complete eradication is very difficult.

Several routes whereby eelworms may enter glasshouses have already been mentioned (pp. 241, 247, 248). The high incidence of potato-root eelworm in some glasshouse districts, in contrast to its rare occurrence in others, is almost certainly related to the extent of local potato growing.

Any soil taken into glasshouses should be disinfested, preferably by steaming. Some exceptions may possibly be made, e.g. when turf soil from old grassland is used for preparing cucumber beds. The risk of introducing eelworm with such soil can be lessened by stacking the soil for at least a year; all risk can be eliminated by effectively fumigating the stack with DD or methyl bromide (see pp. 261–264).

Bought-in plants should be examined carefully for eelworm symptoms. Those of potato-root eelworm on young tomato plants have been described (p. 244), and in the case of *Meloidogyne* spp. root-galls ('knots') may be evident. With some eelworms, or with light infestations, the effect of the pest on the growth of young plants may not always be clearly shown.

It is almost impossible to kill eelworms within roots of growing plants without plant injury, although dormant plant material can sometimes be disinfested by heat treatment. *M. hapla* in rose roots can be controlled in this way.

Soil steaming, chemical treatment, or a combination of these, is the usual method of controlling soil eelworms. Before such measures are taken the roots of an infested crop should be forked out as completely as possible. The superstructure of the house should be cleared of soil, e.g. on ledges, etc. Careful disposal of infested plant material is most important. Ideally, this should be taken direct from the glasshouse, and burnt or steamed. Burial, where danger of spread will not arise (e.g. by drainage water, or from eelworm which persists on host weeds colonizing the grave), is often a practical method.

Steaming

Eelworm control by steaming depends, mainly, on the depth to which the soil is sufficiently heated and on the thoroughness of the work. The soil in the paths should always be steamed. Heating to a depth of 18 in. is usually sufficient with heavy soils, but penetration to even 24 in. may fail to eradicate eelworms in deep, light soils, where root-knot eelworms have been found at a depth of 4 ft. Steaming to a depth of at least 2 ft is often recommended but this is both difficult and expensive as it necessitates the use of perforated pipes (p. 17) placed in trenches about 20 in. deep, or, alternatively, as superimposed pipes at 20 in. and 10 in. depths. The bottom of the trench should be broken up to aid steam penetration. Steam-air treatment may prove as satisfactory as, or even superior to, 'normal' steaming with Hoddesdon pipes, as undesirable side effects are less at the lower temperature attained (Baker, 1962).

Particular care must be taken to adequately heat the soil adjacent to footings of walls, and other underground structures, and to prevent poor heating due to steam escaping through 'blow-holes' in very loose, dry soil, e.g. near heating pipes.

Annual steaming to a depth of 15–18 in. (steaming pipes at 10 in.) will usually enable economic crops to be grown, but, except on very heavy or highly organic (peaty) soils, chemical treatment (preferably combined with steaming) has advantages since the lethal effects penetrate deeper than those of steam.

'Balloon' or 'sheet' steaming is often inadequate owing to the shallow and uneven heat penetration, and the difficulty of dealing with obstructions such as pipe and purlin-post dollies, or footings etc. The use of tougher sheets, or of nets to hold down the ballooning sheets, is said to give better penetration of heat into the soil.

Chemical nematicides

The nematicides at present used to control root-invading eelworms in soil are either volatile liquids, or powders or solutions which decompose in soil to produce the volatile chemical methyl isothiocyanate.

The choice of the most suitable nematicide is influenced by several factors: the species of eelworm, the period which can elapse between its application and subsequent planting, and the presence of other soil pests or diseases. The chemical control measures for root-knot eelworm and for potato-root eelworm, are dealt with separately because of their differing susceptibilities to nematicides.

ROOT-KNOT EELWORM—*Control*

Control by DD

DD, a mixture of dichloropropane and dichloropropene, is widely used for the control of root-knot eelworm. It is injected into the soil as described on p. 33 after removing roots and then digging, or deep rotary cultivation, to thoroughly break up the soil. It is advisable to keep the soil slightly moist until the DD is injected, to promote decomposition of any remaining root fragments. The usual dosage is 400 lb/acre (9·2 lb or $\frac{3}{4}$ gal/1000 ft^2), normally applied by injecting 3·5 ml. at 1 ft intervals to a depth of 6 in. Although 15 in. spacing of the injections is sometimes advocated for light, open soils, better vapour distribution is obtained with the closer spacing. As infestations tend to be most severe, and occur at greater depth in open soils, and also (except in peaty soils) the disappearance of the vapour from the soil is rapid, it may be advisable to increase the dosage in such soils to 500 lb/acre.

The slow release of injurious residues may be troublesome when DD is applied to peaty or clay soils, or to soils with a high water table. Unless heavy, clay soils are cultivated to give a deep, uniform, fine tilth, there may be poor vapour distribution and, so, unsatisfactory control. The regular use of DD on heavy soils has occasionally caused tainting of tomatoes through persistence of the chemical. Lower dosages of 200–300 lb/acre (1·8–2·6 ml. at 12 in. spacings) may have to be used to lessen these difficulties and it becomes preferable to steam-sterilize.

The nematicidal activity of DD increases with increasing temperature, but so does the rate of its elimination from the soil. Satisfactory results can be obtained at soil temperatures around 40 °F (5 °C) for although the chemical is less active the eelworms are subjected to it for a longer period. The most convenient time to apply DD depends on the cropping programme: in tomato and cucumber houses, autumn applications are usual.

After injecting the DD, the surface of the soil should be sealed by light watering (p. 34). To prevent rapid loss of DD vapour, watering should be repeated if necessary to keep the surface moist for 7 days. To prevent survival in the top inch or so of soil dilutions of 'solubilized' xylenols or emulsions of cresylic or other tar (phenolic) acids, at about 0·4% active ingredient, are recommended instead of water as a 'soil seal'. However, satisfactory results are usually obtained without their use; they are not needed when DD is used in conjunction with steaming.

Growing plants are injured by the vapour of DD. Dormant, hard-wooded plants, however, are usually uninjured by the vapour in the glasshouse atmosphere—there must, of course, be no DD near their roots. No residues must remain when plants are set out. To hasten the loss of DD, soil should be forked over about 3 weeks after treatment (2 weeks in summer), and this 'opening-up' should be repeated after a further 1 or 2 weeks if the odour of DD persists. When DD is applied in the autumn to loam soils, it is advisable to allow an interval of at least 8 weeks before planting; in the summer 5 weeks is usually sufficient. These intervals should be extended by several weeks with peaty or clay soils.

DD has little fungicidal action and, as the control of root diseases is usually as important as that of eelworms, DD injection may be combined with treatment with steam, metham-sodium, dazomet or formaldehyde. If steam is to follow DD, the latter should be deeply injected (8 in.) to destroy deep-seated eelworms. Steaming should follow about 3 weeks later, using either 'Hoddesdon pipes' inserted 8–10 in. deep, or spiked pipes or grids. Balloon (sheet) steaming may be satisfactory on light, well-tilled soils.

Dazomet or metham-sodium both have fungicidal and nematicidal properties (p. 267) and can also be used with DD to control certain soil-borne diseases and potato-root eelworm. It should be remembered that metham-sodium stimulates hatching of *Heterodera* cysts but that mixtures with nematicides containing DD destroy this property. Where dazomet is used, or metham-sodium is applied by the rotary-incorporation methods (p. 267) DD may be injected as soon as the atmosphere of the house is free from the irritating vapour they produce. If metham-sodium is applied as a drench, the DD should not be applied until the soil has thoroughly drained. Formaldehyde may be applied either before, or after, treatment with DD. In all cases, the period which must elapse between the first treatment and planting will be increased; the increase is greater when dazomet or metham-sodium are used (p. 267) than with formaldehyde.

The necessity for separate applications of chemicals to give cumulative control of root-knot eelworm, certain soil-borne fungi, and potato-root eelworm, may, in future, be avoided by injecting mixtures of DD and methyl isothiocyanate—such mixtures are now available in some countries.

Inhalation of DD vapour is poisonous as is absorption through the skin, as well as ingestion. Rubber boots and gloves should be worn when handling the fluid and the glasshouse needs full ventilation when injecting is in progress, or when persons enter the house within a period of several days after treatment. Any splashes of DD on the skin should be promptly removed by rinsing with paraffin or kerosene, and by washing with soap and water.

Control by ethylene dibromide

Ethylene dibromide (EDB) is a more effective nematicide than DD, but it is also more expensive. Although now seldom used in Britain, it is applied extensively elsewhere, often in mixtures with other nematicides, or fungicidal fumigants.

To control root-knot eelworm, a dosage of $5\frac{1}{2}$ lb/1000 ft^2 is usually applied as a proprietary solution in an organic solvent. EDB is not approved for use in glasshouses in Britain, and no recommendations for its safe use have been issued. As EDB is less volatile than DD it requires higher soil temperatures for optimum results, although it is more satisfactory at lower temperatures than dibromo-chloropropane (DBCP) ('Nemagon'), a nematicide which is sometimes unsatisfactory when applied in the autumn in Britain.

Control by methyl bromide

Methyl bromide is volatile and a penetrating gas at temperatures above 39 °F (4 °C). It is highly poisonous. Although its nematicidal properties were demonstrated by Goodey in 1945, and its value as a soil fumigant to control root-knot eelworm was shown by Newhall & Lear in the same year, its commercial use in glasshouses in Britain has only recently been appreciated. Its application to border soils was considered impractical, although it had long been used to treat soil on glasshouse benches in America. The availability of small, measured quantities of methyl bromide and supplies of cheap, gas-proof sheeting, have led to renewed investigations of its use as a soil sterilant. However, because it is an insidious, lethal poison, methyl bromide should be applied only by suitably equipped, trained personnel.

The odour of methyl bromide is slight and may be imperceptible to some people, particularly after a few minutes exposure, consequently it is usual to add a small proportion of chloropicrin, a powerful lachrymator ('tear-gas'), as a warning substance. Provided the fumigation is conducted by qualified operators, the application of methyl bromide containing 2% chloropicrin from cylinders should be safe.

The procedure adopted when applying methyl bromide from canisters is to use specially constructed trays, fitted with a spike to puncture the canisters and release the methyl bromide into the tray. To obtain deep penetration the soil should be well cultivated and the second spit broken up; it should be slightly moist (not wet) for methyl bromide is less effective in dry soil.

The required number of trays, one per canister, is placed on the area to be treated, the canisters placed on them, and the whole area and canisters covered by gas-proof polythene sheeting, the edges of which are inserted in a narrow trench about 8 in. deep surrounding the treatment area, and made gas-tight by pressing the back-filled soil firmly in the trench. In some cases it may be more convenient to secure the sheet by weighing down the edges with moist sand, trodden to effect a good seal. The methyl bromide is released under the plastic sheet, by forcing down the canisters so that they are punctured by the spike on the tray. The methyl bromide rapidly vaporizes, and the vapour diffuses into the soil.

Equally good results can be obtained by discharging the methyl bromide from cylinders, especially by the 'hot-gas method'. The methyl bromide is passed through a vaporizer ($\frac{1}{2}$-in. diam. copper coil in water bath at 170° F) and distributed under the gas-proof sheet through perforated plastic pipes.

In this procedure it is usually necessary to weigh the cylinder to determine when the correct dosage has been applied.

The dosage of methyl bromide required to control root-knot eelworm, under late autumn and winter conditions in Britain, has not been satisfactorily established. The currently accepted rate is 1 lb/100 ft^2. However, work by Hague *et al.* (1964) suggests that considerably smaller doses may suffice to control potato-root eelworm, provided fumigation is continued for 2 days. They found that dosages of 1 lb/100 ft^2 can give C.T.P.'s (p. 34) at 18 in. considerably exceeding 400 mg hr/l., which Hague & Sood (1963) consider necessary to control eelworms. The dosage used in tomato houses may, however, be governed by the amount required to give adequate control of brown root rot (a fungal disease), for which purpose fumigation should continue for 4 days at the highest possible temperature.

It is usual to remove the plastic sheet 3 or more days after treatment. This must also be done by trained persons wearing respirators, who will employ a halide-detector lamp, or other device, to determine when it is safe for unprotected persons to enter the glasshouse. (N.B. Special respirator canisters are required for protection against methyl bromide.) Meanwhile the house should be well ventilated, the doors locked, and warning notices affixed. The precautionary measures required when methyl bromide is used are given in a Home Office (HMSO) leaflet entitled *Fumigation with Methyl Bromide. Precautionary Measures.* See also Anon. (1968b) and Flegg (1968).

The advantages of methyl bromide over some other soil fumigants are that phytotoxic residues rapidly disappear from the soil, that it is a fungicide at high doses, and also a weedkiller. Methyl bromide treatment results in increased available soil nitrogen, and some adjustment of the fertilizer programme may be necessary, particularly where the soil has not been steamed or treated with 'sterilizing' chemicals for several years.

Residual methyl bromide vapour is less injurious to plants than that of methyl isothiocyanate. Nevertheless, Hague *et al.* (1964) found that small seedlings, and seeds, may be harmed by residues in soil, although larger plants may be unaffected. Carnations are apparently especially sensitive. Plants should not be set out in methyl-bromide-treated soil less than a week after treatment ends.

Propagating or potting soil may be freed of eelworm by treatment with methyl bromide, but other means should be used to ensure that all pathogenic fungi are also eliminated.

POTATO-ROOT EELWORM—*Control*

Deep, thorough steaming gives the best results, but eradication of the pest is rarely achieved; on some nurseries the high cost of steaming may make this method of control uneconomic, and recourse must be made to chemical methods.

The most effective chemicals are methyl bromide, methyl isothiocyanate, metham-sodium and dazomet, the last two decomposing to methyl isothiocyanate in the soil. The use of methyl bromide has been described in connection with the control of root-knot eelworm (p. 263).

Control with methyl isothiocyanate

Solutions of methyl isothiocyanate have seldom been used, so far, in Britain for control of nematodes, although elsewhere their use is frequent. This chemical is, however, the active nematicide produced when metham-sodium and dazomet are applied to moist soil; hence it is mentioned first. The precautions necessary to prevent injury to plants are common to all three substances. The safety precautions necessary when injecting methyl isothiocyanate are similar to those when injecting DD.

The value, under British conditions, of injected mixtures of methyl isothiocyanate with other volatile nematicides and fungicides, has not yet been established. Injection methods may provide a means of treating deep, open soils to a depth impractical by rotary cultivation or drenching methods. The recommended dosage for controlling potato-root eelworm is 2·4 ml. (12 ml. of a 20% solution) injected at intervals of 9 in. and 6 in. deep into deeply cultivated soil. This is equivalent to approximately $2\frac{1}{4}$ pints of 20% solution/100 ft^2. Methyl isothiocyanate is also an effective fungicide and controls brown root rot of tomatoes at the above rate. After injection the soil surface should be 'sealed' (p. 34). The alternative and originally recommended method of application (in which the solution is sprinkled into the furrows when digging), is unsuitable in glasshouses, as the operator is excessively exposed to the highly lachrymatory vapour.

Methyl isothiocyanate is highly phytotoxic, particularly when present as a vapour in the atmosphere; the soil must be free from residues before plants are taken into the glasshouse. Tomatoes are particularly susceptible to damage, which is expressed as severe growth abnormalities. Suppression of the growth of the main stem, and the development of abnormal flower trusses may occur after three or more over-night exposures to atmospheres containing as little as 0·01 g/1000 ft^3. On several occasions tomato seedlings have been seriously injured by the entry of methyl isothiocyanate vapour through ventilators from nearby, detached, houses treated with metham-sodium. Mullions and partitions, although apparently sound, may permit the passage of injurious amounts of vapour.

The elimination of residues is hastened by forking over the soil to a depth of about 10 in. at least once after the application, and by heating the glasshouse. The first forking should take place not less than 2 weeks after treatment, and if rapid dispersal is required, a second forking should be given about 1 week later. Rotary cultivation is considered to be less effective than forking and leaving the surface rough.

Residues in the soil, sufficient to injure plant roots or give rise to phytotoxic atmospheric concentrations, can be detected by the 'cress test' devised by Messrs. Schering. This test is made by placing about 1 lb of soil in a screw-topped jar. The soil sampled should be taken, without delay, from mixed samples removed down to a depth of 10 in. from a number of distributed sites in the treated house. Cress seed is placed on a wetted pad of cotton wool, or other absorbent material, on top of the soil, the jar closed tightly and kept in a warm place. If appreciable methyl isothiocyanate residues are present, the germination of the seed will be inhibited, or the growth of the seedlings will be abnormal. As a check some cress seed should be sown in jars containing untreated soil.

Practical experience suggests that the elimination of residual amounts of methyl isothiocyanate may be quicker when solutions of the compound are applied to the soil, than when it is derived from metham-sodium or dazomet applications.

The time taken for the elimination of phytotoxic residues depends on the nature of the soil and the temperature. The rate of volatilization may be 50% higher at 61 °F (15 °C) than at 50 °F (10 °C). An interval of 9 weeks between application and planting tomatoes should be allowed for loam soils where methyl isothiocyanate was applied in the late autumn and 4–5 weeks for summer applications. With peaty soils these intervals should be 12 weeks, and 6 weeks respectively. The high absorptive properties of peaty soils may reduce the effectiveness of the fumigant.

The prevention of skin contamination with methyl isothiocyanate is important. Solutions of it penetrate rubber. Any splashes on the skin should be wiped off immediately with a dry cloth, followed by washing the contaminated parts with soap and water. A respirator should be worn if the vapour causes discomfort.

Proprietary mixtures of methyl isothiocyanate with other fumigants such as methyl bromide, DD or chloropicrin, are available in some countries. They have advantages where simultaneous control of other nematodes or of certain fungi is required.

Control with metham-sodium

Commercial, aqueous solutions containing 32–35% of metham-sodium are extensively used to control potato-root eelworm. They are applied at the rate of 2 pints/100 ft^2 by several alternative methods. The most effective method is to dilute the concentrate and drench the soil. Although the usual recommended dilution is 1 in 100 (2 pints in 25 gal), better control of both potato-root eelworm and tomato brown root rot has been obtained where dilutions of 1 in 160 or 1 in 200 were used. Investigation of the distribution of the nematicidal methyl isothiocyanate produced in soil showed that there were higher concentrations at a depth of 1 ft using these greater volumes of more dilute solutions. Where the soil is free draining, or where application can be made by methods which avoid flooding, the use of 2 pints of the concentrate in 50 gal/100 ft^2 is recommended. For heavy, retentive soils, the dilution should be 2 pints in 25 gal/100 ft^2 so that the interval before it is safe to plant tomatoes is not unduly long. The soil should be well cultivated to obtain uniform penetration.

An alternative satisfactory method is to water the soil evenly with 2 pints of the concentrate in 4 gal of water/100 ft^2, followed immediately by deep thorough rotary-cultivation to a depth of at least 8 in. Only relatively small areas should be treated at one time to avoid discomfort due to excessive vaporization of methyl isothiocyanate. The surface of the soil should then be 'sealed' by lightly watering, and this sealing repeated if the surface dries within 7 days.

Injection of the undiluted concentrate, a method frequently employed in early trials, may give unsatisfactory results due to the slow conversion of the metham-sodium to methyl isothiocyanate. Better results may be obtained by injecting 11 ml. of concentrate 6–8 in. deep at 12 in. intervals, followed by rotary cultivation and surface sealing.

Although metham-sodium has been applied annually on many nurseries without any falling-off in its efficiency, there have been instances where nematicidal and fungicidal activity has declined after several consecutive yearly treatments. This decline may result from changes in the catalytic activity normally occurring in soil (Ashley *et al.*, 1963). It is known that the breakdown to methyl isothiocyanate does not occur in pure sand or peat, and may do so only to a very limited extent in aqueous extracts from 'normal' soils. Another explanation, in some cases, may be a build-up of micro-organisms which metabolize methyl isothiocyanate so that nematicidal concentrations cannot form. The biological destruction of methyl isothiocyanate in soil has been demonstrated by Ashley *et al.* (1963).

The disadvantage of metham-sodium is the long period before phytotoxic residues disappear. This period depends on the soil temperature, texture and structure. Metham-sodium, however, gives good control of some root-invading fungi, and weeds.

To avoid phytotoxicity metham-sodium should be applied before the soil temperature falls, although it is advantageous to allow the residual roots of removed tomato plants to start to rot. Provided that the soil is forked over a fortnight after treatment, and again 3 weeks later, the normal practice of winter flooding can be undertaken not less than 8 weeks later, if the application was made before the end of October, or not less than 10 weeks later if applied in November. Tomato plants can be taken into the glasshouse in a further 2 weeks' time. These periods may be insufficient for very peaty soils unless at least four cultivations are given. In all cases it is advisable to conduct a 'cress test' (p. 265) a week before planting is intended.

Bulbs, boxed for forcing, can be housed 3–4 weeks after the soil has been treated, provided it is turned by forking on two occasions after a 2-week period of undisturbed fumigation.

Control with dazomet

The application of metham-sodium solutions by hose, or watering can, often causes irritation of the eyes and nose by the liberated vapour of methyl isothiocyanate, and under some adverse conditions a respirator may have to be worn. This disadvantage rarely arises when dazomet, another methyl isothiocyanate-producing compound, is mixed with the soil.

Commercial, powder formulations containing about 85% of this compound are first evenly distributed on the soil surface and, when intimately mixed with the soil, give results equal to those obtained with metham-sodium. The admixture must be thorough; it can be achieved by the somewhat unconventional method of slow rotary cultivation. Thus, Hams & Collyer (1964) recommend that the rotovator blades be buried to their full depth, rotated at 120 r.p.m., and that the forward speed be limited to 12 ft/min. 'L'-shaped rotor blades should be used. Admixture to a depth of 9 in. or more is necessary. The rate of application to control potato-root eelworm is 400 lb/acre (18 oz/100 ft^2). Moisture is necessary for the decomposition of dazomet, so dry soils should be watered and allowed to drain thoroughly before treatment. The surface of the soil should be 'sealed' (p. 34) after rotary cultivation, and the postfumigation precautions and cultivations mentioned in connection with metham-sodium must be observed.

Rubber gloves and boots should be worn when distributing dazomet, and inhalation of the dust must be avoided by wearing a mask. Recently, dazomet has been offered in a 'prill' formulation with virtually no dust hazard (Anon., 1968a).

SECONDARY EFFECTS OF SOIL DISINFESTATION

Steaming, the application of methyl bromide, and use of methylisothiocyanate-type fumigants causes an increase in the mineralization and availability of nitrogen to plants. Such a phenomenon is known as the soil-amendment effect. This is most pronounced when glasshouse soils are treated following several years without treatment, and reduction in the usual amount of nitrogen given in base fertilizers may be needed to prevent unbalanced early growth of tomatoes. The increased availability of manganese which follows steaming (which may induce phytotoxicity with some soils) does not occur when soils are treated with metham-sodium, or, by inference, with dazomet. The effect of methyl bromide on the availability of manganese is unknown.

Cultural and biological control

The measures needed to prevent infestation with potato-root eelworm are evident from the description of its biology (p. 243), as is the long period needed to starve out the pest. Re-soiling to the necessary depth of at least 2 ft is uneconomic commercially, and there is always the danger that some cysts will remain. On small nurseries growers may find it convenient to grow tomatoes in containers of eelworm-free soil, isolated from the infested soil by polythene sheets or other means.

The grafting of tomatoes on to vigorous rootstocks which are resistant to certain diseases can partly overcome the eelworm problem. This, however, is only a palliative, as the rootstocks merely tolerate the eelworms which may then increase to seriously affect future plantings. (See also p. 248.)

Although the exudates from tomato and potato roots stimulate eelworm hatching (p. 247), the variable rate of eelworm response prevents control by growing these plants for a short period to encourage larvae to emerge and be either removed with the plants or starved in the soil. The use of synthesized hatch-stimulants also seems unlikely to be successful in practice. There seems little likelihood of developing a practical control based on the use of 'enemy' plants such as *Tagetes* spp. or *Crotalaria* spp., for these have little effect on potato-root eelworm in the soil; predaceous (eelworm trapping) fungi have never been shown to exert a useful effect on a commercial scale.

CONTROL OF OTHER ROOT-ATTACKING EELWORMS

Peachey & Brown (1965) obtained control of the dagger nematode, *Xiphinema diversicaudatum*, in a glasshouse soil from which infested roses had been removed, by injecting 7 ml. of DD, 6 in. deep and at 12 in. intervals (800 lb/acre). No living eelworms were found in samples of soil taken to 24 in. depth. It is probable that DD, on account of its deep penetration, will give better control of this eelworm than normal steaming. The DD was applied in November, when the soil temperature was 50 °F (10 °C). Roses were planted 13 weeks later (after one rotary cultivation at 4 weeks, and digging in farmyard manure at 5 weeks) and appeared to be uninjured even though the soil smelt strongly of DD. As glasshouse roses are usually cropped for 10 years, the ultimate value of the treatment is, as yet, unknown. Methylisothiocyanate-type nematicides are not sufficiently effective against *Xiphinema*. Steaming is normally undertaken before replanting with roses, and it is possible that satisfactory control would be obtained by injecting 400 lb/acre of DD at 10–12 in. depth, combined with steaming to destroy eelworms in the upper soil.

Although useful control of dagger nematodes resulted from the application of 5·8 gal/acre of dibromochloropropane (Nemagon) as an emulsified drench to soil in which roses were growing, the ultimate success of this procedure is not reported. This compound, although toxic to herbaceous plants, is safely used at 1–2 gal/acre in the U.S.A. on some woody nursery stock, e.g. against *Pratylenchus* spp., *Tylenchorhynchus* spp. and *Trichodorus* spp. It may be worth trying on infested woody stock plants required for propagating purposes. Maggenti *et al.* (1967) showed that Telone (1,3-dichloropropene and related chlorinated C3 hydrocarbons) was effective against *Pratylenchus penetrans* when used in a split application at 30 gal/acre followed after 2 weeks by 10 gal/acre.

Stem cuttings from root-infested plants will be eelworm free if any soil splashes are removed by thorough washing. Some stock plants may be freed from root-invading eelworms by heat treatment for 30 min at 118–120 °F (47·8–48·9 °C) after washing all soil from their roots. Soil used for propagation and potting should always be heat sterilized to ensure freedom from both nematode and fungal pathogens.

REFERENCES

ANON. (1962). Soil sterilization. Bull. No. 22. *M.A.F.F., Lond.* 49 pp.

— (1965). Fumigation with methyl bromide. Precautionary measures. H.M.S.O., London. 8 pp.

— (1968a). *Boots Field and Glasshouse Soil Steriliser, Dazomet.* Boots Farm Sales Ltd., Nottingham.

— (1968b). Safe and efficient fumigation practice. M.A.F.F., Lond. 5 pp.

ASHLEY, M. C., LEIGH, B. L. & LLOYD, S. (1963). Action of metham-sodium in soil. II. Factors affecting the removal of methyl isothiocyanate residues. *J. Sci. Fd Agric.*, **14**, 153–61.

BAKER, K. F. (1962). Principles of heat treatment of soil and planting material. *J. Aust. Inst. agric. Sci.*, **28**, 118–26.

FLEGG, P. B. (1968). Methyl bromide and mushrooms. *M.G.A. Bull.* **228**, 582–4.

GALLEY, D. J. & HAGUE, N. G. M. (1967). Some factors affecting the efficacy of methyl bromide fumigations of glasshouse soils. *Proc. 4th Brit. Insecticide & Fungicide Conference*. pp. 56–62.

GOODEY, T. (1945). *Anguillulina dipsaci* on onion seed and its control by fumigation with methyl bromide. *J. Helminth.*, **21**, 45–59.

GRAINGER, J. & KENNEDY, R. R. (1964). Temperature distribution and performance in balloon-sheet soil steaming. *Hort. Res.* **4**, 27–41.

HAGUE, N. G. M., LUBATTI, O. F. & PAGE, A. B. P. (1964). Methyl bromide fumigation of nematodes in soil under gas-proof sheets. *Hort. Res.*, **3**, 84–101.

HAGUE, N. G. M. & SOOD, U. (1963). Soil sterilization with methyl bromide to control soil nematodes. *Pl. Path.* **12**, 88–90.

HAMS, A. F. & COLLYER, J. (1964). Methods of incorporating dazomet into soils in relation to biological efficiency. *Proc. 3rd Brit. Insecticide & Fungicide Conference* (Brighton, 1963). pp. 225–39.

HUGHES, J. T., READ, W. H. & SMITH, R. J. (1962). Soil sterilization studies. *Rep. Glasshouse Crops Res. Inst.*, (1961), 83–6.

MAGGENTI, A. R., HART, W. H. & LENZ, J. V. (1967). Soil treatments for the control of root-lesion nematode (*Pratylenchus penetrans*) on Easter Lily. *Pl. Dis. Rptr*, **51**, 549–52.

MARTIN, G. C. (1968). Control of *Meloidogyne javanica* in potato tubers and *M. hapla* in the roots of young rose bushes by means of heated water. *Nematologica*, **14**, 441–6.

PEACHEY, J. E. & BROWN, E. B. (1965). Ridding a glasshouse soil of dagger nematode before planting roses. *Expl Hort.*, **13**, 45–9.

PEACHEY, J. E. & CHAPMAN, M. E. (1966). Chemical control of plant nematodes. *Technical Communication No. 36. Commonwealth Agric. Bureaux. Farnham Royal.* 119 pp.

PEACHEY, J. E., COLLINGWOOD, C. A. & RAO, G. N. (1963). Control of potato-root eelworm in a glasshouse soil with a formulation of methyl isothiocyanate. *Pl. Path.*, **12**, 84–7.

PEACHEY, J. E., GREEN, C. D. & GREET, D. N. (1965). Effects of DD and Nemagon on eight-year-old rose bushes infested with dagger nematode. *Expl Hort.*, **13**, 49–50.

PIEROH, E. A., WERRES, H. & PASCHKE, K. (1959). Trapex—a new nematicide for soil disinfection. *Anz. Schädlingsk.*, **32**, 183–9.

READ, W. H., HUGHES, J. T. & SMITH, R. J. (1962). Some recent investigations with chemical soil sterilants. *Proc. 2nd Brit. Insecticide & Fungicide Conference* (Brighton, 1961). pp. 361–9.

BUD AND LEAF EELWORMS

General

Several species of *Aphelenchoides* are included under the general title of bud and leaf eelworms (Table 8.2) and collectively they attack hundreds of plant species. They are often of considerable nuisance value, and one species, the chrysanthemum eelworm, *Aphelenchoides ritzemabosi*, can be a very serious pest. All have a similar, simple life history. They commonly live as endoparasites in the natural spaces between cells within the parenchymatous tissue of aerial parts of the plant; they rarely attack roots. In buds, shoots and leaf axils some species live ectoparasitically, sheltering between the tightly folded or adpressed leaflets, or in the depths of the angle between a leaf petiole and plant stem. All these habitats are similar, for the eelworm is always surrounded by plant cells whose contents are sucked for food, and it is enclosed in a protected, moist, micro-environment. Here, the females lay eggs which hatch, after a few days, to release larvae which closely resemble the parent except in size and absence of reproductive organs, and in about 2 weeks the larva matures and the newly adult females commence egg laying. Female bud and leaf eelworms are each believed to lay only 20–40 eggs, but the short life cycle compensates for this low fecundity and enormous numbers of nematodes—e.g. 10,000 in one leaf—are commonly found. Some bud and leaf eelworms feed only on the contents of plant cells, but others are also fungal feeders.

While the plant surfaces are dry, bud and leaf nematodes are restricted to their enclosed plant habitat, but when a plant is wetted (as by rain, dew or irrigation water) so that water films persist, eelworms emerge from open stomata or wounds and swim actively over the plant surfaces in the films.

TABLE 8.2. Bud and leaf eelworms

Aphelenchoides besseyi	Christie, 1942
**A. blastophthorus*	Franklin, 1952
A. fragariae	(Ritzemabos, 1890); Christie, 1932
A. lilium	Yookoo, 1964.
A. ritzemabosi	(Schwartz, 1911) Steiner & Buhrer, 1932
A. subtenuis	(Cobb, 1926) Steiner & Buhrer, 1932

* This species is only important as a pest of Scabious (*Scabiosa caucasica*)—an outdoor crop not within the scope of this book.

Rapid eelworm spread is achieved both in this way and by the transport of eelworms in excess water dripping from the plant; as the plant dries eelworms re-enter and infest healthy tissue. Eelworms washed from a plant may invade the basal growth of that and other lower growing herbs; at least one species (*A. ritzemabosi*) tends to move upwards in water films and will migrate, from soil, up host-plant stems to invade lower leaves.

Aphelenchoides ritzemabosi (Schwartz, 1911) Steiner & Buhrer, 1932 (chrysanthemum eelworm) is widespread in the temperate zones of the world. Its importance in modern chrysanthemum growing has perhaps declined with the advent, and increase, of year-round (YR) bloom production under glass, for, on the whole, YR plants are grown under glass where conditions are less favourable to eelworms than are those outdoors. The incidence of eelworm is also very low, or non-existent, in the cuttings supplied to the YR grower. Nevertheless, those who raise cuttings, or who grow chrysanthemums of 'early outdoor' and 'late' varieties in the traditional manner, often find that eelworm is a serious problem—especially when expensive stocks of new varieties are devastated. The eelworm problem is always worse in a wet, cool season, for eelworm spread is then encouraged by water films in which the eelworms migrate.

Chrysanthemum eelworm is a bisexual species which does not reproduce parthenogenetically. The sex ratio averages approximately 1 male to 5 females. The female is 0·77–1·2 mm long, the male 0·71–0·93 mm; both sexes are about 20 μ wide. The long slender adult is active, it swims with a wide sinuous movement and moves rapidly in water films on the leaf surface where leaf hairs assist its progress. The eelworm is colourless and translucent and cannot be seen in plant tissue with the unaided eye; it is just visible when suspended

FIG. 8.6. Chrysanthemum eelworm 'leaf-blotch' (healthy leaf, *centre*). (J. Hesling.)

in water in a transparent container and held against the light (Figs. 8.7 and 8.9).

The female eelworms lay eggs where they feed, and breeding occurs not only within leaves but also in leaf axils, growing shoots and flower buds. Eggs hatch in 3 days, the larvae mature and egg laying begins in a further 10–15

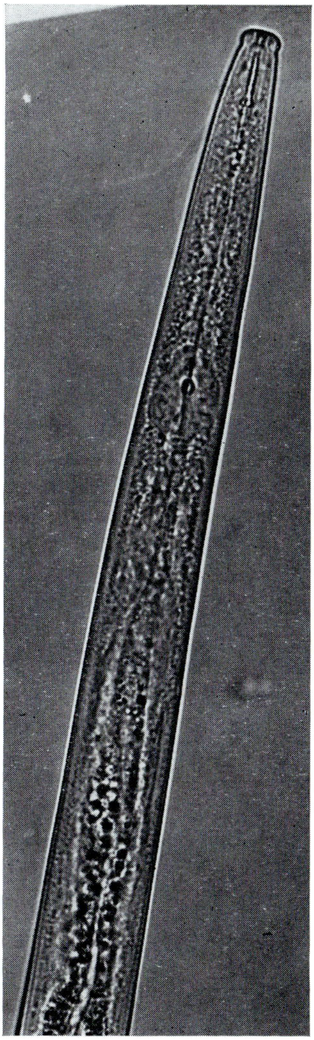

FIG. 8.7. Head of *Aphelenchoides ritzemabosi* (× 750). (J. Hesling.)

days; development times are dependent on temperature and other factors. Because female eelworms tend to migrate into undamaged tissue for egg laying the larvae are afforded a better chance of successful development.

Symptoms may be found on plants at all stages of growth. Damage to 'resistant' varieties (no variety is *completely* immune) is sometimes not recognized as due to eelworm. Some of the eelworm symptoms on cuttings and young plants are insufficiently known.

The 'classical', best-known symptom of eelworm attack is interveinal leaf-blotching (Fig. 8.6). This is caused by eelworm feeding within the leaf and is usually first noticed on the lower leaves of older plants. It begins as yellow areas on the leaf (which at first are often dotted with irregularly shaped and spaced minute brown spots), demarcated by the main veins. (In the leaf, the cells in and around the veins have thicker walls and there is less intercellular space so that this region is not so readily penetrated by the nematode

FIG. 8.8. Terminal bud of wallflower seedling damaged by *Aphelenchoides ritzemabosi*. (J. Hesling.)

and its spread within the leaf is thereby hindered.) In time, and more quickly in wet weather, the yellow areas turn brown or black as the interveinal areas die, while the disease follows a similar course in the neighbouring sectors to which the nematodes migrate via water films on the leaf surfaces. Ultimately the entire leaf dies; such leaves do not drop off the plant but hang down in ragged clusters; as each leaf may contain thousands of eelworms their prompt destruction is advisable.

Eelworms from infested leaves tend to move upwards and may migrate up the plant stem as much as 6 in. overnight; in this way the disease appears to spread upwards from the base of the plant. In severe cases the eelworm reaches

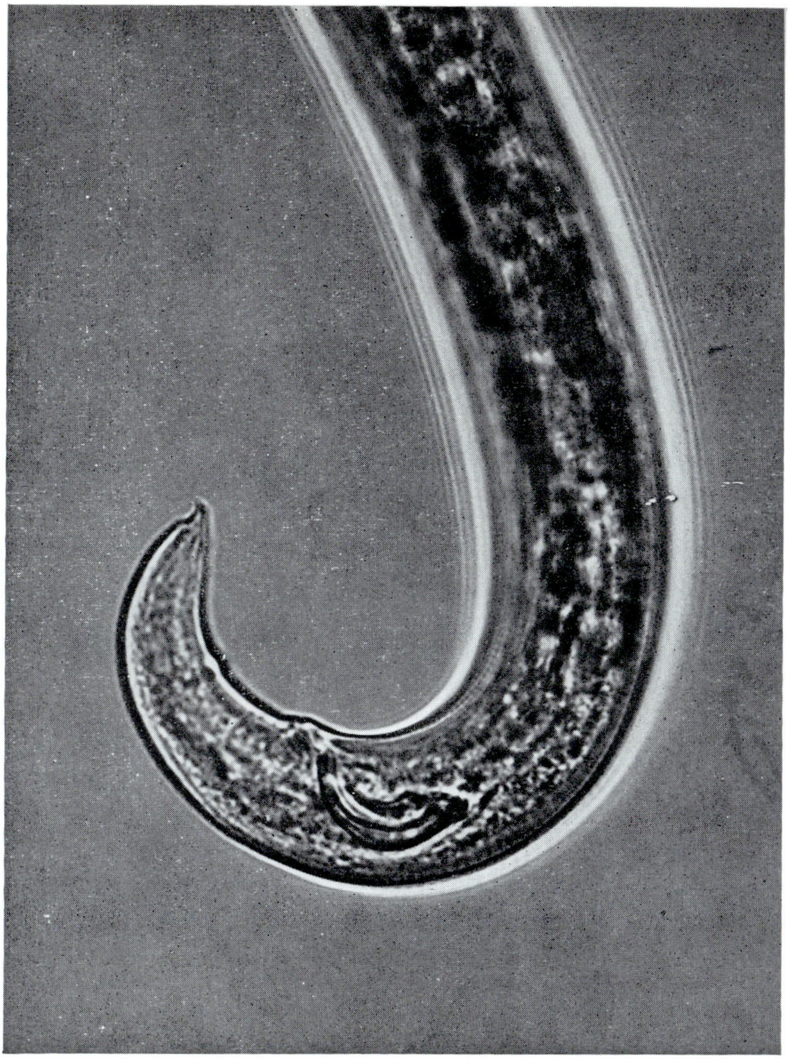

FIG. 8.9. Tail of *Aphelenchoides ritzemabosi* male (×1000).

the shoot; here, in the protection of the closely folded leaflets, it feeds externally as well as internally. At its worst, shoot damage results in a blind plant; more commonly the apical leaves are undersized and misshapen, often being discoloured, lobe-shaped and unusually thick. The flower buds and their stems may be attacked, brown necrotic patches appearing on the sepals; the flower bud may be undersized or distorted and fail to develop—in severe cases it will shrivel and die.

Any cuttings taken from an infested stock which has had no eelworm control treatment will probably be infested. Much of the eelworm damage to cuttings and young plants is the result of ectoparasitic feeding by eelworms in the shoot. Therein the nematodes browse on the epidermis of the embryo leaves and stem. This often results merely in the death of patches of superficial cells; as the shoot extends and grows the damage is revealed as brown, scarred tissue. On the expanding leaf the scars are usually on the upper leaf surface and, in more severe cases, the leaves emerge misshapen—narrow, spoon-shaped, distorted, asymmetrical, and abnormally thick. Eelworm damage to the stems is manifest as superficial ladder-like scars of brown tissue extending upwards from the base of a leaf petiole. These scars result from the extension and splitting-up of an eelworm feeding area originally in, and superior to, the leaf axil when this was enclosed within the developing shoot.

In severe cases eelworms feeding in the shoots of cuttings may kill them or completely 'blind' the young plants. 'Blinding' has a similar effect to 'stopping' the plants to induce 'breaking', and such eelworm-blinded plants may produce precocious side growth before it is required and so upset the timing of flowering. Lower leaves on young plants may show leaf-blotch symptoms. Eelworm damage to young plants and cuttings may pass unnoticed if the attack is slight; furthermore, the symptoms (apart from leaf-blotch) are not eelworm-specific and so could be confused with, or masked by, damage attributable to other pests, diseases, frost, faulty management, and phytotoxicity of insecticidal or fungicidal sprays. As few as 30 eelworms in a young plant may result in heavy damage when the plant matures. In general, any cuttings that are stunted, with short internodes and distorted leaves, should be suspect and the presence of eelworm demonstrated by immersing a leaf, shoot, or entire cutting (after washing off any adhering soil, peat, etc., and removing the roots) in the minimum quantity of water for about 12 hr. The undisturbed sediment or water from the bottom of the container should be carefully removed with a pipette, examined under a microscope and any eelworms identified. Chrysanthemum eelworm is easily distinguished from harmless eelworm contaminants of the plants by its elongate form and active swimming: its differentiation from other *Aphelenchoides* is more difficult (see p. 279).

When infested cuttings are transplanted their rapid growth masks the eelworm symptoms and such plants remain 'healthy' while temperatures are high and the plants not wetted for long periods. These conditions frequently prevail under glass and perhaps explain why YR plants appear free from eelworm. But if the plants are grown outdoors the disease will manifest itself without fail with the return of the long damp nights of autumn.

Chrysanthemum varieties vary in their susceptibility to eelworm but there are little quantitative data on the degree of resistance in any variety and no variety is completely immune. Leaf-browning symptoms appear more rapidly on resistant than on susceptible varieties and the eelworms find themselves isolated in a necrotic and unfavourable environment in which they cannot thrive, thus, 'resistant' varieties seem hypersensitive to eelworm (Wallace, 1961).

Chrysanthemum eelworm is an obligate parasite, i.e. it must feed on living plant tissue to survive and breed. Apart from chrysanthemum 160 species of ornamentals, crop plants and weeds are known to be hosts (Fig. 8.8); the eelworm is therefore common and persistent. It is not believed to feed on fungi, and if soil is completely free of all plant growth the eelworm is unlikely to persist for more than 3–4 months. If infested plant tissue dies and dries the eelworms within coil up like watchsprings and, if kept dry, will remain viable for up to 3 years. On re-wetting they resume activity and may then find their way to other host plants. Adult eelworms withstand desiccation better than larval stages. *A. ritzemabosi* does not survive in rotting or buried chrysanthemum debris.

In outdoor chrysanthemum beds the eelworm may spread up to 30 in. in a dry season and several feet in a wet one; most of this spread is passive, and results from splashing of infested water during rain or irrigation. At the season's end all parts of the plants may be infested, including embryonic basal and stem cuttings.

Control. See p. 281.

REFERENCES

FRENCH, N. & BARRACLOUGH, RUTH, M. (1961). Observations on the reproduction of *Aphelenchoides ritzemabosi* (Schwartz). *Nematologica*, **6**, 89–94.

GREET, D. N. & WALLACE, H. R. (1962). Diagnosis of eelworm attack in chrysanthemums. *Pl. Path.*, **11**, 43.

HESLING, J. J. (1961). Chrysanthemum eelworm. *Gdnrs' Chron.*, **151**, 34–5, 50–1.

SANWAL, K. C. (1961). A key to the species of the nematode genus *Aphelenchoides* Fischer, 1894. *Can. J. Zool.*, **39**, 143–8.

SOUTHEY, J. F. (1952). Unusual chrysanthemum eelworm symptoms. *Pl. Path.*, **1**, 48–9.

STURHAN, D. (1962). Über neue Wirtspflanzen der Blattälchen *Aphelenchoides fragariae* und *A. ritzemabosi*, mit Bemerkungen zu den Wirtspflanzenkreisen beide Nematodenarten, *Anz. Schädingsk.*, **35**, 65–7.

WALLACE, H. R. (1961). The nature of resistance in chrysanthemum varieties to *Aphelenchoides ritzemabosi*. *Nematologica*, **6**, 49–58.

— (1962). Chrysanthemum eelworm. *Yb. natn. Chrysanth. Soc.*, 107–28.

'Cauliflower' disease of strawberry. 'Cauliflower' disease of strawberry is the result of the combined effects of *Aphelenchoides ritzemabosi* and certain strains of a bacterium *Corynebacterium fascians* (Tilford) Dawson. In cauliflower disease the plant crowns continue to produce axillary buds, and in extreme forms the crown resembles a rosette of leaf tissue of which the greater part is short, enlarged petiole. 'Cauliflower' is predominantly a bacterial disease that is modified by eelworms. Although the nature of the disease has been clearly demonstrated in the laboratory using *A. ritzemabosi* only, *A. fragariae* may also be implicated. *A. besseyi*, and *Ditylenchus dipsaci*, (both pests of strawberry) have not been unequivocally associated with 'cauliflower'.

REFERENCE

PITCHER, R. S. & CROSSE, J. E. (1958). Studies in the relationship of eelworms and bacteria to certain plant diseases. II. Further analysis of the strawberry cauliflower disease complex. *Nematologica*, **3**, 244–56.

Aphelenchoides fragariae (Ritzemabos, 1890) Christie 1932 (strawberry eelworm; fern eelworm). In the U.S.A. this species is also known as 'spring crimp' or 'spring dwarf' nematode.

A. fragariae ranges more widely than chrysanthemum eelworm, *A. ritzemabosi*, (p. 271) for it is found not only in the temperate but also in the tropical zones of the world. Its common names suggest that it is the only eelworm which attacks strawberry and fern, but this is not so, for strawberries are also commonly attacked (sometimes simultaneously) by *A. ritzemabosi* and *Ditylenchus dipsaci*. Seventy ferns are hosts of *A. fragariae*, one (*Struthiopteris orientalis*) is also a host of *A. ritzemabosi* (Sturhan, 1962). Over 250 plants in 47 families are hosts of *A. fragariae*, and many of these are also hosts of *A. ritzemabosi*. Correct identification of the eelworm (see p. 279) is necessary for some control treatments.

A. fragariae (Fig. 8.10*b*) is a bisexual species. It is a small elongate eelworm; males range in length from 0·48–0·65 mm, females from 0·45–0·8 mm; both adults are from 10–15 μ wide.

The symptoms of attack depend on whether the eelworms are endoparasitic, as within the leaf lamina, or ectoparasitic, as when feeding on and between the tightly folded growth in a plant bud or shoot. Endoparasitic feeding within leaves usually results in leaf-blotch symptoms—yellow and necrotic areas of leaf demarcated by main veins. The form and pattern of leaf damage and its clarity varies with the leaf venation and plant species. Attacked tissues are commonly paler green, changing colour through yellow to brown as the disease progresses. Attacked plants are often dwarfed and distorted; in severe cases the leaves, or the whole plant, die. Leaves, petioles, and stems expanding from buds attacked by ectoparasitic eelworms may be misshapen, crinkled, stunted, and show brown scars—eelworm 'feeding-areas'; sometimes the bud is killed and the plant 'blinded'. In general, damage by *A. fragariae* and *A. ritzemabosi* is very similar.

There are reports of damage by *A. fragariae* to many glasshouse plants but the attacks are rarely sufficiently common, or severe, for them to rank as a serious crop problem under glass. Nevertheless, isolated cases may cause financial loss. Franklin (1965) records serious damage to mint brought into glasshouses for forcing; on this plant the eelworm was endoparasitic in lower leaves, where it caused blackened patches, and ectoparasitic in the terminal buds so that internodes were shortened and young leaves rendered distorted, brittle and easily shed.

Strawberry is the main horticultural crop attacked by this nematode. In the U.S.A. the disease it causes is known as 'spring dwarf' or 'spring crimp'. 'Red plant' is another name for the disease in Europe and is so called because foliage of some varieties turns red when attacked. The symptoms which give the disease these names have been recorded wherever strawberries are grown; they are stunted plants with deformed stems, leaves, and flowers. Damaged plant parts range in size from rudimentary to almost normal. Leaf edges may be crinkled, the petioles less hairy than normal; red plant symptoms may be present. Runners from affected plants are usually infested.

On some ferns *A. fragariae* produces striking and characteristic leaf-blotch symptoms. Sharply defined dark-brown or black areas appear on the fronds; on *Pteris* sp. the leaf-blotches are almost chevron-like. These symptoms are most marked in the winter months; when plant growth is active the symptoms are largely masked.

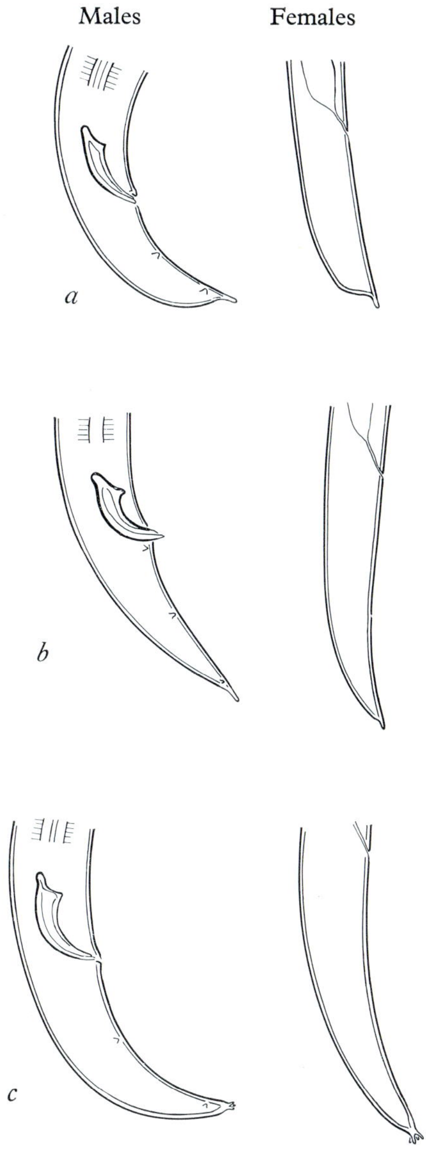

FIG. 8.10. Tails of: (*a*) *Aphelenchoides subtenuis*; (*b*) *A. fragariae*; and (*c*) *A. besseyi*.

A. fragariae breeds within the tissues it inhabits; many thousands of eelworms may be found in plant buds. The life cycle from egg to adult occupies 10–14 days. *A. fragariae* can also feed on some fungi, though it does not thrive well on them; nevertheless, this faculty may enable it to persist in soil. It is said not to survive in dead leaves.

Little is known about the extent of its spread but this is, no doubt, enhanced in wet, cool weather, when the eelworm moves in water films on foliage, and is passively transported in water draining from foliage. *A. fragariae* becomes endoparasitic by entering leaves through a stoma. Apart from plants already mentioned, it has been reported damaging ornamentals such as begonia, ferns, gloxinia, Croft or Easter lily (*Lilium longiflorum* cvs) Bellingham hybrid lilies (*L. aurelianense* hybrids), peony, *Primula denticulata* and violet. *A. fragariae* is clearly polyphagous so its appearance on almost any plant should not give cause for surprise.

REFERENCES

BALL, E., MANN, C. E. T. & STANILAND, L. N. (1927). Strawberry investigations at Long Ashton, *II. J. Minist. Agric. Fish.*, **34**, 627–41.

COURTNEY, W. D. (1945). Nematode infection of Croft Easter Lilies. *Phytopathology*, **66**, 572.

FRANKLIN, M. T. (1965). *Tech. Bull. Minist. Agric. Fish. Fd*, **7**, 2nd ed., pp. 131–41. H.M.S.O., London.

STRÜMPEL, H. (1967). Beobachtungen zur Lebensweise von *Aphelenchoides fragariae* in Lorraine-Begonien. *Nematologica*, **13**, 67–72.

STURHAN, D. (1962). Uber neue Wirtspflanzen der Blattälchen *Aphelenchoides fragariae* and *A. ritzemabosi*, mit Bemerkungen zu den Wirtspflanzenkreisen beider Nematoden. *Anz. Schädlingskunde*, **35**, 65–7.

Aphelenchoides besseyi, Christie, 1942 (summer crimp nematode of strawberry). This species thrives best in a warm climate, as in the Atlantic Seaboard of the southern U.S.A., where the strawberry disease it causes is known as 'summer crimp' or 'summer dwarf'. The nematode is also an important parasite of rice, and seems widespread in the world. It is not known in Britain, but if accidentally introduced it will most likely be found under glass where the environment should favour its survival.

The nematode (Fig. 8.10*c*) is small and elongate; males range in length from 0·6–0·9 mm, females are 0·45–0·74 mm long; both sexes are approximately 13–15 μ wide. Development from egg to adult occupies about 2 weeks.

On strawberry plants *A. besseyi* is most commonly ectoparasitic in buds; it thrives well in hot weather but its numbers decline under cool conditions. Like other bud and leaf nematodes *A. besseyi* probably moves in moisture films on foliage. It is a persistent eelworm, for the pre-adult larva is able to survive about 2 years in dry leaves and the species can be cultured on the fungi *Alternaria*, *Curvularia*, *Helminthosporium* and *Fusarium*. Christie (1959) has suggested that it might survive on fungi in plant-free soil. On strawberry, symptoms of *A. besseyi* are a plant with a spidery appearance, this results from damage to developing leaves within the plant crown. Leaves are deformed and undersized with crinkled margins and a reddish colour cast on the veins, petioles and leaf serrations. Leaves are brittle, generally darker green than normal and the petioles almost denuded of plant hairs.

Runners from affected plants may be infested. Unlike *A. fragariae* and *A. ritzemabosi*, *A. besseyi* does not seem to be associated with cauliflower disease of strawberry (p. 277).

Knowledge of the host range of *A. besseyi* seems limited; chrysanthemum, hydrangea, orchid (*Vanda* spp.), coleus, dahlia, *Ficus* sp. and zinnia are amongst its hosts.

REFERENCES

CHRISTIE, J. R. (1959). *Plant nematodes. Their bionomics and control.* Gainesville, University of Florida. xi, 256 pp.

SHER, S. A. (1954). Observations on plant parasitic nematodes in Hawaii. *Pl. Dis. Reptr*, **38**, 687–9.

Aphelenchoides subtenuis (Cobb, 1926) Steiner & Buhrer, 1932 (=*A. hodsoni* Goodey, 1935). This species is known principally from narcissus but has also been recorded from *Colchicum autumnale*, crocus, iris, phlox, scilla, and tulip. It attacks all parts of the plant, including roots. In infested narcissus, leaves are yellow and may be split, and the bulbs show brown-ring symptoms (see p. 290). Very large eelworm populations may be found in infested tissue.

A. subtenuis is also a fungal feeder and so might be expected to be more common than appears the case. In fact, cases of damage seem relatively rare and the eelworm cannot be rated as a serious pest. It is included here because it occurs in bulb stocks and its damage might be confused with that caused by *Ditylenchus dipsaci* (see p. 289).

A. subtenuis (Fig. 8.10*a*) is a bisexual species, the males are 0·75–1 mm in length, and slender, being 16 μ wide. The female is 0·8–1·3 mm long and about 20 μ wide, see p. 279.

REFERENCE

GOODEY, T. (1935). *Aphelenchoides hodsoni* n.sp., a nematode affecting narcissus bulbs and leaves. *J. Helminth.*, **13**, 167–72.

Aphelenchoides lilium Yookoo, 1964. This nematode was found in diseased bulbs of *Lilium speciosum*. It can be distinguished from *A. fragariae* and *A. subtenuis* by its stouter body, shorter post-vulval sac, the position of the excretory pore and shape of the mucro on the female tail.

REFERENCE

YOOKOO, T. (1964). On a new species of *Aphelenchoides*, parasite of bulb of lily from Japan, *A. lilium*, n.sp. *Agric. Bull. Saga Univ.*, **20**, 67–9.

Control of bud and leaf eelworms

In addition to the older method of hot-water treatment, systemic nematicides provide another method of control in some plants. So far, systemic compounds have been tried on relatively few plants, but there is promise that they can be used on a wide range of species not amenable to heat treatment. Although leaf and bud eelworms usually occur under glass only because plants have been propagated from diseased stock, it is nevertheless advisable to sterilize all composts because the eelworms may be expected almost anywhere. Where a susceptible crop is to be planted into probably infested ground, the soil should be disinfested by steaming (p. 260) or treatment with DD or methyl bromide (pp. 261, 263).

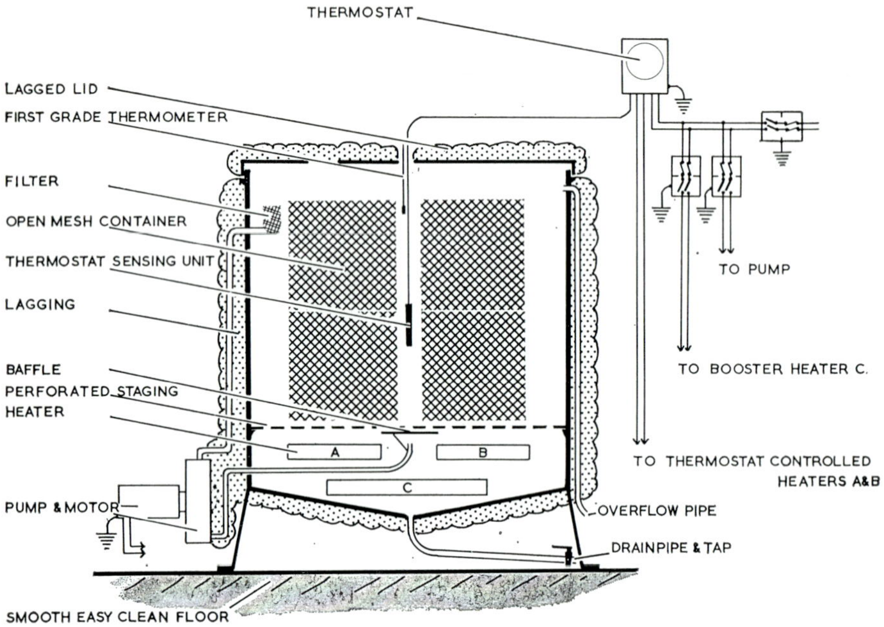

FIG. 8.11. Diagram of electrically-heated, temperature-controlled water bath for hot-water treatment to control eelworm in bulbs and other plant material (see pp. 282, 295). The pump should circulate the water four to five times an hour to ensure even temperature distribution. The booster heater is used only to raise the bath temperature quickly between treatments. All electrical equipment must be safely earthed.

Hot-water treatment

Aphelenchoides ritzemabosi can be controlled in chrysanthemum stools and healthy cuttings obtained, by immersing the stools in hot water as soon as the plants are dormant. As this method ideally requires large baths with efficient water circulation and precise temperature control (Fig. 8.11), it is now being superseded by chemical control techniques, particularly on smaller nurseries. Furthermore, the small stools produced in modern 'direct-planting' cultural procedures are less able to withstand hot-water treatment than those grown by traditional methods.

Before immersion the stools must be carefully washed free of soil and debris, and any green shoots removed. The stems should be cut back to about 3 in. unless stem cuttings need to be taken, when about 9 in. should be left. Washed stools should be allowed to drain thoroughly before immersion. The standard temperature (which should be maintained by an accurate thermostat and be checked by a thermometer of known accuracy) and time of immersion is 115 °F (46·1 °C) for 5 min, although the older treatment of 20–30 min at 110 °F (43·4 °C) is recommended for 'Loveliness' and its sports, which are liable to damage. There is only a small margin between the heat exposures lethal to eelworms and those damaging to plants, so there is always a risk of damage if the treatments are extended. To ensure even heat distribution, large mesh containers should be used, and the bath should not be over-loaded (if possible do not exceed 3 medium-sized stools/gal). Immediately on removal from the bath the stools should be plunged in clean cold water, allowed to drain, and boxed or potted, using a clean compost with a neutral pH and low pC. Treated stools should be grown at about 50 °F (10 °C) and watered only lightly until shoots appear. This will reduce both the slight retardation in cutting production resulting from the treatment, and minimize losses due to fungi. Those unable to obtain the highly poisonous nematicides mentioned later can obtain satisfactory results by immersing the cleaned stools in water at 118 °F (48 °C) and removing them 5–5½ min later at about 115 °F (46·1 °C). This entails some experimentation to ensure that the temperature of the water (which is not heated during the treatment) falls by the required amount in the stated time. Chrysanthemum stool miner is also controlled by these treatments (p. 170).

Some other plants, or plant material, can also be freed of leaf and bud eelworms by hot-water treatment but in most cases it is better to obtain plants free from eelworm. With runners of mint (*Mentha villosa-nervata*) for forcing there are advantages from heat treatment since not only is complete control of mint rust (*Puccinia menthae*) obtained but also an increase in shoot production, which reduces the number of runners required. Immersion for 10 min in water maintained at 115 °F (46·1 °C), followed by immediate rapid cooling, will eliminate *A. fragariae*. Treatment should be applied to the washed, dormant runners shortly after they have been lifted.

Strawberry runners infested with *Aphelenchoides* spp., or with the stem and bulb eelworm *Ditylenchus dipsaci*, can be freed from these by hot-water treatment. Treatment of plants destined for forcing is undesirable but eelworm-free stocks can be obtained by immersion of runners in water maintained at 115 °F (46 °C) for 10 min or 7 min if infested with *Aphelenchoides* or *D. dipsaci* respectively. Plants should be treated shortly after lifting; runners which have been lifted some time may be severely checked. They must be immediately plunged into cold water after treatment and planted out into eelworm-free soil as soon as possible. Firm planting at the correct depth is essential.

Begonias in small pots are said to withstand immersion for 3 min in water at 117–119 °F (48 °C), while ferns such as *Asplenium*, and similar species, can be immersed for 10–15 min at 110 °F (43·5 °C). Most plants, however, are seriously injured or killed by hot-water treatments of sufficient severity to destroy eelworms.

The reinfection of treated plants by eelworms from debris, containers, soil or other plants must be prevented by strict attention to hygiene.

As leaf and bud eelworms move in films of water on the leaves and stems, cultural conditions influence the attack. Keeping the foliage dry greatly restricts eelworm movement.

Chemical control

Repeated applications of parathion (0·01%) on infested foliage have some value but they very seldom eradicate eelworm and are unsatisfactory on stock plants. Applications should be made at intervals of 2–3 weeks. Where strawberry runners for forcing need to be taken from an infested stock it is advisable to dip them in 0·1% parathion. Protective clothing requirements must be strictly observed: it is necessary to wear rubber gloves when handling runners, even when they have dried.

The systemic nematicide thionazin is more effective than parathion, but little is yet known about its action on most glasshouse plants. It has proved a simpler and equally effective alternative to hot-water treatment for the control of chrysanthemum eelworm. Thionazin is absorbed through the roots of plants and kills eelworms within the plant tissues.

Infested cuttings, when well-rooted, are drenched with 0·03% thionazin. Two applications 3 weeks apart are recommended, but trials by Moreton (1966) have shown that control may be obtained by a single application of 0·02% thionazin applied to cuttings at the rate of 10 pints/yd^2, (1 pint/standard seedbox). Drenching the stools has not given such good results. Drenches of thionazin can be applied if the plants are found to be infested after planting-out. Leafminer, red spider mite and *Myzus persicae* are also susceptible to this material. Thionazin is extremely poisonous and the recommended precautions must be observed.

When no safe (to the plant) alternative method is available for controlling leaf and bud eelworm on an ornamental pot-plant, drenching the soil with 0·02% thionazin may be a rewarding experiment, but it must be remembered that plants susceptible to injury by drenches of other organophosphorus compounds are likely to be injured.

REFERENCES

ANON. (1965). Eelworms on strawberries. *Advis. leaflet No. 414 M.A.F.F. Lond.* 8 pp.

— (1967). Hot-water treatment of plant material. *Bull No. 201 M.A.F.F.* H.M.S.O., London. 42 pp.

HESLING, J. J. (1961). Problems in the routine hot-water treatment of chrysanthemum stools. *Pl. Path.*, **10**, 139–41.

MORETON, B. D. (1966). Personal communication.

STANILAND, L. N. (1950). Experiments on the control of chrysanthemum eelworm (*Aphelenchoides ritzemabosi*, Schwartz) by hot-water treatment. *Ann. appl. Biol.*, **37**, 11–18.

Ditylenchus dipsaci (Kühn, 1857) Filipjev, 1936 (stem nematode, stem (and bulb) eelworm). This eelworm is an important plant pest. In some of its four hundred host plants it increases to astronomical numbers, it is able to persist for many years even under adverse conditions, and only a few eelworms are needed to set up an infestation in a crop. Stem eelworm is distributed throughout the temperate zones of the world and also occurs in some warmer countries. Its world-wide distribution has doubtless been aided by man.

Different isolates of the eelworm thrive on different host plants, and so it is accepted that the eelworm exists in perhaps twenty or more 'biological races' or 'strains'. For example, the 'narcissus race' thrives on narcissus, the 'hyacinth race' on hyacinth; eelworm from narcissus does not thrive on hyacinth and vice versa. Distinct common names, such as narcissus eelworm, tulip eelworm and phlox eelworm are frequently used. But no distinguishing features, other than host range, have yet proved reliable for race differentiation. Recently, increased knowledge of the eelworm's biology, and of the interbreeding and the host range of the races, has suggested that races may arise by selection from a heterogeneous background population, that host preferences are genetically controlled, and that new races with slightly differing food preferences may be constantly in evolution. If this is so, it perhaps explains the many records of apparently spontaneous outbreaks of this eelworm, and also increases the difficulty of giving sound advice on crop rotations.

Stem eelworm is elongate with a pointed tail. Larval and adult stages of the eelworm are active, all carry a slender mouth-spear with well developed basal knobs. Adults are 1·0–1·3 mm in length and about 30 μ wide; being colourless and translucent they are invisible in soil or within plants. The male has a well-developed bursa which extends almost to the tail (Figs. 8.12, 8.19); in the female the post-vulval sac extends about half-way to the anus (Figs. 8.14, 8.21).

Stem eelworm feeds and breeds only on living plants. In some hosts breeding is very rapid; often, these are the 'type-host plant' of a particular race and the presence of eelworm in them is easily noticed because of characteristic symptoms (see below). In many plants, however, the eelworm breeds slowly, and symptoms are not outstanding; such host plants are easily overlooked.

The nematodes live almost entirely as endoparasites—eggs being laid where the female eelworms feed. Males are necessary for breeding and, in narcissus, the sex ratio is roughly 3 females:2 males. Each female may lay several hundred eggs, development to adult takes about 21 days at 65 °F (18 °C); the rates of breeding and development vary with temperature.

In suitable plants the annual eelworm increase may be 10–15 thousandfold, and when plant senescence or eelworm-induced necrosis occurs it is common to find that over 90% of the eelworms are pre-adult larvae. These are active forms, and resistant to adverse conditions. They can withstand frost, desiccation and summer temperatures; in the laboratory, under cool conditions 36 °F (2 °C), they may persist in an active state without food for at least 2 years. In *genuine* fallow field soil, however, persistence for over a year is unlikely. Few field soils under fallow obtained by mechanical cultivation, however, are devoid of some plant life and there can be little doubt that small residual populations of eelworm may persist for several years after an outbreak. *D. dipsaci* is said to persist longer in heavy clay than in light sandy soils but the reason for this is not known.

Stem eelworm is spread in soil contaminating footwear, machinery, etc., and is also carried in infested foliage even if this be dry and dead. The eelworm may also spread in draining surface water or in contaminated irrigation systems. These passive modes of spread contrast with the eelworms' own active spread which is probably limited to only a few feet each year. Stem eelworm moves

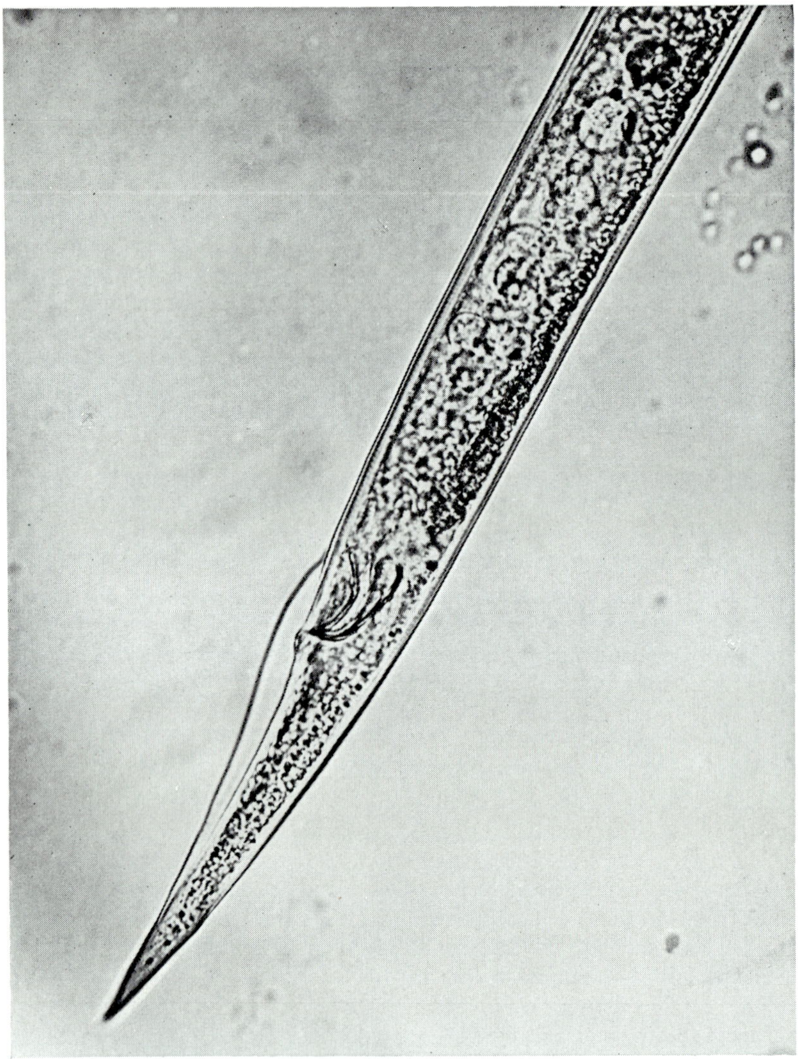

FIG. 8.12. Tail of male *Ditylenchus dipsaci* (× 600). (J. Hesling.)

both horizontally and vertically in soil; in the spring it tends to move from depth to upper soil layers. Deep-seated eelworms could therefore survive nonpersistent control treatments effective only in superficial soil layers.

The general symptoms of attack are outlined below, followed by descriptions of symptoms, and other details, of infestation by *D. dipsaci* on some important glasshouse crops.

Stem eelworm enters plants from the soil by penetrating through epidermal cells or by passing through a plant stoma or into a wound. Basal parts of plant shoots, and germinating seedlings in soil, are most often attacked. Roots are

FIG. 8.13. 'Spikkels' caused by *Ditylenchus dipsaci* on narcissus. (Courtesy of Glasshouse Crops Research Institute.)

occasionally entered but this is unusual and there seems to be no record of root damage by *D. dipsaci*. The root/shoot junction of a plant is often a preferred site; frequently cotyledonary tissue seems to be suitable even when true stems and leaves are not. Within some plants, especially the 'type-hosts' of the races, a marked plant reaction occurs. In stems, cells enlarge and become loosely attached to each other, or separate; local swelling follows and tissues assume a mealy, whitish appearance due to entrapped air in intercellular spaces, they may also be abnormally brittle or fragile. There is proliferation of side and basal shoots, while stem growth may be asymmetrical so that twisting, curling and other distortions occur. Despite local enlargement, plant growth is stunted

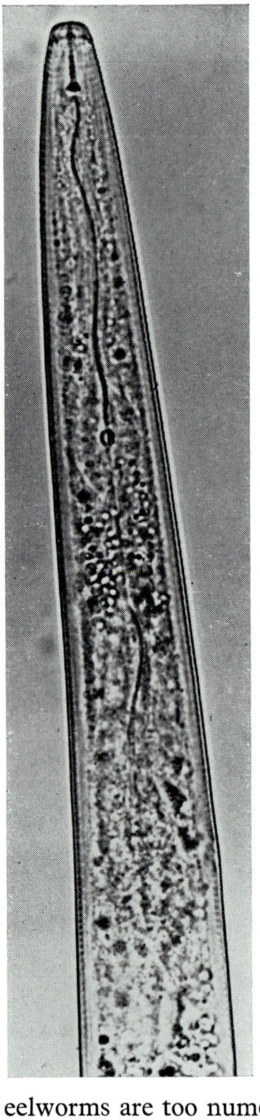

FIG. 8.14. Anterior region of female *Ditylenchus dipsaci* (×600). (J. Hesling.)

overall, and if the eelworms are too numerous, or the plant sensitive, tissues will turn yellow and die. Occasionally, leaves and petioles are attacked, they become wrinkled, curled, and brittle, and in some plants (*Lysimachia clethroides*, and annual aster) leaf-blotch symptoms appear which resemble those caused by *Aphelenchoides* spp. (see p. 272).

Symptoms of *Ditylenchus dipsaci* on narcissus

The characteristic foliage symptom is the 'spikkel'—a blister-like elongate area on the leaf which is readily detected if the leaf is passed between finger and thumb (Fig. 8.13). Spikkels are often yellow at the centre; sometimes several coalesce, then, the leaf is often unusually broad and short. Spikkels

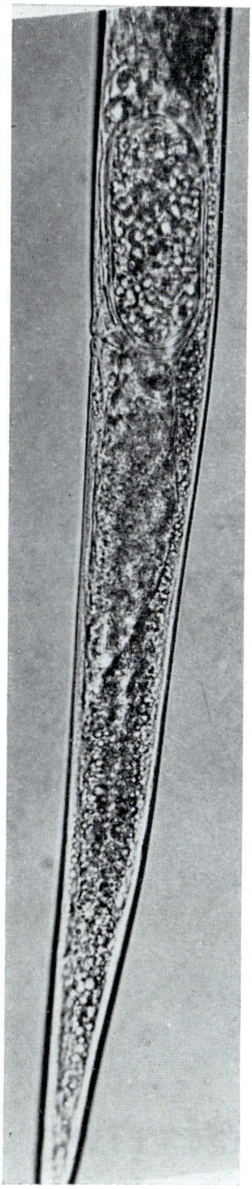

FIG. 8.15. Posterior region of female *Ditylenchus dipsaci* (× 500). (J. Hesling.)

may also be found at the edge of a leaf as a short yellow section on its margin; leaves with edge spikkels are often sickle-shaped—a symptom which can also be attributed to bulb scale mite (p. 211) or virus attack. With severe eelworm attack the foliage is dwarfed, distorted and necrotic, and the tissue may be so mealy that it almost falls apart on touch. Flowers from infested plants often have short deformed stalks and emerge at an unusual angle.

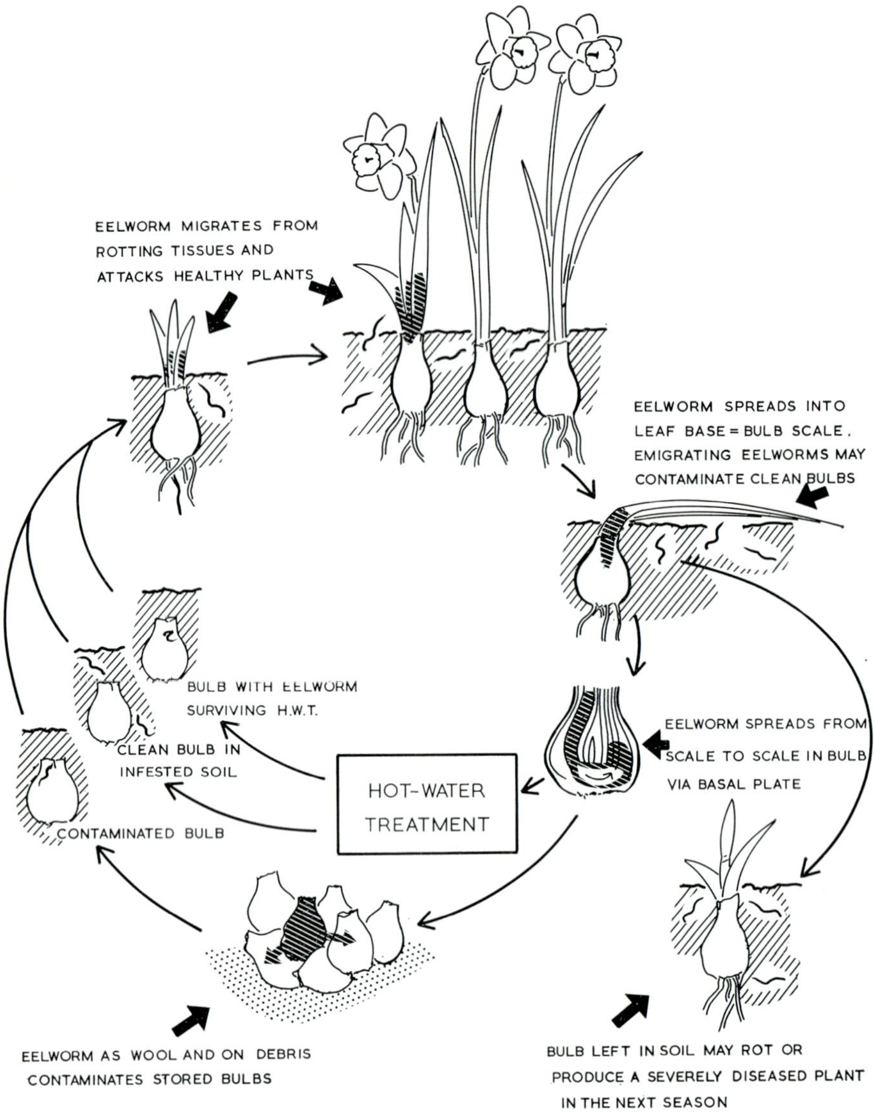

FIG. 8.16. *Ditylenchus dipsaci* and the narcissus-growing season.

If an infested bulb is lifted and cut across traversely one or more scales will be found to be chocolate to pale brown in colour, the colouration extends ring-like, in the scale, around the bulb. This 'brown-ring' symptom is a means of diagnosing eelworm in dry bulbs bought for forcing. Eelworm-infested, dry bulbs also often lack lustre and are soft near the neck; such bulbs should be cut to find out if 'brown-ring' is present. 'Brown-ring' is not a specific symptom so eelworm should be confirmed by microscopic examination of suspect tissue, selected particularly from the edge of brown area. Soon after lifting, infested bulbs may show no symptoms of softness or dullness and if eelworm is suspected

FIG. 8.17. *Ditylenchus dipsaci* 'wool' at the neck of narcissus bulb (× 4). (Courtesy of Glasshouse Crops Research Institute.)

a sample should be examined for 'brown-ring'. If eelworm is confirmed it is likely that many bulbs with no symptoms whatever may be contaminated, and it is risky to use such a stock for forcing purposes. 'Clean' bulbs cannot be sorted from infested stocks. The risk of economic loss is great with forced, infested bulbs for the eelworm thrives well in the warmer conditions. Eelworm is likely to be widely dispersed in the stock because bulbs for forcing have rarely had recent control treatment, since this sometimes incurs risk to the all-important flower crop. The bulb forcer is advised to purchase stocks that have regularly been hot-water treated and also examined while growing during previous years and found to be free of eelworm. The relationship between eelworm and the narcissus-growing season is shown in Fig. 8.16.

All trash from infested bulbs, foliage, diseased bulbs, soil from forcing boxes, and the boxes themselves will harbour thousands of eelworms. All should be burnt, or sterilized either by heat or by several hours immersion in 5% formalin. Eelworm-diseased bulbs very often are attacked by lesser bulb flies (p. 162) and by other saprophages such as some mites (p. 225), eelworms, insects and fungi. These quickly rot the bulb and may mask the original disease, for few *D. dipsaci* remain in the rotting mass; the eelworms die in the foul conditions or escape from the bulb into soil or to contaminate others. When eelworms migrate from stored bulbs they often accumulate at the exit point—usually at the bulb neck or at cracks at the base of the scales—and the creamy-white masses of eelworms so formed are termed 'wool' (Fig. 8.17). 'Wool' is comprised almost entirely of pre-adult and adult eelworms.

Narcissus is also attacked by *D. dipsaci* from tulip (see below); and the symptoms are similar to those just described. Host plants of narcissus *D. dipsaci* include broad, runner and french beans, garden pea, onions, shallots, *Brassica* seedlings, chickweed, beet, turnip.

Ditylenchus dipsaci on hyacinth

Symptoms of stem eelworm on hyacinth are very similar to those on narcissus. Leaves carry spikkels, yellow flecks or blotches, and are often dwarfed and twisted and may split. 'Brown-rings' are found in bulbs cut transversely, and swelling of the infested bulb tissue sometimes leads to a splitting of the outer scales. Flower spikes are often stunted or deformed. Forced hyacinths usually show eelworm disease most severely.

Modern, specialist, careful propagation and preparation of hyacinth bulbs has made eelworm outbreaks less common than hitherto; bulb forcers must avoid trouble by ensuring that they buy only healthy stock and do not expose it to eelworm during growth. The eelworm from hyacinth is known to breed in bluebell and *Scilla siberica*.

Ditylenchus dipsaci on Hydrangea

D. dipsaci attacks young growth of hydrangea, especially basal shoots. The first indications are slight swellings, lighter green than their surrounds, on lower leaves. Leaves, buds, and stems are invaded, and later symptoms are swollen (sometimes split) dwarfed, distorted, curled stems and leaf petioles, and curled or crinkled leaves. Infested leaves often show necrotic, interveinal patches, especially near the petiole.

When diseased shoots are taken as cuttings, propagated in moist warm conditions, and grown on under glass, the eelworms become generally dispersed and may persist to damage flowers. The stems of the florets are dwarfed and the flower 'head' is distorted or asymmetrical and may even abort (Fig. 8.18). Sometimes, stems of the florets show brown scars—this, and other flower damage, is probably the result of eelworm activity when the flower was embryonic in the shoot. Hydrangea sometimes seems sensitive to attack by *D. dipsaci*, and infested parts may simply wither and die. When infested growth

FIG. 8.18. Hydrangea damaged by eelworm *Ditylenchus dipsaci*, showing distorted, crumpled side shoot and stem lesions. (Courtesy Glasshouse Crops Research Institute.)

becomes woody, eelworms tend to disappear but may find their way into young basal growths so that the infestation is perpetuated.

It is not clear which 'race' of *D. dipsaci* attacks hydrangea. In an outbreak in Denmark an isolate from narcissus was believed to be responsible; in England the 'oat-onion race' has been blamed (French & John, 1965, *in litt.*), while in other countries the origin of the infestation was not established. Thus no reliable advice can be given on safe, alternative crops to prevent eelworm carry over. Little experimentation has been done on control; shoots produced well above ground from old woody branches are often eelworm free but some risk is still entailed if these are used for propagation. Chevalier & Scotto la Massese (1966) found that repeated applications of parathion gave some control, but with some phytotoxicity.

Ditylenchus dipsaci on strawberry

Symptoms are most marked in spring and autumn when the effect of eelworm is not masked by the luxuriant growth made in glasshouse or summer conditions. Leaflets are corrugated; crumpled and curled backwards (downwards), their marginal serration is less clear and the leaf outline irregular. Infested leaves change in colour, varying from reddish to paler green than normal, dependent on cultivar. The leaf may be brittle, its midrib wider, thickened, uneven or puffy. Leaf petioles are swollen and stunted, the swellings are almost gall-like and may show a brown core. Infested flower stalks are dwarfed, swollen, and almost sessile; attacked ripe fruits are pale in colour, loose in texture and easily squashed. Where the attack is severe the crown of the plant may die and the lateral outgrowths result in a multi-headed plant. Infested stolons are thick, stunted and distorted; runners may be infested as they grow from the mother plant, or later, from infested soil.

Races of *D. dipsaci* from onion, oat, red clover, narcissus and tulip are known to attack strawberry. Growers are advised to buy runners from producers whose stocks have been regularly inspected and found to be free from eelworm (and other pests and diseases).

Ditylenchus dipsaci on tulip

On tulip, stem eelworm symptoms are more obvious on foliage and flowers than on bulbs. Flower stems show pale lesions, often just below the flower, the stem may be thick, distorted and curved with the flower bent over towards the damage. Affected flowers fail to colour completely, the petals remain more or less green, perhaps with distortion and a silvery appearance. Infested leaves are fragile, often showing transverse and longitudinal splits; their bases may be deformed and swollen, and they may shrivel and die prematurely.

Eelworm infestation in tulip bulbs is not so easily detected as is that in bulbs of narcissus or hyacinth for no distinct 'brown-ring' symptoms are evident. Slightly infested bulbs show little sign of eelworm at first, but in storage the damage increases and soft greyish-brown patches with indistinct margins appear on the bulbs. In time, badly infested bulbs are killed; in store they rot, dry up and shrivel. Stocks containing such bulbs are likely to be generally infested.

The tulip race of *D. dipsaci* is potentially a very serious pest. It not only renders flowers and bulb stocks unsaleable but it is difficult to control because tulips may be seriously affected by hot-water treatment (see p. 295) and need careful pretreatment if they are to survive it. Tulip eelworm will thrive on hyacinth, narcissus, phlox and strawberry as well as on several common weeds.

Conversely, races from lucerne, white clover, narcissus and oats will breed on tulip (Webster, 1967).

Tulip, by its extreme reaction to *D. dipsaci*, by the necrosis of infested parts, and by its method of forming a new bulb, seems not to be a host plant ideally suited to the perpetuation of its eelworm race. It has been demonstrated (Cooper, *in litt.*) that careful inspection of tulip stocks and the removal of diseased plants can almost eliminate eelworm, and in this way, coupled with control treatments, the incidence of tulip eelworm has been reduced to a very low level in some parts of England. Growers who buy bulbs for forcing should insist that they are from healthy stock which has been found free of eelworm during previous growing seasons.

REFERENCES

ANON. (1968). Stem and bulb eelworm on tulips. *Advis. Leafl. 461. Minist. Agric. Fish. Fd*, H.M.S.O., London. 4 pp.

— (1967). Stem and bulb eelworm on narcissi. *Advis. Leafl. 460. Minist. Agric. Fish. Fd*, H.M.S.O., London. 7 pp.

CHEVALIER, M. & SCOTTO LA MASSESE, C. (1966). Essais dirigés contre l'anguillule de l'hortensia, *Ditylenchus dipsaci*. *Phytiatrie-Phytopharmacie*, **4**, 243–50.

FRENCH, N. & JOHN, MARGARET (1965). Personal communication.

GOODEY, J. B. (1951). Observations on the attack by the stem eelworm, *Ditylenchus dipsaci* on strawberry. *Ann. appl. Biol.*, **38**, 618–23.

HESLING, J. J. (1965). Narcissus eelworm—yesterday and today. *Daffodil Tulip Yb.* **1966**, 76–94.

— (1966). Biological races of stem eelworm. *Rep. Glasshouse Crops Res. Inst.*, **1965.** 132–41.

SCOTTO LA MASSESE, C. (1964). L'anguillule de l'hortensia. *L'hortensia; Journées d'études de l'Horticulture et des Pépinières, no. 1, Versailles*, **1963**, 43–60.

THORNE, G. (1961). *Principles of Nematology*. McGraw-Hill, New York and Maidenhead. pp. 115–38.

WEBSTER, J. (1967). The significance of biological races of *Ditylenchus dipsaci* and their hybrids. *Ann. appl. Biol.*, **59**, 77–84.

Control of stem and bulb eelworm

Stem and bulb eelworm may cause loss of flowers from forced plants; losses can be avoided by using only bulbs from eelworm-free stock. Such bulbs are usually obtained from stocks either hot-water or chemically treated a year or two previously, and expertly and thoroughly rogued several times during their seasons of growth. The application of sprays of thionazin (0·35%) to infested narcissus plants immediately after flowering may avoid the necessity, during roguing, of removing and destroying infested plants and treating the soil in their vicinity with DD but it is still advisable to control eelworm in the stock when it is lifted. Foliar sprays of thionazin give insufficient eelworm control in tulips.

Narcissus—hot-water treatment

The hot-water treatment of narcissus bulbs incurs some risk of flower malformation, and should only be carried out in emergency on bulbs for forcing; it should preferably be done at the right stage of development of the embryonic flowers, and after warm storage (see below). This is immediately after the paracorolla (cup and trumpet) has formed. If a bulb is dissected and examined with a hand lens the paracorolla can be found inside the base of the longer perianth segments and surrounding the more readily identified stamens. Beyer's (1942) scale shows the various stages of flower development in bulbs, and is illustrated by symbols and schematic diagrams, while excellent photomicrographs are given by Preece & Morrison (1963). The time of year when the correct stage is reached depends on the variety and district. Hot-water treatment at other growth stages may seriously affect flower quality and production. Flower damage can be reduced, but with some risk of slightly inferior eelworm control (Winfield, 1968), if the bulbs are stored for a week at 85–86 °F (29·5–30 °C) immediately before treatment. Storage below 65 °F (18 °C) will aggravate damage. In view of these considerations, hot-water treatment is more suitable where the bulbs are both grown, and later forced, by the same organization, with the necessary equipment available. Under many circumstances chemical control, as described later, is becoming acceptable. Heavy infestations so reduce the vigour of bulbs that they are unfit for forcing even though the eelworms are satisfactorily controlled.

To kill eelworms, cleaned, sorted bulbs should be immersed for 3 hr in water, plus wetter, at 112 °F (44·4 °C) (Fig. 8.11). It is very advantageous to add a nematicide and fungicide such as formaldehyde to give a bath concentration of 1 in 200 of commercial 38–40% formaldehyde (formalin). Ensure that this concentration is maintained in successive baths. Alternatively, a proprietary organomercury fungicide may be added to the bath; the precautions required when using mercurial substances must be closely observed and the procedure carried out only in registered premises. Immediately after hot-water treatment the bulbs should be spread out to dry and cooled naturally. When cool they can be planted or boxed, or dried and prepared for early flowering. The hot-water treatment will also control bulb scale mite (p. 209), bulb mite (p. 225), bulb flies (p. 160) and most other contaminant eelworms.

Treated bulbs must be strictly isolated from infested bulbs, debris and contaminated containers. Bulb trays may be disinfected by immersion in 1% cresylic acid (prepared from cresylic acid as an emulsifiable concentrate or as 'solubilized' cresylic acid), drying and airing them before re-use.

The hot-water treatment of 3 hr at 112 °F (44·4 °C) may not kill all eelworm 'wool'. Heavily infested stocks should not be forced, but should be lifted early (before wool forms) and treated as soon as possible, irrespective of the stage of flower development. They should then be planted into clean ground, carefully examined during growth and given further treatment as necessary.

Other plants—hot-water treatment

Although hot-water treatment controls eelworms in tulips it seriously affects flower production and is quite unsuitable as a pre-forcing procedure. Chemical control offers more prospect of successful application (see below). Eelworms cause little damage to the flowers from lightly infested bulbs, so these are frequently forced as an economic means of getting rid of diseased stock. Where this is done care must be taken to prevent infestation of other stocks through the careless disposal of spent bulbs and soil, or the use of contaminated boxes.

When the flower spike has formed in the bulb, hyacinths may be subjected to 4 hr immersion in water at 110·5 °F (43·6 °C) to control stem and bulb eelworm, but this treatment should be used for the production of clean stock for forcing in future years, rather than on bulbs for immediate forcing.

Sometimes valuable propagating stock of other plants may be freed from *Ditylenchus dipsaci* by hot-water treatment, even though this results in plant injury (see also p. 283). The use of chemical drenches is likely to be less injurious to many plants and much more convenient to the grower.

Chemical treatments

Immersion of infested narcissus and tulip bulbs in a bath containing the systemic organophosphorus compound thionazin ('Nemafos') has recently been shown to give good control of stem and bulb eelworm, and is officially approved for such usage. The chemical is used as a dip, at a concentration of 0·23%, and must be constantly agitated. *Cleaned* and *dry* narcissus bulbs should be immersed in this very poisonous dilution for 2½ hr, and tulips for 2 hr. The manufacturer's safety precautions must be observed. The bulbs may be temperature-conditioned for early flowering before immersion but must not be treated if root growth has commenced (cold storage promotes rooting), or be pre-soaked before dipping (see also below). It is advisable to include a fungicide (e.g. formaldehyde or organomercury compound) in the bath, and also aldrin if there is likelihood of narcissi being attacked later by bulb flies. Thionazin dips must be 'topped up' from time to time as recommended by the chemical manufacturers. Accumulation of sludge in the dip tank should be avoided because this may impair the agitation of the solution and also, some thionazin may be adsorbed on the sludge, so lessening the amount available for the bulbs. After use, thionazin must be neutralized as recommended by the manufacturers and the spent substance safely disposed of. Bulbs contaminated with eelworm 'wool', or which are soft or damaged, should be destroyed; indeed, badly infested stocks are better hot-water treated. Thionazin persists for a relatively short time in plant tissues, and sufficient may not remain long enough to kill eelworms which invade the plants later. There is evidence, however, that it may persist in some soils; it disappears more rapidly in alkaline conditions. Owing to the recent introduction of thionazin the responses of different narcissus and tulip species and cultivars to immersion treatments are not well known. Slight checking of the growth of some tulips, and spotting of leaves of some daffodils, may occur. In some cases, therefore, the treatment should only be applied as an emergency measure, e.g. if bulbs destined for immediate forcing are found to be infested. Normally, the treatment should be used to lessen infestation in stocks due for forcing in future years. Treated bulbs should be planted within a week of dipping to take advantage of the systemic action which results from root uptake as well as by absorption by the bulb tissues. Also, the dipping promotes root growth from the bulbs, and such bulbs would be damaged if not planted at once.

Drenching sprays of thionazin, shown by Oliff (1966) to effectively control eelworms in narcissus but not in tulips, are unlikely to be of value in preventing damage to forced narcissus since injury will have occurred by the time the presence of the pest is observed. Their use before flowering might constitute a hazard to workers picking flowers. It is likely, however, that safer and more effective modifications of thionazin treatments, or the use of other systemic chemicals, may result from further investigations.

Soil drenches of thionazin, or H.V. foliar sprays of the compound, may control stem and bulb or other eelworms in other hosts; the intolerance of plants to the chemical is likely to be the main factor restricting its use. Moreton (1966–*in litt.*) found that stem and bulb eelworm in some *Hydrangea hortensis* cultivars being forced under glass was almost completely controlled by drenching the soil in the pots with 0·01–0·02% thionazin, but that the conditions necessary for its safe use required further investigation as severe phytotoxicity occurred in one test. He also found that hydrangeas will not tolerate the hot-water treatment necessary to kill eelworms. It must be remembered that thionazin and similar substances are highly poisonous and are officially approved for limited use only.

Cultural control

Ditylenchus dipsaci does not usually infest roots, consequently, where root cuttings provide a means of propagation, eelworm-free plants can be obtained even from infested plants. All top growth must be removed and the root portions should be thoroughly washed before insertion in clean compost.

REFERENCES

ANON. (1969). Hot water treatment of Narcissus bulbs. *Short term leaflet* (*S.T.L.*) *No. 1. Minist. Agric. Fish. Fd*, H.M.S.O., London.

— (1967). Hot-water treatment of plant material. *Bull. No. 201 M.A.F.F.* H.M.S.O., London. 42 pp.

BEYER, J. J. (1942). De Terminologie van de Bloembolgewassen. [The terminology of the flower development in bulbous plants]. *Meded. LandbHoogesch., Wageningen*, **46**, 3–16.

CHEVALIER, M. & SCOTTO LA MASSESE, C. (1966). Essais dirigés contre l'anguillule de l'hortensia. *Phytiat.-Phytopharm.*, **4**, 243–50.

HAGUE, N. G. M. & PURNELL, R. E. (1965). The control of plant parasitic nematodes by water dispersed nematicides. Field experiments on the control of *Ditylenchus dipsaci.* (Kühn) Filipjev in narcissus bulbs. *Hort. Res.*, **5**, 65–75.

HESLING, J. J. (1965). Narcissus eelworm—yesterday and today. *Daffodil Tulip Yb.*, **1966**, 76–94.

OLIFF, K. E. (1966a). The control of plant parasitic nematodes by water dispersed nematicides. *III.* Foliar application of thionazin to narcissus and tulip to control the stem eelworm *Ditylenchus dipsaci* (Kühn) Filipjev. *Hort. Res.*, **6**, 79–84.

— (1966b). Ibid. IV. The effect of dipping narcissus bulbs infested with *Ditylenchus dipsaci* (Kühn) Filipjev in thionazin. *Hort. Res.*, **6**, 85–90.

PREECE, T. F. & MORRISON, J. R. (1963). Growth stages of the narcissus flower within the bulb. Illustrations of the Beyer Scale. *Pl. Path.*, **12**, 145–6.

WINFIELD, A. L. (1968). Control of *Ditylenchus dipsaci* on Narcissi in Eastern England. *C.r. VIIIeme symp. int. nematologie Antibes 1965.* p. 128.

WOODVILLE, H. C. (1964). Lethal times and temperatures for two species of eelworm. *Expl. Hort.*, **10**, 90–5.

Ditylenchus destructor Thorne (1945) (the potato-rot nematode; tuber nematode; iris nematode). This species has a world-wide distribution in temperate zones. Although it is probably best known as a pest of stored potatoes, it has a large host range in both outdoor and glasshouse crops as well as in weeds and so its appearance as a pest under glass should give no cause for surprise. In addition, there is good evidence that some isolates feed and thrive on many soil fungi as well as on higher plants. It is possible, however, that more than one species is involved and further research is needed to confirm this.

The inclusion of fungi in the host range precludes the use of 'fallowing' as a control—even if this were possible under glass where economics usually dictate that cropping be continuous.

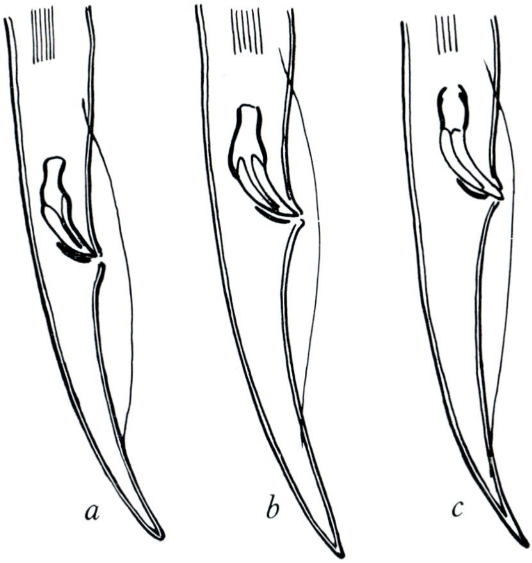

FIG. 8.19. Posterior region of males of: (*a*) *Ditylenchus myceliophagus*; (*b*) *D. destructor*; and (*c*) *D. dipsaci*.

D. destructor is essentially a parasite of bulbs, corms, tubers, roots, rhizomes and underground stems. Amongst other plants attacked are potato (often brought into greenhouses for sprouting), dahlia tubers, gladiolus corms, bulbous iris, tigridia, tomato, and tulip bulbs. Soil populations of *D. destructor* seem, often, to be very small and it has been suggested that the species over-winters in the egg stage. However, Henderson (1958) found that the nematodes accumulate in the region very close to the root (rhizosphere) where they possibly overwinter, feeding perhaps on root-pathogenic and other fungi, and Andersson (1967) showed that the eelworm can overwinter in the decaying remains of infested plants. In some cases the eelworm enters necrotic areas of root, but this is perhaps only when feeding on endoparasitic fungi.

The life history of *D. destructor* is probably very similar to that of *D. dipsaci*. Breeding seems to be continuous when conditions are favourable. The eelworm is bisexual and does not reproduce by parthenogenesis. Unlike *D. dipsaci*, there seems to be no stage when a plant's eelworm population comprises a very high percentage of pre-adult and adult eelworms. No resistant 'wool' stage has been recorded. The eelworm seems unable to resist desiccation and its long persistence in a cryptobiotic (dormant) state seems unlikely.

D. destructor probably leaves the plants it has destroyed and persists in the soil before attacking further hosts. Little is known about its spread. Draining water, and infested soil carried on implements, etc., must be responsible for

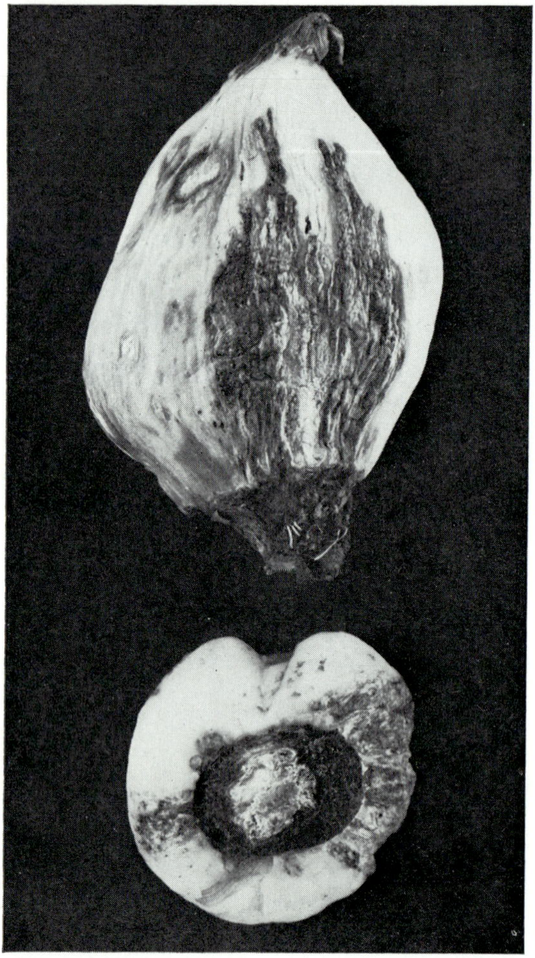

FIG. 8.20. Iris bulb showing symptoms of attack by *Ditylenchus destructor*. (Courtesy of Glasshouse Crops Research Institute.)

some eelworm dispersion, but their importance in this and in disease aetiology cannot be assessed with a species that seems polyphagous, indigenous, and already widespread in many countries. Duggan & Moore (1963) showed that the eelworm spread from infested to healthy potatoes in store, and the same can be expected in stores of bulbs, tubers, etc.

D. destructor damages a wide range of plants and no general description of symptoms seems adequate to cover those found on crop plants. Necrotic lesions on roots and stems, and soft, dry patches with rotting centres on tubers are a frequent result of *D. destructor* attack. Bulbous iris spp., which are sometimes forced under glass, are often damaged by *D. destructor*, and it seems useful to describe the attack and the symptoms in this case (Fig. 8.20).

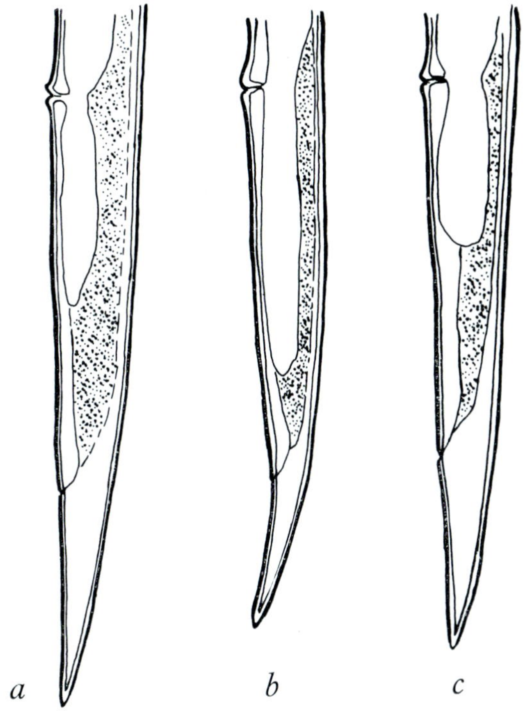

FIG. 8.21. Posterior region of females of: (*a*) *Ditylenchus dipsaci*; (*b*) *D. destructor*; and (*c*) *D. myceliophagus* (diagrammatic).

From soil, the eelworms often penetrate iris bulbs via loose tissue around the edge of the basal plate, where roots emerge. The basal plate is invaded and some eelworms pass upwards into the bulb scales while others may move downwards into the roots where they cause a grey-black rot. Heavy attack on the roots is very damaging, so that the leaves are poorly developed and yellow-tipped. Damage within the bulb is dependent both on the degree of attack and on the variety of iris; the bulb may be killed or survive to yield daughter bulbs which are often infested. Daughter bulbs are commonly invaded through the basal plate shortly before lifting, and the invading eelworms cause brownish-cream streaks of damaged tissue in the scales. The streaks darken to blackish, rotten areas which coalesce so that tissues are ultimately destroyed. The initial colour of the streaks varies with plant variety, thus, in English iris 'White Excelsior' the colour at first is light orange.

Any iris bulbs with a crevice between the basal plate and the scales, and with internal streaks or rot, should be regarded as suspect, and a microscopic examination made of tissues at the edge of internal lesions. At this edge is the eelworm 'front' from which *D. destructor* will most likely be found for identification. More-rotten tissues may contain few *D. destructor*, for secondary agents of rot—bacteria, fungi, mites, fly larvae, etc.—frequently follow attack by this eelworm.

D. destructor closely resembles *D. dipsaci* (p. 284) and *D. myceliophagus* (p. 365) in size and form. These three species are distinguishable by differences visible under the microscope (Figs. 8.12, 8.14, 8.15, 8.19 and 8.21). *D. destructor* varies greatly in size according to both its host (plant or fungus) and feeding conditions; care is needed in identification.

Control

Iris bulbs ('Wedgewood') infested with *Ditylenchus destructor* may be immersed for 3 hr in a 1 in 200 formaldehyde solution at 110 °F (43·3 °C). Pretreatment bulb storage at 73 °F (22·8 °C) for 3–14 days immediately after lifting is beneficial.

After treatment the bulbs should be spread thinly in a cool, clean, well-ventilated place for drying. Treated bulbs should be kept away from contaminated stocks and planted in uninfested soil. There is little information, if any, on chemical control methods for use on the dry bulbs. Soil fumigation, by one of the methods outlined in the section on the control of soil-infesting eelworms (pp. 259–269), may be necessary when glasshouse soils become contaminated.

REFERENCES

ANDERSSON, S. (1967). Investigations on the occurrence and behaviour of *Ditylenchus destructor* in Sweden. *Nematologica*, **13**, 406–16.

ANON. (1967). Hot-water treatment of plant material. *Bull. No. 201. M.A.F.F.* HMSO, London. 42 pp.

DUGGAN, J. & MOORE, J. F. (1962). Host-range of tuber-rot eelworm. (*Ditylenchus destructor*). *Irish J. Agric. Res.*, **1**, 109–14.

— (1963). Observations on tuber-rot eelworm (*Ditylenchus destructor* Thorne, 1945). *Ir. J. agric. Res.*, **2**, 75–86.

FAULKNER, L. R. & DARLING, H. M. (1961). Pathological histology, hosts, and culture of the potato-rot nematode. *Phytopathology*, **51**, 778–86.

GOODEY, J. B. (1951). The potato-tuber nematode, *Ditylenchus destructor* Thorne, 1945; the cause of eelworm disease in bulbous iris. *Ann. appl. Biol.*, **38**, 79–90.

— (1952a). The influence of the host on the dimensions of the plant-parasitic nematode *Ditylenchus destructor*. *Ann. appl. Biol.*, **39**, 468–74.

— (1952b). Investigations into the host ranges of *Ditylenchus destructor* and *D. dipsaci*. *Ann. appl. Biol.*, **39**, 221–8.

GOODEY, T. (1949). Tuber-rot eelworm of potato and its weed hosts. *J. Helminth.*, **23**, 89–90.

HENDERSON, V. E. (1958). Relationship between some clovers and *Ditylenchus destructor* Thorne, 1945. *Nature, Lond.*, **181**, 59–60.

KUIPER, K. & SILVER, C. N. (1959). Een proef met *Ditylenchus destructor* van verschillende herkomsten. *T. plZiekt.*, **65**, 64.

SAFYANOV, S. P. (1965). [Nutritional specialization of the potato stem eelworm] *Vest. Sel'-khoz. Nauki, Alma-Ata.*, No. 8, pp. 121–6. (In Russian).

MUSHROOM PESTS

Bionomics and Control

Chapter 9

Insect Pests of Mushrooms

Insect pests of mushrooms are often referred to under the comprehensive term 'fly'. This anonymity is due to the close relationship, and hence similar appearance, of all the species involved. However, since sound control must be based on accurate identification both growers and entomologists should learn to appreciate quite minute differences.

Order COLLEMBOLA: Springtails (see Fig. 5.1)

Springtails are not normally pests of mushrooms except where the beds are made up on the floor. Catch crops grown in glasshouses are a notable example. There, enormous numbers may develop, so many in fact, that they may appear to colour the surface of the bed. Since the commonest species are slate-blue they are known to many growers as 'gunpowder mites'.

The main characteristics of these insects, especially their habit of leaping when disturbed, are described on p. 65, as are suitable references for their identification.

A number of species have been found in mushroom beds including *Achorutes armatus* Nicolet, *Proisotoma minuta* (Tulberg), *Xenylla mucronata* Axelson, *X. welchi* Folsom, *X. corticalis* Börn, *Ceratophysella granulata* Stack, *Hypogastrura manubrialis* Tulberg, *Lepidocyrtus lanuginosus* Smelin and *L. cyaneus* Tulberg.

The true 'gunpowder mite' is *A. armatus*, a tiny (1·5 mm long), dark greenish-grey to bluish-black insect (Fig. 9.1), countless numbers of which occasionally form 'carpets' on the floor and even piles on the bed surface. Close examination shows that the dorsal surface of the body is commonly marbled while the ventral surface is yellow with dark round spots. Between the third and fourth antennal segments is a large, ventral, bilobed sac. The female lays almost white, slightly granular, eggs in groups of about 30. When laid, they are only about half normal size but, within a few minutes, they absorb water and become about 0·16 mm in diameter. At 75 °F (24 °C) the eggs hatch in 8 days and the larvae continue to moult throughout life, as many as 24 moults having been recorded in a single individual. At 75 °F (24 °C) *A. armatus* becomes sexually mature about 20 days after hatching, adults living for at least 4 months. It feeds both on the mycelium and the sporophores; minute, open pits are eaten in the stems and caps, leading to dry, branched tunnels within the tissues. Tunnelled sporophores often split along the stipe. Attacks, on pinhead or button mushrooms, prevent growth and may cause premature death.

H. manubrialis oviposits at intervals of 12–14 days producing spherical, white eggs, usually laid singly but occasionally in small groups of up to 30. At 71 °F (22 °C) these hatch in 8–10 days but may take 3 weeks at cropping

FIG. 9.1. 'Gunpowder Mite', the collembolan *Achorutes armatus* on mushroom beds (× 6). (J. Hesling.)

temperatures. Where food is abundant *H. manubrialis* is gregarious and, in ideal situations providing plenty of food, the females become sexually mature only 18 days after hatching though they do not normally reproduce for 5–7 weeks. Individuals may live for 5–10 months moulting almost every week.

Where springtails are present they are attracted to newly implanted spawn and, if only a few portions of manure spawn are used to inoculate the beds may, at least temporarily, inhibit the normal 'run' of mycelium. Since springtails have no pupal stage they feed continuously from hatching till death so that high populations may cause severe damage involving complete hollowing out and collapse of button mushrooms.

Control

As springtails favour damp situations with an abundance of decaying vegetable matter, they are particularly common in glasshouses used for cucumbers, or where large quantities of organic matter have been 'dug in' to improve tilth. Catch crops of mushrooms in glasshouses should therefore be safeguarded either by raising the bottom trays at least 6 in. above the floor, or making the beds on sound polythene sheet. Where mushrooms are grown in special houses or adapted buildings the floor should be made with concrete surfaced with a smooth finish.

Many species abound in the surface layers of manure during the composting process and it is, therefore, essential to ensure that the pasteurization in the peak-heat process is adequate. Springtails are readily killed by heat so that, if 135–140 °F (57–60 °C) is maintained for 3–4 hr, there is almost no chance of survival. However, temperature of the lowest beds and trays is often less than desired, but fans may be used to achieve a better heat distribution.

The best control is achieved by mixing 50 p.p.m. of gamma-BHC in the compost (1 pint 10% emulsion, in a convenient quantity of water, per ton) at the last turn. Where control becomes necessary after the beds have been made up, BHC dusts (1 lb 5% dust/1000 ft^2 of bed) and malathion dusts ($1\frac{1}{2}$ lb 4% dust/1000 ft^2 of bed) may be applied to the surface.

REFERENCES

BRITT, N. W. (1951). Observations on the life-history of the Collembolan *Achorutes armatus*. *Trans. Am. microsc. Soc.*, **70**, 119–32.

RIPPER, W. (1930). Champignon-Springschwarze. Biologie und Bekampfung von *Hypogastrura manubrialis* Tulberg. *Z. angew. Ent.*, **16**, 546–84.

VAIL, P. V. (1965). Colonization of *Hypogastrura manubrialis* (Collembola: Poduridae) with notes on its biology. *Ann. ent. Soc. Am.*, **58**, 555–61.

Order DIPTERA: Flies

Family PHORIDAE

The most serious pest problems in mushroom culture are posed by species in this family.

Megaselia nigra Meig. (=*albidihaltaris* Felt) has been a pest ever since mushrooms were grown commercially and is also responsible for much of the tunnelling frequently found in wild mushrooms.

The black adults are frequently seen scuttling over the sporophores of field mushrooms; their rapid, jerking run and hump-backed appearance are quite characteristic. Closer examination will reveal a large, rounded third antennal segment and a unique wing venation (Fig. 9.2*a*). The female *M. nigra* lays elongate, ovoid (0·5 × 0·2 mm) eggs devoid of surface sculpture; these are usually attached to mushroom gills although they may also be laid on the surface of soil, or casing, immediately adjacent to developing pinheads. At 64 °F (18 °C) these eggs hatch within 3 days into white, almost transparent larvae. The larval body is narrowed anteriorly to a vestigial head bearing a pair of black 'mouth-hooks'—heavily chitinized, mandibulary sclerites—bearing seven teeth (Fig. 9.4). At the posterior end the body is truncated with a pair of small, but prominent, terminal pupillae.

Newly emerged larvae tunnel into the sporophore tissues and develop rapidly, passing through the three larval instars within 5 days at 64 °F (18 °C).

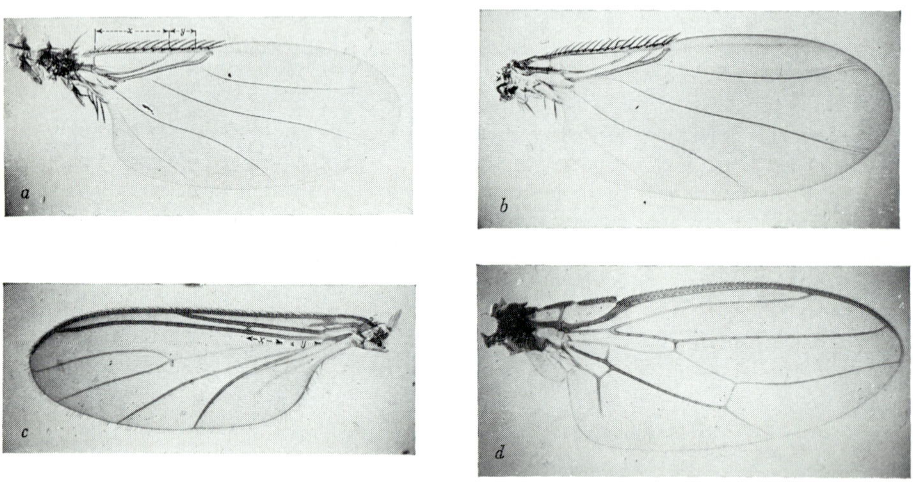

FIG. 9.2. Typical wing venations: (*a*) *Megaselia nigra*; (*b*) *M. halterata*; (*c*) *Lycoriella solani*; and (*d*) *Leptocera heteroneura*. (Courtesy of Glasshouse Crops Research Institute.)

The cap and stipe are often tunnelled by scores of larvae (Fig. 9.3) and, in severe cases, rapidly deteriorate as bacterial infection spreads from the site of larval damage. Just prior to maturation, the larvae vacate the mushroom and form a characteristic pupa in the adjacent compost. The pupa develops within an ovoid, but dorso-ventrally flattened, puparium, white at first but soon becoming pale yellow and then dark brown. In all phorid pupae a pair of black, curved prothoracic respiratory horns develop after about 2 days (Fig. 9.4). At 64 °F (18 °C) the fly emerges after 5 days. Outdoors, in heavily infested sporophores, the developmental cycle proceeds more slowly because of the lower temperature and intense competition for food between the larvae. Developmental periods as long as 26 days between egg and adult have been recorded. In nature the adult females live at least 16 days but the males only 10.

In the early days of outdoor, commercial, mushroom growing the species caused severe damage but in recent years it has become much less important. *M. nigra* lays eggs only where daylight falls on the beds. In one case, where beds had been made in a well-lit cowshed, almost every mushroom was tunnelled. When the windows were blacked out, however, attacks ceased despite the large numbers of *nigra* present.

M. nigra is troublesome in late summer, and light-trap studies have confirmed that it flies from early June to December.

Many pupae, especially outdoors, become dark brown and develop white, subcutaneous spots. The prothoracic horns never develop in such pupae for, instead of the phorid, they contain the parasite, *Aspilota concolor* Nees (Hymenoptera: Braconidae) (Fig. 9.5). Heavy parasitism (more than 90%) occurs but, since *Aspilota* lays its eggs in young larvae of *M. nigra*, and parasitized individuals continue to feed and develop normally until pupation, damage to

FIG. 9.3. Mushrooms tunnelled by *Megaselia nigra*. (Courtesy of Glasshouse Crops Research Institute.)

the mushrooms continues. Each female phorid lays at least fifty eggs so, to prevent damage, the parasite would need to reduce the phorid population over a wide area.

Control

As the eggs are laid on, or near, developing mushrooms chemicals cannot be applied without serious risk of taint, or even toxic residues, in the harvested crop. It is, therefore, fortunate that darkness prevents oviposition. In most purpose-built houses with adequate ventilation the light intensity is low enough to inhibit oviposition but, in small houses, where the doors are opened to cool them in hot weather, serious losses may occur unless light-tight screening is used.

REFERENCES

BRAUNS, A. (1950). Zur kenntnis der Schadinsekten an Champignonkulturen. *NachrBl. dt. PflSchutzdienst, Berl.*, **2**, 153–6.

HUSSEY, N. W. (1959). Biology of mushroom phorids. *Mushr. Sci.*, **4**, 260–70.

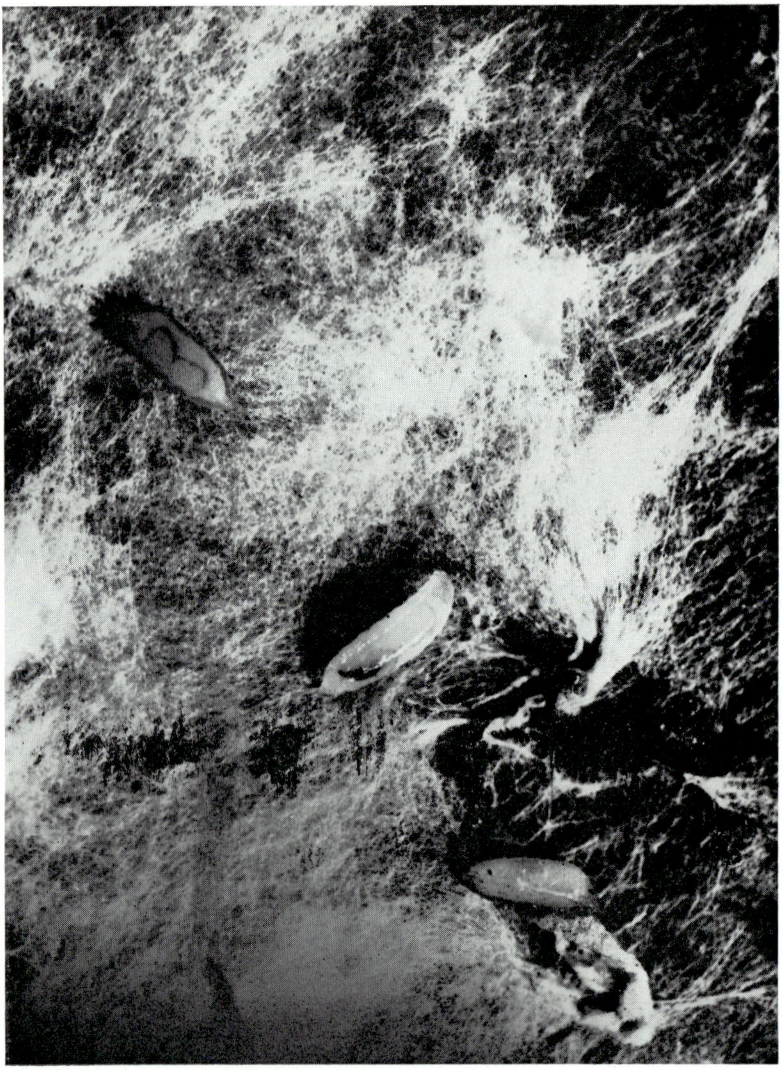

FIG. 9.4. Larvae and pupae of *Megaselia halterata* (×10). (Courtesy of Glasshouse Crops Research Institute.)

Megaselia halterata Wood (the Worthing phorid) so-called because of a vast outbreak in the Worthing district in 1953, occurs on most mushroom farms. It occurred in small numbers before this but was not regarded as a pest nor was its biology known. Since 1953 it has remained the most serious pest problem in mushroom culture throughout Western Europe. *M. halterata* is smaller and lighter than *M. nigra*, the dorsum of thorax and abdomen varies from yellowish to reddish-brown (Fig. 9.6). Individuals bred within mushroom farms tend to be deeper brown with the halteres, which are usually yellow, darkened, at least distally. The adults have the characteristic jerky movement associated with other members of the family and are common in summer and early autumn. They congregate on the surface of boxes and trays, especially near the doors and around lights within the buildings.

Within 24 hr of emergence the females are attracted to swarms of males which form at the doors of cropping houses, spawn rooms and other buildings as well as at prominent objects such as trees up to several miles from mushroom farms. The females can only be trapped while making directional flights to

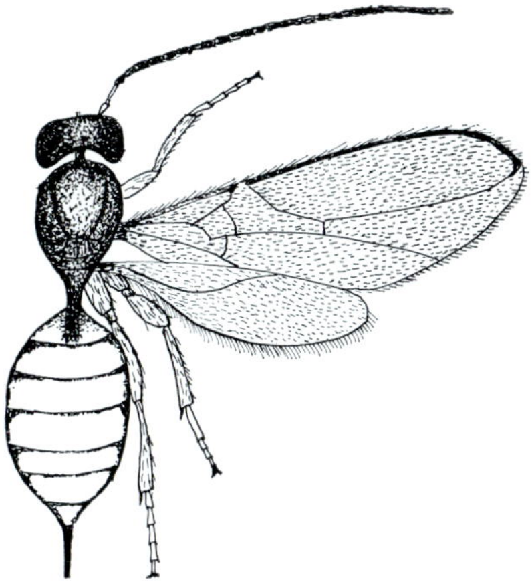

FIG. 9.5. Female *Aspilota concolor* Nees, parasite of the phorid *Megaselia nigra*.

these swarms while males tend to fly at random. Presumably, sexually excited males produce a chemical stimulus which is carried downwind to females several miles away.

Once mated, the females become sensitive to an attractant odour emanating from growing mushroom mycelium. They can detect small volumes of spawned compost at considerable distances. Data, collected from suction trap catches, have shown that a mushroom crop increases in attractiveness to phorids throughout the spawn-running period, reaching a maximum at 'casing'. Within 1 week of 'casing' the mycelial growth is normally complete and the flies are no longer attracted in any numbers. It is interesting to record that, while the sex ratio in *M. halterata* is almost 1:1, collections *outside* spawn rooms produced 62 males to every female while *inside* the females were in preponderance in the order of 80 to 1. Each female phorid lays about 50 eggs, placing each within a millimetre or so of the tips of growing hyphae. Individual females produce their eggs within a few days though they may live several weeks after egg laying. Indeed, they may survive for as long as 10 weeks under winter conditions.

Adult phorids shelter from inclement weather in dense bushes and in rank vegetation, at ground level. This behaviour is especially obvious in windy weather. Flight is encouraged by temperatures above 60 °F (16 °C) though the minimum threshold is 55 °F (13 °C). The flies are, therefore, usually on the wing from about 9 a.m. and continue to fly in suitable weather either until

FIG. 9.6. Adult female *Megaselia halterata* (phorid) (× 40). (Courtesy of Glasshouse Crops Research Institute.)

the temperature falls below 60 °F (16 °C) or sunset reduces the light intensity below a critical level. Even when the temperature and light intensity are suitable for flight, wind and rain may prevent the flies reaching the spawn-running rooms. Rain, however, has to be quite heavy to prevent activity for, in still conditions, flight movements may occur even in drizzle or light rain.

The temperature differential between the spawn-running and cropping rooms has a profound effect on the development of a phorid population. Eggs laid in the spawn-running rooms hatch after 2 days at 75 °F (24 °C) and the larvae mature to pupae in another 4 or 5 days. The adult fly emerges 7 to 8 days later. Early infestation ensures that a generation will mature at, or soon after, casing. Eggs laid after casing, however, experience temperatures between 60–62 °F (16–17 °C) and take 4 days to hatch, the larvae feed for at least a fortnight while the pupae do not mature for a further 4 weeks. This slower development accounts for the swarms of flies commonly found after 4 to 5 weeks' cropping.

While adult phorids may play some, as yet unidentified, role in the transmission of mushroom diseases, and certainly have an important nuisance value, they do not normally affect cropping. Occasionally, exceptionally heavy and early infestation may prevent the mycelium colonizing the compost. This serious damage was more prevalent when beds were spawned with relatively few clumps of manure spawn. The widespread adoption, in recent years, of through-spawning with grain spawn provides a large number of foci for mycelial growth and so prevents the concentration of phorid larvae at a few vital centres.

Even when the compost is colonized normally, growers often report a loss of mycelium after casing. This damage usually occurs below the interface of casing and compost, the mycelium dying back progressively for 2 or 3 in. from the surface. Such conditions largely prevent sporophore production. However, only large larval populations are capable of inflicting such damage; experiments have suggested that the critical number of larvae is about 100/30 g sample (=handful) of compost. This is roughly equivalent to 3 million larvae/ton. Fortunately attacks on this scale are rare, and almost unknown where the crop is free of larvae at casing. Where progressive loss of mycelium occurs in the absence of numerous larvae other causes such as 'die-back' virus should be suspected.

Vast numbers of adult *M. halterata* may, however, develop from compost that is lightly infested but it is important to appreciate that these do not lay eggs in the compost within which they have developed because the mycelium is no longer in a state of active growth attractive to the females.

The key to the control of phorid outbreaks is, therefore, attainment of an insect-free, fully-run compost at casing. Since there are numerous difficulties in attaining efficient chemical control of this species the obvious preventative is to make the spawn-room fly-proof. As adult *M. halterata* have a remarkable propensity for squeezing through minute cracks and crevices the sealing must be exceptionally thorough. This applies especially to the doors which should, preferably, be double. The inner doors should be screened with a mesh of less than 40 apertures per inch. Ideally, the outside doors should be seated against a rubber seal, like that on a commercial refrigerator.

Within the spawn-room rapid, and complete, mycelial growth must be encouraged either by extending the spawn-running period beyond 14 days, or increasing the rate of spawning above the normal one carton per 25 ft^2.

It was suggested in Chapter 1 that parasites exert little effective control in mushroom houses. However, there is an exception, for *M. halterata* is attacked by a parasitic eelworm which might, with artificial manipulation, become an important commercial control. The eelworm is a species of *Bradynema*

(Nematoda: Allantonematidae) which, although recognized as a distinct species, has not yet been described. If a parasitized, adult *M. halterata* is dissected, the abdominal cavity will be found tightly packed with adult female *Bradynema* eelworms, together with their eggs and larvae. This parasitism often entirely inhibits the maturation of the phorid eggs; the fecundity of parasitized *M. halterata* averages less than 1 egg per female compared with the normal 50. Despite the fact that flies carry almost no ripe eggs they still attempt to oviposit but, instead of eggs, groups of 20–40 *Bradynema* larvae are passed down the oviduct. Any eelworms which do not escape during 'oviposition' remain and die within the female fly. Male phorids are parasitized as frequently as females but the *Bradynema* is unable to escape at any time, even when the fly dies and decomposes.

The eelworms usually escape from the phorid host as second instar larvae. They quickly moult, the females being fertilized in a pre-adult, infective stage, by males which have matured in the compost. Once fertilized, the *Bradynema* seek out the larval stages of their host. They penetrate through the cuticle of first, or young second, instar *M. halterata*. Once inside, they mature and undergo some remarkable transformations. A swelling appears in the middle of the pre-adult eelworm which gradually becomes lemon-shaped. During the pupal phase of the host this lemon-shaped object moults and extends to a flaccid, elongate adult, often with a pronounced 'foot'. The eggs are produced within this unusually shaped eelworm and, at about the time that the adult phorid emerges from the pupa, laid into the haemocoele of the host. Multi-parasitism is common, as many as eighteen *Bradynema* females having been found in a single phorid larva.

The main factor limiting the proportion of phorids parasitized is the density of parasitized, female flies ovipositing in the compost. Observations on commercial farms have shown that the proportion of parasitized flies increases steadily through the fly season from June to November. This increase may range from less than 10% to over 80% of the population. Since this represents a decline in the reproductive potential of about 450% it would obviously be advantageous if the effect could be advanced to the beginning, both of the fly season and individual crops. This could, in theory, be accomplished by breeding the eelworm artificially and watering it on to the beds as required. Because the larval *Bradynema* do not feed on the mycelium and merely seek out the host larvae, there would be no deleterious effect on mushroom production. In this way, the phorid population in any crop would be restricted to a single generation. The proportion of phorids parasitized would also be increased so that the number of larvae developing, even in this one generation, would be drastically reduced. The practical effect might, however, be less than expected if a large proportion of the flies attracted to the spawn-run came from non-parasitized field populations.

Chemical control

Successful control of *Megaselia halterata* depends on preventing oviposition by flies which invade houses during the first 3 weeks of spawn growth. Where flies have access to the crop within spawn-rooms the larvae must be killed as they hatch. There are many methods for killing flying phorids which, due to the tight packing of boxes within the rooms, must largely be based on aerosols, fogs or smokes. BHC and sulfotep smokes, thermal fogs of pyrethrin and aerosols of diazinon, malathion and pyrethrin are all effective but only at, or soon after, application—a serious limitation if the flies are continuously entering the building in suitable weather. Where reliance is placed on such methods it may even be necessary to apply the insecticide twice daily. The first application should be made as soon as the outside air temperature rises above 60 °F (18 °C) and the second in mid-afternoon. No treatment is necessary when the wind is strong, or if there is moderate to heavy rain, because the flies are not on the wing. A material which remains in the air as a toxic vapour is likely to achieve the most efficient control. Dichlorvos theoretically fulfils this demand, and, where correctly used, has given outstanding results. Tests have shown that a concentration/time product of 0·07 μg/hr/l. is effective. In practice, as the flies spend only a short time in the air before entering the compost for oviposition, an aerial concentration of 0·3 μg/l. is the minimum target. While dichlorvos vaporizes readily from aerosols and PVC-resin strips it is hydrolysed rapidly in high humidities. Tests have revealed that, when applied as an aerosol at a rate equivalent to 3 g (6 cc 50% e.c.)/1000 ft^3, the aerial concentration falls below 0·3 μg/l. in 5–6 hr. Daily applications are therefore essential though the residual vapour gives dichlorvos a distinct advantage over other materials. Since it is possible to calculate the input of pure material necessary to give any desired concentration under different environmental conditions it may, in the future, be possible to introduce metered amounts of dichlorvos automatically, at set intervals.

Where it is impossible to maintain a lethal concentration of dichlorvos during the critical spawn-running period it may be preferable to apply insecticides, either to reduce oviposition or to kill larvae as they hatch from the eggs. The only chemical which deters ovipositing females is pyrethrin and, if applied at a rate equivalent to about 0·01 g/ft^2 of bed area, a beneficial effect should be obtained. However, as the flies enter the compost through cracks and crevices in the sides and bottom of the trays it is never fully effective. Close stacking of the trays within the spawn-room prevents efficient coverage of all surfaces so that, in practice, only superficial dusting of the compost surface directly after spawning is possible.

The incorporation of insecticides within compost, is, theoretically, preferable, but, as pointed out in Chapter 1, there are difficulties in achieving a uniform admixture of small quantities of insecticide in large stacks of compost. Chemicals lethal to newly hatched larvae include diazinon and chlorfenvinphos. Concentrations of 10 p.p.m. and 50 p.p.m. respectively are completely effective. Where a spawning-machine is available the insecticide may be applied to the surface of each tray and mixed through the compost by the rotating tines. This is a most economical, convenient and efficient technique. Efficiency is essential as phorids will develop normally in any untreated pockets.

The target concentration of 10 p.p.m. diazinon may be conveniently obtained by making up, for each 1000 ft^2 of tray surface, a stock solution of 50 gal of water containing 10 fl. oz of a 20% diazinon emulsion and applying this solution at a rate of 2½ pints to each 6 ft^2 tray. The emulsifying agents in commercial emulsions, can, if the treatment is made in a confined space, create rather unpleasant vapours which become objectionable during a long working day. It is, therefore, desirable to conduct the spawning and mixing operations in the open air. Those operators handling treated compost during the tamping process should wear rubber gloves. Wettable powder formulations may be more convenient for mushroom growers (1 lb 40% w.p./100 gal of water—apply 3 pints/6 ft^2 tray).

Not all growers have spawning machines, and then the only convenient time for incorporating insecticides is during the composting process. To avoid interference with bacterial action, it is customary to add the diazinon during the last turn of the compost stack. A continuous flow of the insecticide in water, at a rate equivalent to 15 gal/ton, is applied to the exposed surface of the stack in front of the advancing composting machine. Further admixture occurs during the filling operations when additional forking is commonly practised. Any insecticide mixed in compost at this time has, of course, to withstand the peak-heat process. Temperatures between 130–140 °F (54–60 °C) are maintained for several days, often with the production of considerable

FIG. 9.7. Hind tibia of male *Megaselia nigra*.

quantities of ammonia. Tests on several different farms showed that, after the different peak-heating processes, the average survival of diazinon was half the amount originally incorporated. Practical experience has confirmed that, where diazinon is incorporated during composting at 4 fl. oz of 20% emulsion per ton of compost, control of *M. halterata* has been satisfactory.

It must be remembered that both methods of incorporating insecticides in compost protect only the spawn-run. As it grows up through the casing, preparatory to the formation of thick mycelial strands before pinning, the flies are still actively attracted to the mycelium. Failure to give protection during this phase of mycelial growth occasionally results in severe damage at the compost/casing interface. Such damage may be prevented either by intensive use of insecticide aerosols or fogs in the first 10 days after casing, or by the application of insecticides to the compost surface. Pyrethrin dusts at 0·01 g/ft^2 (5 lb 0·2% dust/1000 ft^2), are normally effective. It may also be possible to incorporate diazinon in the casing material, at 2 fl. oz, 20% emulsion per ton, by mixing the chemical in the water used to soak the peat. Operators handling treated casing should wear rubber gloves. Even when incorporated at many times the recommended rate, diazinon has no effect on mushroom production. No residues have been detected in mushrooms where 10 p.p.m. of diazinon is incorporated at, or before, casing. The protection afforded lasts at least a month after treatment.

Despite the intensive insecticidal programmes commonly adopted, resistance has not become a serious problem. However, resistance to fogs and aerosols of pyrethrin has been detected and it is, therefore, wise to avoid excessive use of related compounds.

Where it is difficult to exclude flies from the early stages of mycelial growth, diazinon or chlorfenvinphos should be incorporated in composts during the 'season' from June to late October. At other times, and in well-constructed spawn-rooms, aerial treatments must be used. Nicotine, pyrethrin and gamma-BHC smokes or aerosols may be alternated from November to June, reserving dichlorvos for the period of greatest threat. In the cropping houses aerosols of diazinon, malathion, gamma-BHC and pyrethrin can be alternated. It is preferable to use one material for 3 or 4 months before switching to the next in the programme which should also be designed to avoid excessive use of a single insecticide during the life of an individual crop.

Because female *M. halterata* are attracted from considerable distances, site hygiene plays only a small part in the control of phorids. Adults seek shelter in rank vegetation during bad weather so that concrete surrounds to the houses and neatly cut grass are advantageous.

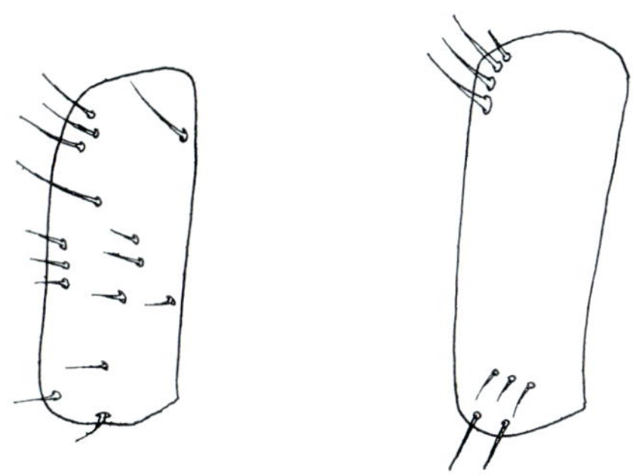

FIG. 9.8. Propleura of adult phorids: left, *Megaselia halterata*; right, *M. nigra*.

Super 'lure' traps baited by chemicals, concentrated from the odours produced by growing mycelium, may possibly be exploited for control in the future.

A third species, *M. bovista*, has occasionally been found tunnelling sporophores.

IDENTIFICATION OF PHORID SPECIES AFFECTING CULTIVATED MUSHROOMS

Adults of the three British species can be separated by the following key:

1. Males (recognizable by the large, black copulatory apparatus at the apex of the abdomen) 2
— Females 4
2. Apical third of dorsal surface of hind tibiae flattened, as if cut away, and dorsal seam therefore abruptly bent anterodorsally . . . . (Fig. 9.7) *Megaselia nigra* Meig.
— Hind tibia not thus 3
3. Hypopygium (terminal copulatory apparatus) without bristles . . . . . . *M. bovista* Grimm
— Hypopygium with at least two strong bristles on each side . . . . *M. halterata* Wood
4. Dorsal seam of hind tibiae abruptly bent anterodorsally at apical third; Propleura with only a few hairs (Fig. 9.8) 5

5. Divisions of costa of forewing $x=3y$ (Fig. 9.2*a*) *M. nigra*
— $x=4y$ ....*M. bovista*
6. Dorsal seam of hind tibiae evenly curved at apical third; Propleura with many hairs, $x=4y$ ....*M. halterata*

REFERENCES

BRAUNS, A. (1950). Zur kenntnis der Schadinsekten an Champignon Kulturen. *NachbrBl. dt. PflSchutzdienst, Berl.*, **2**, 153–6.
COLYER, C. N. (1956). Phorid flies damaging cultivated mushrooms: *Megaselia plurispinosa* (Lundbeck, 1920) (Diptera, Phoridae) a synonym of *M. halterata* (Wood, 1910). *J. Soc. Br. Ent.*, **5**, 166–71.
HUSSEY, N. W. (1959). Biology of Mushroom Phorids. *Mushr. Sci.*, **IV**, 260–9.
— (1960). Identification of Phoridae (Diptera) attacking cultivated mushrooms in Britain. *Ann. Mag. nat. hist.*, **13**, 599–603.
— (1965). Economies in pest control suggested by a study of phorid flight activity. *M.G.A. Bull.*, **182**, 71–83.
MORETON, B. D. & JOHN, M. E. (1955). The mushroom phorid fly epidemic of 1953. *Pl. Path.*, **4**, 9–11.

Family LYCORIIDAE: Fungus Gnats or Sciarids

Fungus gnats are small, delicate, gnat-like flies with long thread-like antennae and large compound eyes which meet above the bases of the antennae in a form of 'eye bridge'. The maxillary palpi consist of one, two or three segments. The distribution of the setae at the apex of the front tibiae of the males are useful for identification (Fig. 9.10). These either take the form of a crest, or comb, or a definite field with a greater, or smaller, number of almost parallel setae. The number of apical tibial spurs is also of diagnostic importance. The head and thorax are generally black but the abdomen, apart from the black dorsal and ventral plates on each segment, is often yellowish. The whole body is covered with fine microtrichia. The iridescent wings have a characteristic venation with the anterior veins more prominent than the posterior and there is a pronounced fork at the apex of the M (Median) vein (Fig. 9.2*c*). The apex of the abdomen is quite pointed and terminates with a pair of small cerci. The males, on the other hand, have prominent genitalia with well-developed claspers (Fig. 9.11).

Under natural conditions fungus gnats inhabit wild fungi, leaf mould, manure piles, and rotting wood, within which they have a predominantly saprophagous habit. Under certain conditions they enter mushroom houses where they are commonly seen running rapidly over the surface of the beds and on the sides of the boxes. They do not move with the characteristic jerking movement of phorids, from which they may be readily distinguished both by the long antennae, held out in front of the head at 45°, and the more elongate body form.

The larvae (see Fig. 5.49), when full grown, are up to 12 mm long and are easily recognized by the black, shining head and long, white body comprising twelve abdominal segments. The cuticle is partially transparent and, as the alimentary canal is frequently dark and the fat body clear white in colour, the internal organs are often clearly visible. The last abdominal segment is frequently rather enlarged and serves as a pseudopod to facilitate movement.

There are eight pairs of spiracles, situated on the prothorax and the first seven abdominal segments.

When ready to pupate the larvae spin a covering of fine silken threads within which are held fragments of faeces and compost. Herein the larva

FIG. 9.9. Adult *Lycoriella solani* (sciarid) from light-trap with attached mites. (Courtesy of Glasshouse Crops Research Institute.)

contracts into a quiescent prepupal stage which lasts for one day before the free pupa is formed. The head of the latter is flat with the antennae curved over the upper margin of the eye (see Fig. 5.51). The wings extend to the third abdominal segment while the three pairs of legs all extend approximately to the fourth abdominal segment. It is not, at the present time, possible to differentiate between the known mushroom-infesting species as far as their biology and potential damage is concerned. More than a dozen species have been associated with mushrooms but the confusion in nomenclature has made some of these records doubtful. In any event, damage has only been linked with a specific determination in a few cases. We shall therefore only discuss the three species about which there is valid evidence of pest status in mushroom culture.

Lycoriella solani Winnertz (=*Neosciara fenestralis* Zett) (Fig. 9.9) is the best-known species in Europe. Each female lays between 150–170 oval, white eggs (see Fig. 5.50), each $0{\cdot}7 \times 0{\cdot}3$ mm, either in the compost or casing, or actually on the sporophores. They may be laid singly but more usually they occur in groups of ten or more. They hatch, under mushroom cropping conditions, within seven days and then develop through four larval instars in 19–27 days—averaging 24 days at 60 °F (15 °C). The pupal phase occupies between 8–14 days. A complete generation takes about 5 weeks.

Copulation normally occurs during the first day after emergence, the female commencing oviposition 2–3 days later. The larvae feed largely on compost which they chew with their strong biting mouthparts. Manure spawn used to inoculate the beds may, occasionally, be completely destroyed. More commonly however larvae break down localized areas of compost into a soggy, foul mass. The pH of this rancid material is usually above 7 and the mycelium, therefore, rarely colonizes such patches. It has been shown experimentally that female *L. solani* prefer to oviposit in unspawned compost and that even limited mycelial growth in compost reduces the number of eggs laid. In general, therefore, larger populations develop where the flies have access to compost before it is fully colonized by mycelium. Larvae hatching from eggs laid at this time are almost full-grown as the first flush of mushrooms appear. Large larvae may tunnel up into the stipes of developing sporophores. If these are small 'buttons' they become brown and leathery and fail to grow further but, in larger sporophores, the stipes and caps may be extensively tunnelled. Since very high larval populations have been found in compost with only minimal damage to sporophores it is difficult to avoid the conclusion that these, chiefly saprophagous larvae, are only attracted to sporophores when these are themselves unhealthy. The association between Lycoriids and the virus disease 'die-back' seems too frequent to be entirely accidental.

There are two other species *L. auripila* Winnertz and *Bradysia brunnipes* Meigen which occur from time to time in mushroom houses. *L. auripila* is more common and, in recent years, there has been some evidence that this species may cause more primary damage to sporophores than *L. solani*. *L. auripila* can be readily detected by the 'slime' trail left by the larvae as they move over, and through, compost. When tunnelling in mushrooms they often cut a 'window' which is covered by silken threads and mucilage and through which frass is ejected. Larvae may be seen hanging from infested trays on silken threads.

All three species seem, however, to become pest problems only after a large

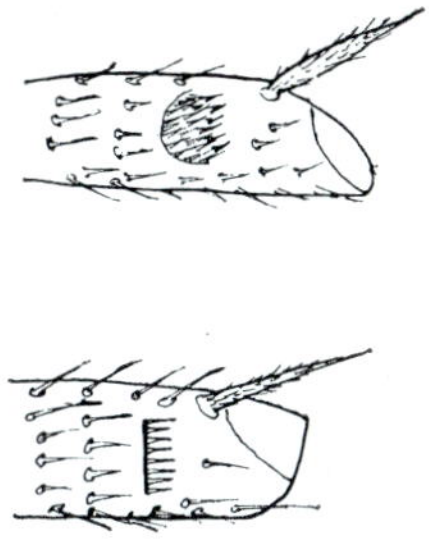

FIG. 9.10. Fore tibiae of male *Lycoriella* (*upper*) and *Bradysia* (*lower*).

population has developed within a single, probably diseased, crop. The adults locate new crops randomly rather than purposefully as in phorids. There is some recent evidence that they may be attracted to composts containing nitrogenous additives. Once established in a crop however, they may develop enormous populations with three to four overlapping generations. Probably their most important role in mushroom culture is disease and mite transmission. Unlike phorids, which seem to spend long periods cleaning themselves, fungus gnats readily pick up mites which breed in the foul patches of compost created by the larvae. These species of mites often have specially adapted forms (hypopae) which cling to passing insects. A single fly may carry as many as 30 mites in this way (Fig. 9.9). Since these mites are usually associated with various forms of bacterial breakdown the host fly may effect undesirable disease transmission. Adult sciarids also carry spores of *Verticillium*, an important mushroom disease.

These Lycoriid species may be attacked by nematode parasites of which three have been identified—*Tetradonema plicans* Cobb, *Attractonema gibbosum* Leuck. (see Flachs, 1941) and *Tripius sciarae* (Bovien).

T. sciarae penetrates the host larva as a mated, adult female by first producing an adhesive substance around its own head by digesting the last, unshed larval cuticle. Adhering to the host with this adhesive it uses its buccal spear, and probably digestive enzymes, to penetrate the host cuticle. The digested larval cuticle of the eelworm seals the penetration hole. Penetration may be accomplished within 10 min but sometimes takes several hours.

Once within the host the eelworm rapidly becomes gravid, increases greatly in size, and lays eggs within a week. These hatch in 3 days, and 2 weeks later the host is full of juvenile eelworms. These larvae develop to the pre-adult stage

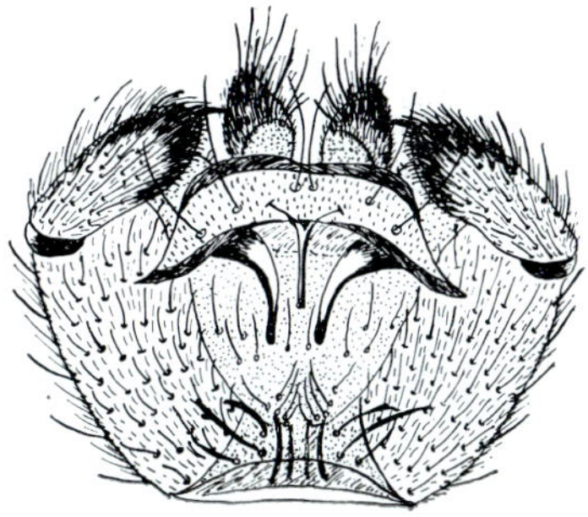

FIG. 9.11. Male genitalia *Bradysia brunnipes* Meig.

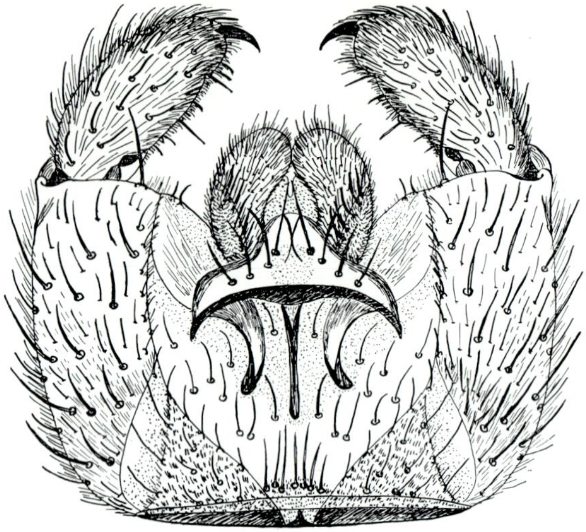

FIG. 9.12. Male genitalia *Lycoriella auripila* Winn.

within the host, which then dies, and liberates the parasites. As even the few parasitized Lycoriid larvae which do mature to flies are sterile, *Tripius* has a potential for biological control which could be exploited if artificial culture methods became available.

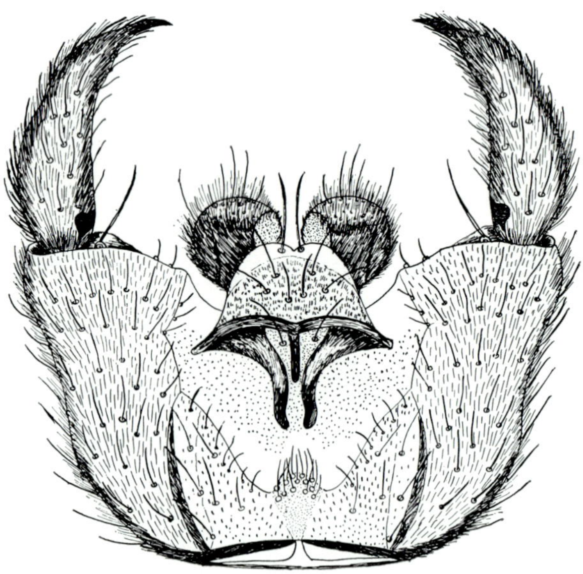

FIG. 9.13. Male genitalia *Lycoriella solani* Winn.

KEY TO MALES OF MUSHROOM-INFESTING LYCORIIDAE

(Males are the smaller of the sexes and carry a large copulatory apparatus at the end of the abdomen.)

1.(2) Fore tibia with crest of stout, long setae (Fig. 9.10). Claspers of genitalia partially constricted, apically rounded with a small tooth (Fig. 9.11). Eyebridge with four irregular rows of facets. Setae on both *x* and *y* (Fig. 9.2) . . . . *Bradysia brunnipes* Meig.

2.(1) Fore tibia with an area of setae (Fig. 9.10) Clasper with one stout tooth apically and two long hairs on inner face (Figs. 9.12, 9.13). . 3

3.(4) Basal lobe of hypopygium indefinite bearing 6–8 long setae (Fig. 9.12) Body length 2·3–2·6 mm. . . . . *Lycoriella auripila* Winn.

4.(3) Basal lobe of hypopygium distinct bearing a tuft of long stout setae (Fig. 9.13) Body length 2·6–3·2 mm . . . . . *L. solani* Winn.

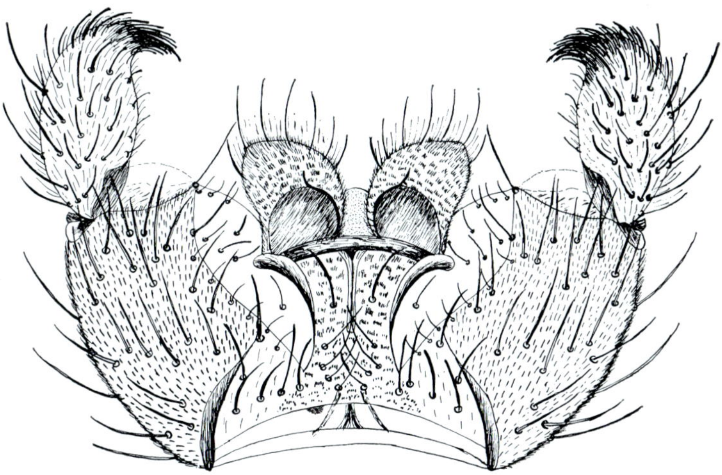

FIG. 9.14. Genitalia of *Bradysia paupera* Tuomikoski.

Control

As with phorids, exclusion of the adults from spawn-running rooms would largely prevent Lycoriid infestations since the compost would be fully colonized by mycelium before the flies could oviposit. Chemical control may be directed against adults and larvae. In the case of the adults, insecticides are most conveniently applied as smokes and aerosols. As every effort should be made to prevent the development of resistance and, bearing in mind that prevention of oviposition up to pinning should prevent the development of a sizable population late in the crop, it is wise to adopt a rotational programme of insecticides during the pre-cropping period. This should be based on chemicals with distinct modes of action (e.g. pyrethrins, nicotine, organochlorine or organophosphorus materials). Pyrethrin fogs, thermally or mechanically induced, applied at 1 fl. oz oil-based pyrethrin per 1000 ft^3 could be followed by gamma-BHC smokes (0·7 g BHC/1000 ft^3) or atomizing concentrates at 1 g BHC/1000 ft^3. The organophosphorus material may be chosen from a wide range including malathion or diazinon at 1 fl. oz (20%) per 3000 ft^3, and dichlorvos at 6 cc (50%) per 1000 ft^3. The frequency of application depends both on the size of the fly population on the farm and the weather at the time. In general, applications should be made twice weekly, between spawning and pinning, as in a similar programme outlined for phorids (p. 316). During cropping, insecticides need be used only if unduly large numbers of flies appear. To avoid danger from toxic residues and discomfort to workers, pyrethrin is to be preferred for treatment of houses with mature mushrooms on the beds. Spraying of the walls, boxes and bedboards with 0·5% DDT or BHC is undoubtedly a useful measure in reducing residual populations. Recently it has been shown that sciarid larvae may be controlled by incorporating diazinon or chlorfenvinphos at 50 p.p.m. in compost at spawning.

Where other measures have been omitted or unsuccessful, control of sciarid larvae in the casing may be necessary though this is more difficult than is the case with phorid larvae as lethal concentrations of insecticides tend to be fungitoxic. As already explained the larvae are seriously injurious only when they are in the superficial layers of compost and casing where they may tunnel into sporophores from below. Drenches of 0·01% malathion are toxic to the larvae within 48 hr; if applied at about 40 gal/1000 ft^2 of bed area these should penetrate sufficiently to reduce infestations when sporophore damage is taking place. As many mushrooms as possible should be removed from cropping beds before treatment and no further sporophores may be harvested for 4 days after the application.

As sciarids are all predominantly saprophagous it is particularly important to ensure that stalks, trashings, spent manure, etc., are either removed as far as possible from the farm or sterilized in some way.

REFERENCES

FERRIS, J. M. & FERRIS, V. R. (1966). Observations on *Tetradonema plicans*, an entomoparasitic nematode, with a key to the genera of the family Tetradonematidae (Nematoda: Trichosyringida). *Ann. ent. Soc. Am.*, **59**, 964–71.

FLACHS, K. (1941). Die Trauermucke *Neosciara solani* Winnertz als Schadling an Champignonkulturen. *NachrBl dt. PflSchutzdienst, Berl.*, **19**, 1–20.

GERBACHEVSKAYA, A. A. (1963). Leaf-midges (Diptera, Lycoriidae) injurious to vegetables and common mushrooms in hothouses of the Leningrad region. *Ént. obozr.*, **42**, 496–511.

HUNGERFORD, H. B. (1919). Biological notes on *Tetradonema plicans* Cobb a nematode parasite of *Sciara coprophila* Lintner. *J. Parasit.*, **5**, 186.

HUSSEY, N. W. & GURNEY, B. (1968). Biology and control of the sciarid *Lycoriella auripila* Winn. in mushroom culture. *Ann. appl. Biol.*, **62**, 395–403.

POINAR, G. D. (1965). The bionomics and parasitic development of *Tripius sciarae* (Bovien) (Sphaerulariidae: Aphelenchoidea), a nematode parasite of sciarid flies (Sciaridae: Diptera). *Parasitology*, **55**, 559–69.

POINAR, G. D. & DONCASTER, C. C. (1965). The penetration of *Tripius sciarae* (Bovien) (Sphaerulariidae: Aphelenchoidea) into its insect host, *Bradysia paupera* Tuom. (Mycetophilidae: Diptera). *Nematologica*, **11**, 73–8.

THOMAS, I. (1930). The structure and life-history of *Sciara nitidicollis* Meig. *Proc. zool. soc. Lond.*, **65**, 1009–26.

TUOMIKOSKI, R. (1960). Zur kenntnis der Sciariden (Dipt.) Finnlands. *Ann. zool. Soc. 'Vanamo'*, **21**, 1–164.

Family CECIDOMYIIDAE

The mushroom cecids are unusual in that reproduction is achieved by *paedogenesis*, a process in which the normal egg-adult development is short-circuited by 'mother' larvae which give birth to 'daughter' larvae.

Five different species are associated with mushroom production in Britain. Two, *Heteropeza pygmaea* Winnertz and *Henria psalliotae* Wyatt, have white larvae while those of *Lestremia cinerea* MaQuart, *Mycophila speyeri* Barnes and *M. barnesi* Edwards are orange or yellowish.

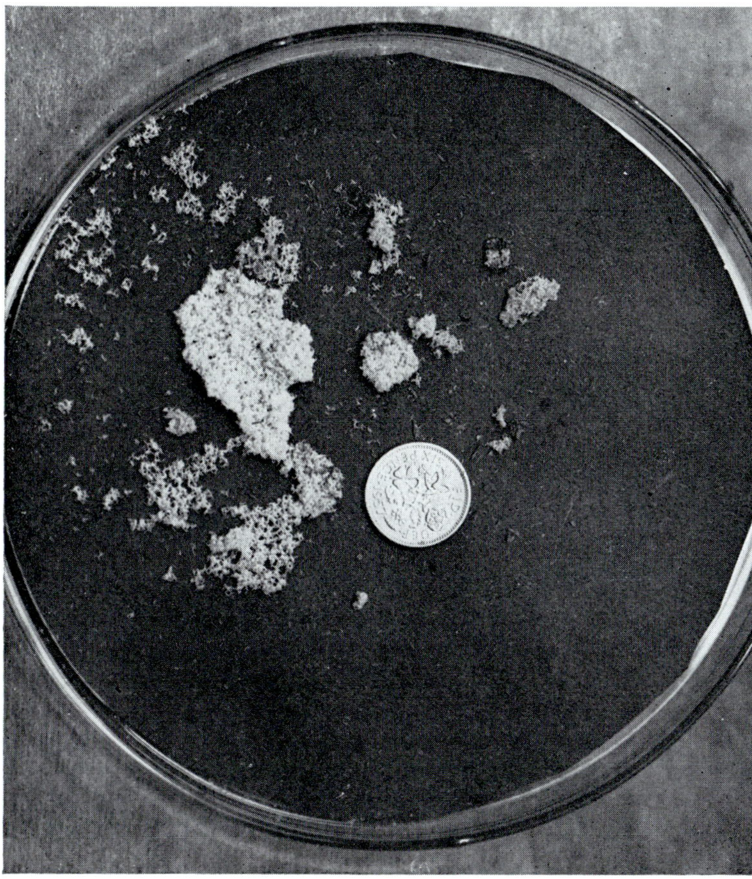

FIG. 9.15. Swarming *Heteropeza* larvae. (Courtesy of Glasshouse Crops Research Institute.)

Heteropeza pygmaea (= *Oligarces paradoxus* Meinert) is the most widespread and commercially important species and its life cycle is described in detail. The newly born, paedogenetic larva, about 1 mm long, has contiguous thoracic eye-spots but no sternal spatula (an anchor-shaped chitinized process on the ventral surface, Fig. 9.19). These young larvae move in search of food and, if this is scarce, they may migrate to the surface of mushroom beds, particularly when these have been watered. If they reach a dry surface, such as the edge of the boxes or the floor, they cannot crawl, but by flexion of the body followed by a sudden straightening they are able to jump several centimetres. In these dry conditions their tacky cuticle causes them to adhere together in writhing masses, often 3 cm across (Fig. 9.15). It is at this stage that they can easily be carried about the farm by workers and their cultural equipment. The positively photokinetic behaviour of the larvae which tend to move towards the better lit portions of the house along the main gangways, to ventilators, doors and other light sources, increases the danger of accidental dispersal.

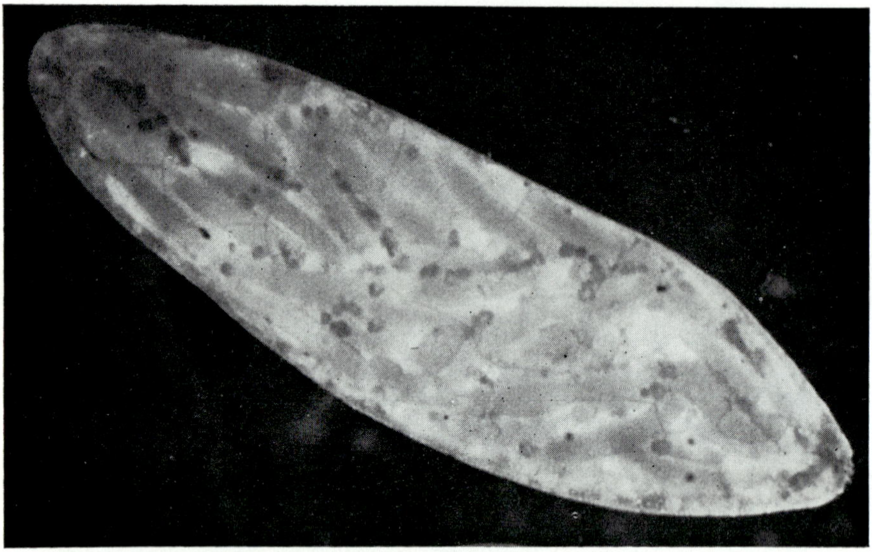

FIG. 9.16. Paedogenetic hemi-pupa of a mushroom cecid (×40). (Courtesy of Glasshouse Crops Research Institute.)

When actively growing mycelium is found the larva pierces the hyphae and sucks out the contents. Older larvae, with violent movements of the head, tear tufts of hyphae and then feed on the exuding sap. In optimal conditions they attain their full size of 2·8 mm in 5 days, growing merely by extension of the cuticle. The mid-gut becomes distended by a brown fluid within the peritrophic membrane (a lining of the mid-gut). Only at this stage is the anus functional and through it this membrane and its contents are voided as a continuous tube up to 10 mm long. Within the larval cuticle, which is not actually shed, the larva then moults to a rigid and immobile 'hemi-pupa' (Fig. 9.16). The ovaries develop rapidly, even in the young larva, and embryos are released into the haemocele to begin growth only a day after feeding begins. By the time the mother larva is full-grown these embryos are almost 1 mm long. After two further days they mature and force their way out at either the head, or anal, end of the mother.

Occasionally, when larvae are undernourished, either because of competition or ageing mycelium, they will only produce one or two embryos during an abnormally long developmental period. At the same time the inner larval cuticle becomes progressively thickened by extra reticulations. The enclosed larva is then able to remain inert, but alive, for long periods—up to 18 months in laboratory observations. The persistence, under adverse conditions, of these 'resting mother larvae' emphasizes the care necessary in the disposal of spent mushroom compost.

About 10 weeks after spawning some physiological change seems to occur within mycelium which induces many of the larvae to become 'imago larvae'. In this form the thoracic eye-spots are widely separated, and a chitinized sternal spatula appears (Fig. 9.19). This structure is variable in shape; the mean width is generally about one-third of its length and the base has broad lateral extensions. This imago larva completes only a single instar before moulting to a normal free pupa from which adults emerge in about 5 days. The adult flies are less than 1 mm long, and orange. Despite the fact that thousands have been caught in light-traps they are rarely seen by growers. Nearly all the flies caught in this way proved to be of one sex. Unless the female has mated, its one or two eggs remain infertile and it is believed that farm infestations rarely arise from adults entering houses. However, *H. pygmaea* occurs widely in decaying wood and rotting vegetation so fertile females may occasionally infest mushroom beds from such natural habitats.

Once larvae have become established in mushroom beds their numbers increase rapidly. Under laboratory conditions each mother larva produces an average of 14 daughter larvae within 6½ days though, in compost, only about 7 larvae are produced every 13 days. Even so, early infestation may result in populations as high as 30,000 in every square foot of bed. This logarithmic population growth is a theoretical concept rarely seen in nature. But, in the mushroom cecids, paedogenetic reproduction permits continuous increase within a favourable environment devoid of both natural enemies or limiting climatic conditions. The food source, mushroom mycelium, is a rapidly, continuously extending and anastomosing network of hyphae which does not suffer so severely from larval feeding as would a slower growing higher plant of finite size. In experiments in which beds were infested with different numbers of *H. pygmaea* larvae at spawning, or at intervals of several weeks thereafter, it was found that population increase proceeded at this constant logarithmic rate irrespective of the time, or numbers, of the initial infestation. This constant rate of increase continued until about 9 weeks after spawning, at which time, irrespective of the population density, it declined sharply. Towards the end of the period of maximum sporophore production there is apparently some alteration in the nutrient value of the mycelium. Even the largest populations do not affect yield unless the initial infestation occurs at, or very soon after, spawning, when sporophore production may be reduced by 10–14%.

The main economic loss is caused, not by a reduction in the numbers of sporophores produced, but by contamination of the mushrooms, especially after the beds have been watered, by random wandering of the larvae. Unlike other cecids, *H. pygmaea* seems to be the vector of a bacterium which produces longitudinal, brown, discoloured stripes on the mushroom stipe as well as minute, black pustules of fluid on the gills which consequently assume a spotted appearance (Fig. 9.17). Early infestation of the beds may result in contamination of as much as 20% of the total crop, though most of this loss occurs in the last 3 weeks.

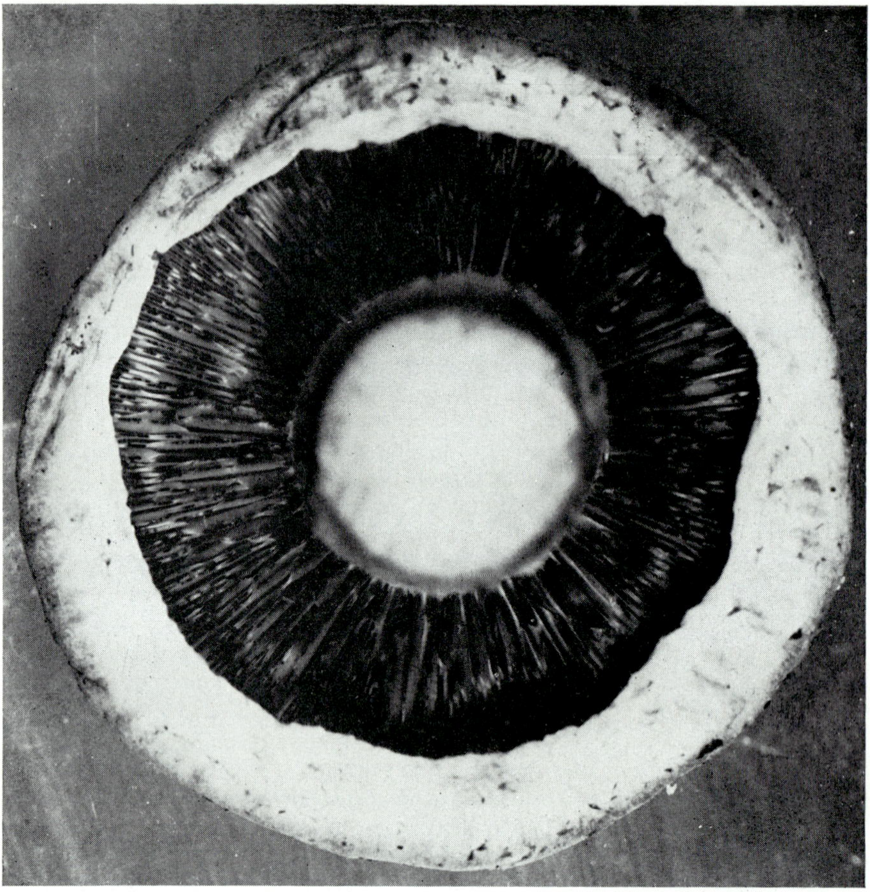

FIG. 9.17. Damage to mushroom gills by *Heteropeza* larvae. (Courtesy of Glasshouse Crops Research Institute.)

Henria psalliotae Wyatt was described from a single outbreak on mushrooms growing in a cellar in Cornwall. Reproduction is again paedogenetic but, before reproducing, each whitish larva completes its development, forms a sternal spatula and then, within a cocoon, becomes a 'hemi-pupa'. This form resembles a pupa but lacks the appendages destined to become the legs and antenna of the adult. Each hemi-pupa produces an average of 56 daughter larvae in about 16 days. This rate of reproduction is much lower than that of *Heteropeza* where a single larva could, theoretically, produce a population of 29 million progeny in 6 weeks whereas, in the same time, *Henria* could produce only 32,000.

This low rate of potential increase suggests that *Henria* is unlikely to become an important pest.

Lestremia cinerea Macquart is a common species which is not paedogenetic and does not occur in such large numbers as other mushroom cecids. It has, in fact, only been twice recorded as a mushroom pest. The earliest record by Speyer (1926) occurred on outdoor grown mushrooms and the second, by Wyatt (1964), was associated with sporophores near the doors of a cropping house. As it is known that *L. cinerea* is a diurnal species it probably lays eggs only where the beds are lit by natural daylight. Further, unlike other mushroom cecids, the white and elliptical eggs are apparently laid on the young sporophores. The entire life cycle takes place on the mushroom rather than in the compost. The young larvae are pale yellow at first, becoming deep orange when full grown (4 mm long). The sternal spatula has a slender central shaft diverging into two long tapering arms. *L. cinerea* is unlikely to be a pest problem under conditions of modern mushroom culture.

Mycophila speyeri (Barnes) has orange larvae which are paedogenetic. They are abundant in a wide range of habitats where suitable fungal mycelium occurs, having been recorded from rotting ash-wood, horse manure, peat, and cucumber beds. Mushroom beds prepared on the soil, within glasshouses, are particularly vulnerable but, as in *Heteropeza*, larvae may also be accidentally introduced into compost on contaminated tools or, of course, in infested peat. As many thousands of adults of both sexes are known to emerge from ageing mushroom crops and infested cucumber beds, flies may also initiate infestations.

Larvae hatching from pale-yellow, flattened, smooth and elliptical eggs ($0{\cdot}3 \times 0{\cdot}1$ mm) are deep orange and about 0·8 mm long. They start feeding on the fungal hyphae at once. Unlike *Heteropeza* the anus is functional so that waste materials are not accumulated in the mid-gut. Growth is rapid, small embryos being visible in the haemocoele within 1 day of birth. When the mother larva is 6 days old, and about 2 mm long, the daughter larvae shed their embryonic cuticle and become active. To emerge they rupture the mother's body either at the head, or anal end. Paedogenesis in *Mycophila*, therefore, takes place in the larval stage rather in the special hemi-pupal stage of *Heteropeza* and *Henria*. Under optimal conditions each mother larva produces 20 daughters in about 6 days. If a larvae is starved it does not, as in *Heteropeza*, form a resting stage capable of surviving many months but merely wanders around for several weeks before dying of starvation. In this species cessation of paedogenetic reproduction seems to be most easily induced by overcrowding, although sufficient food must be available to sustain life. Under these conditions, a mother larva produces only one or two very large, deep-orange embryos. Within 3 days these daughter larva moult to a final instar in which the sternal spatula rapidly becomes pigmented. This spatula consists of a slender shaft, tapering posteriorly and diverging at each end into triangular lateral extensions; there are normally three triangular anterior prongs.

After 3 days this larva excavates a pupal chamber, open at one end, within the compost. It is only at pupation that the eye-spots separate widely. The adult fly emerges 1 week after pupation. Males and females are found with equal frequency, both sexes being produced by a single mother larva. Experimental evidence suggests that sex determination is under environmental control.

The bright orange *M. speyeri* larvae, which may swarm even on to the first flush of mushrooms, cause obvious contamination (Fig. 9.18). However, unlike *Heteropeza,* they do not induce bacterial breakdown of the mushroom. This difference is possibly due to the fact that *Mycophila* larva excrete throughout life while the anus in *Heteropeza* is functional only at maturity; when voided the excreta probably supports a distinctly different bacterial flora.

FIG. 9.18. *Mycophila* larvae swarming on mushroom stalk. (Courtesy of Glasshouse Crops Research Institute.)

M. speyeri spreads to new habitats, in response to population pressure rather than to any change in mushroom metabolism, either by mass swarming of the larvae or by the development of imago larvae. The timing of the initial infestation is critical in determining the amount of economic loss. Where larvae enter the beds within the first 2 or 3 weeks after spawning the maximum population (about 20,000/ft^2) is reached at, or during, the first flush. Since the greater portion of a mushroom crop is produced in the earliest flushes losses through larval contamination can be severe, sometimes exceeding half the total production. Unlike *Heteropeza,* larvae of this species rarely swarm from the beds to form writhing piles on the floor.

Mycophila barnesi Edwards. Like *M. speyeri*, this species has also been found in many habitats other than mushroom beds, including rotting roots and bulbs, peat and even the puff-ball, *Scleroderma verrucosa*. Infestation may therefore arise from a number of local sources though once within spawned

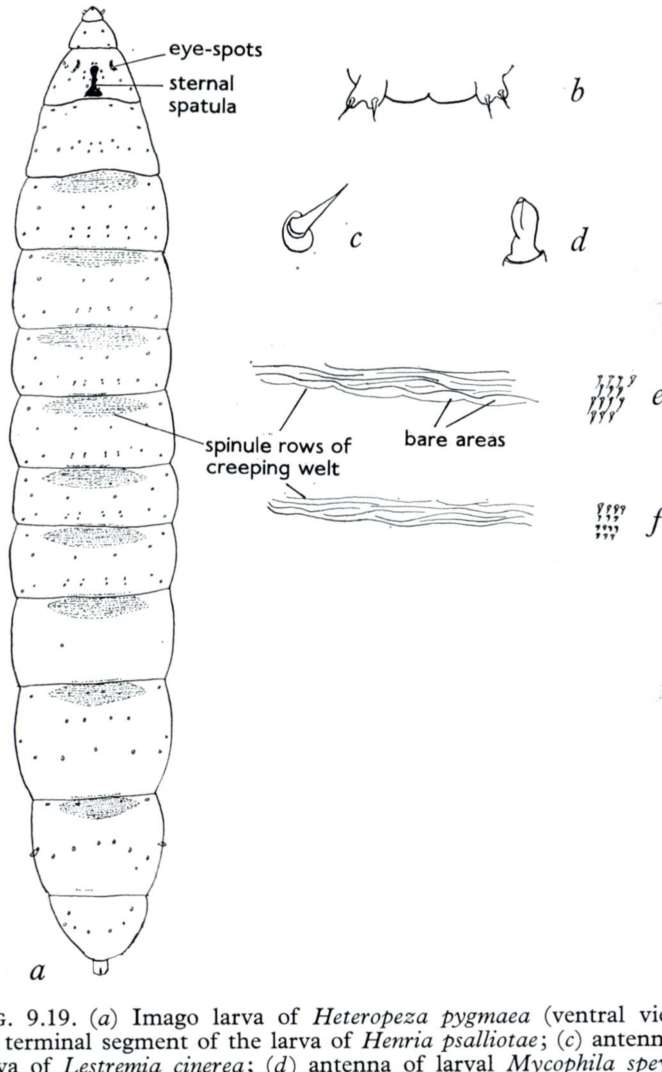

FIG. 9.19. (*a*) Imago larva of *Heteropeza pygmaea* (ventral view); (*b*) terminal segment of the larva of *Henria psalliotae*; (*c*) antenna of larva of *Lestremia cinerea*; (*d*) antenna of larval *Mycophila speyeri*; (*e*) and (*f*), posterior spinules of creeping welt of *M. speyeri* and *M. barnesi*, respectively (diagrammatic).

compost the larval population, as in other paedogenetic cecids, develops exponentially. Each mother larva produces about 19 young which mature in 8 days. This slower rate of development (cf. *M. speyeri*) delays population increase so that, even after early infestation, the second flush of mushrooms is rarely attacked, though the third and subsequent flushes may be severely contaminated. Almost 60% of the total production has been damaged in experimental crops.

Though the larvae are a much deeper orange than *M. speyeri* the two species are difficult to distinguish except by the detailed keys developed by Wyatt (1964):

KEY TO LARVAE OF CECIDOMYIIDAE INFESTING CULTIVATED MUSHROOMS

1. Abdominal creeping welts (Fig. 9.19) with separate spinule rows at posterior border of previous segment. Larvae white or cream........ sub-family HETEROPEZINAE.... 2
— No separate posterior spinule rows. Larvae usually orange or yellowish sub-family LESTREMIINAE..... 3
2. Terminal lobes on each segment bearing pairs of spines (Fig. 9.19b). Mesothorax with one or more spinule rows. Anterior ventral papillae among abdominal spinule rows................*Henria psalliotae*
— Anal segment without terminal lobes, Mesothorax without spinule rows. Anterior ventral papillae behind abdominal spinule rows...... *Heteropeza pygmaea*
3. Antennae tapering to a fine point (Fig. 9.19*c*)....*Lestremia cinerea*
— Antennae short with an obtuse apex (Fig. 9.19*d*)....................4
4. Abdominal creeping welts with sinuous hind margins (Fig. 9.19*f*).... *Mycophila barnesi*
— Abdominal creeping welts with posterior spinule row enclosing ten bare areas (Fig. 9.19*e*).......................*M. speyeri*

KEY TO ADULTS (WYATT, 1959)

1. Wings with at most three simple veins, first tarsal segment shorter than second.....................................HETEROPEZINAE 2
— Wings with at least four veins, one being branched, first tarsal segment longer than second..............................LESTREMIINAE 3
2. Wings with three veins and a cross vein (Fig. 9.20*b*), five tarsal segments, no sclerotized spermathecae present......*Henria psalliotae*
— Wings with only two indistinct veins, two tarsal segments, sclerotized spermathecae present (Fig. 9.20*a*).............*Heteropeza pygmaea*
3. Median vein forked (Fig. 9.20*d*), no sclerotized spermathecae present *Lestremia cinerea*
— Median vein not forked (Fig. 9.20*c*), sclerotized spermathecae present 4
4. Empodium distinct (Fig. 9.20*e*)................*Mycophila barnesi*
— Empodium rudimentary (Fig. 9.20*f*)..........*M. speyeri*

Control

The economic importance of the mushroom cecids arises from contamination of the crop rather than reduction in the number of sporophores. Mushrooms are lost only when infestation occurs soon after spawning so larvae must be excluded at this stage. Adult flies rarely lay viable eggs so the paedogenetic larvae constitute the main threat. As they are easily transported on equipment, clothes and hands every care should be taken to ensure that workers and their tools are not moved from maturing to newly spawned crops. Where equipment and overalls are allocated to separate houses new crops rarely become accidentally infested. 'Cooking-out' or fumigation of matured crops should be regarded as an essential in the hygiene procedure against cecids, especially *Heteropeza*. In the absence of thorough decontamination of the building the 'resting mother' formed by this species can readily survive from an infested crop and contaminate the next. There is also evidence that *Heteropeza* larvae can penetrate

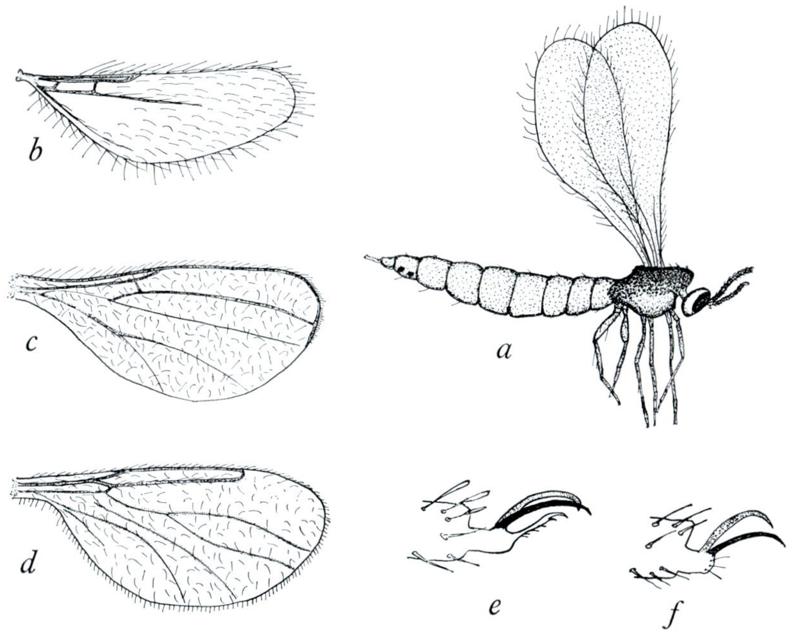

FIG. 9.20. (*a*) Adult *Heteropeza pygmaea*. Forewings of: (*b*) *Henria psalliotae*; (*c*) *Mycophila speyeri*; and (*d*) *Lestremia cinerea*. Tarsi of: (*e*) *M. barnesi* and (*f*) *M. speyeri*.

up to 12 mm into the timber of trays or the supporting structures so that it is important to use only sound wood. Treatment of boxes, either with a dip such as 2% sodium pentachlorophenate or by steam sterilization, is therefore a wise routine precaution. Recent evidence suggests that dips using disinfectants based on mixtures of cresols and phenols will provide the most effective and convenient control.

Where infestation occurs despite routine hygiene procedures, the practicability of chemical treatments must be considered. While it has proved possible experimentally to reduce larval populations by incorporating chemicals in the compost before spawning, the concentrations usually required (in excess of 100 p.p.m.) are fungitoxic and may cause reductions in cropping. Although not yet approved, 10 p.p.m. of thionazin in compost has been shown to be a most effective control.

Since physical contamination of the mushrooms is the main damage caused, control would be economically effective if migration by the larvae could be halted. Unfortunately cecid larvae are extraordinarily tolerant to chemicals. *Heteropeza* larvae have been immersed in 20% parathion for as long as 16 hr without any deleterious effects and, in routine laboratory culture of paedogenetic cecids, the larvae are not harmed by surface sterilization with 7% formalin. Despite their relative invulnerability to chemicals it has been found that insecticides may form a barrier through which the larvae are reluctant to crawl. While this barrier is ineffective if, as in the case of dusts, it is merely a surface layer, incorporation within the casing layer provides a barrier in depth. To be completely efficient the chemical must remain effective for at least 8 weeks, from the application of the treated casing layer to the end of cropping.

Public opinion has hardened considerably against the use of persistent pesticides since the deleterious effects of organochlorine compounds, in wild-life foodchains, were demonstrated. However, within the organochlorine group of persistent insecticides, gamma-BHC is least objectionable and when mixed in the casing layer, at rates between 10–50 p.p.m. of casing, sporophore contamination is reduced. The yellow larvae of *Mycophila* are more easily deterred than *Heteropeza*. The former is satisfactorily controlled by only 10 p.p.m. gamma-BHC (4 fl. oz 10% e.c. per ton of casing) but even 50 p.p.m. (1 pint 10% e.c. per ton) is only partially successful against *Heteropeza*. Higher concentrations tend to delay cropping and reduce yield.

The principal weapon against cecid larvae is sound hygiene and efficient cook-out and peak-heats. As *Mycophila* are killed by 99 °F (37 °C) and *Heteropeza* by 113 °F (45 °C) normal efficiency in both these operations should eliminate all cecid larvae.

REFERENCES

HUSSEY, N. W. & WYATT, I. J. (1959). Cecid control by incorporation of insecticides in composts. *Mushr. Sci.*, **IV**, 280–6.

—— (1960). Insecticidal control of paedogenetic cecid larvae in mushroom beds. II. Incorporation of insecticides in the casing layer. *Ann. appl. Biol.*, **48**, 347–51.

WYATT, I. J. (1959a). Some aspects of cecid biology of importance in mushroom culture. *Mushr. Sci.*, **IV**, 271–9.

— (1959b). A new genus and species of Cecidomyiidae (Diptera) infesting mushrooms. *Proc. R. ent. Soc. Lond.*, (B) **28**, 175–9.

— (1961). Pupal paedogenesis in the Cecidomyiidae (Diptera) Pt. 1. *Proc. R. ent. Soc. Lond.*, (A) **36**, 133–43.

— (1963). Pupal paedogenesis in the Cecidomyiidae (Diptera) Pt. 2. *Ibid.*, **38**, 136–44.

— (1964). Immature stages of Lestremiinae (Diptera: Cecidomyiidae) infesting cultivated mushrooms. *Trans. R. ent. Soc. Lcnd.*, **116**, 15–28.

Family SPHAEROCERIDAE (BORBORIDAE)

Some species of this family occur in manure during the composting process. They resemble phorids which accounts for the common belief that these important pests are introduced into mushroom houses with the compost.

However, examination with a hand lens shows that the arrangement of veins in the forewing is quite distinct. Phorids have no cross veins (Fig. 9.2*a*) but sphaerocerids have several, some forming a rectangular 'box', in the middle of the wing (Fig. 9.2*d*).

One species, *Leptocera heteroneura* Hal., is common in mushroom houses. Austin (1937) thought that it arose from larvae infesting mushrooms but it breeds in compost unsuitable for mushroom mycelium. The species is probably a saprophage and therefore lays eggs only where bacterial decomposition is in progress.

The eggs, 0·2 × 0·6 mm, which have a delicately reticulated surface, are inserted into the surface film of moisture on decomposing straws within the compost. Incubation takes about 3 days. The larvae (Fig. 9.21) resemble house-fly maggots in tapering to a slender head which has prominent mouth hooks; the blunt posterior has two prominent spiracles. These spiracles are surrounded by a circle of bristles which resist the surface tension of liquids, so permitting breathing even when the larvae is itself almost submerged. In suitable conditions the larvae mature in about 14 days though, when there is intense competition for food, they may take as long as 4 weeks. When mature,

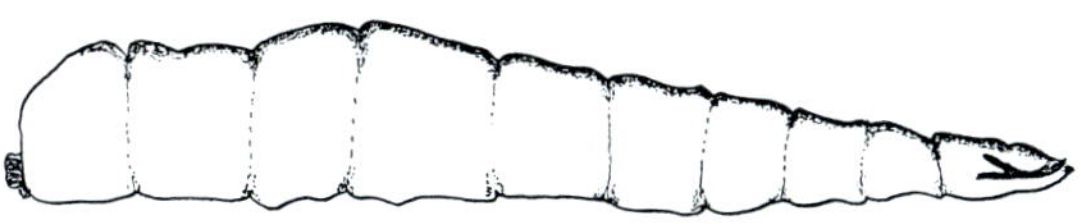

FIG. 9.21. Larva of *Sphaeroceridae*.

the larvae crawl to dry sites and form golden-brown barrel-shaped puparia from which the adults emerge in about 10–14 days.

L. heteroneura, feeding as it does at sites where there is intense bacterial activity, often carries mites which also feed in such habitats. It can therefore be a vector of organisms which are undesirable in healthy crops.

Routine insecticidal treatments for other pests normally prevent increase of Sphaeroceridae but, as with other saprophages, healthy crops and properly prepared compost alone should entirely eliminate them.

REFERENCES

AUSTIN, M. D. (1937). Notes on the status of flies of the family Sphaeroceridae (Borboridae) in the economy of cultivated mushrooms. *Proc. R. ent. Soc. Lond.*, (A), **12**, 15–16.

RICHARDS, O. W. (1930). The British species of Sphaeroceridae (Borboridae: Diptera). *Proc. zool. Soc. Lond.*, **18**, 261–345.

Family DROSOPHILIDAE

Mushrooms are occasionally extensively tunnelled by larvae of *Drosophila funebris* F., and possibly other related species, which are normally associated with decomposing vegetable matter. The damaged mushrooms are reduced to an almost fluid state by the larvae. Where attacks on healthy mushrooms have been noticed, eggs, each with four terminal hairs, are laid below the surface of the cap tissues. Each oviposition site is marked by a brown spot which expands until it resembles an incipient stage of bacterial blotch.

Family SCATOPSIDAE

The larvae of these flies are usually found in manure or decaying vegetable material. In pre-war days, when compost was not peak-heated, they were common in mushroom houses but, under modern conditions of culture, they are seldom seen. However, light-trapping within houses often shows that a few adults are present—these individuals breed in small pockets of compost where bacterial decomposition is in progress. As many as seven different species have been recorded but none have been associated with any form of damage to the mushroom crop. The female (Fig. 9.22*a*) lays about 100 white, elongate

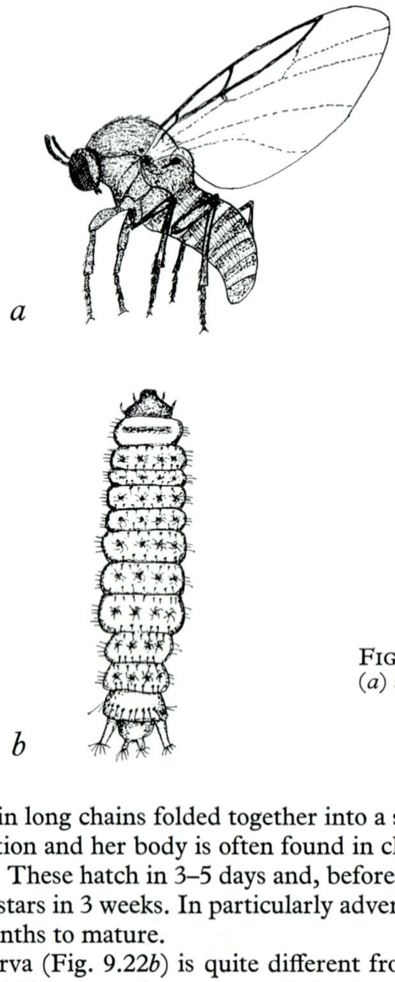

FIG. 9.22. *Scatopsidae:* (*a*) adult; (*b*) larva.

eggs (0·3 × 0·9 mm) in long chains folded together into a single mass; she dies shortly after oviposition and her body is often found in close contact with her complement of eggs. These hatch in 3–5 days and, before pupation, the larvae pass through four instars in 3 weeks. In particularly adverse conditions, larvae may take up to 7 months to mature.

The first instar larva (Fig. 9.22*b*) is quite different from the older instars. The head capsule and biting mouthparts are well developed and the cylindrical body has twelve segments. Whereas older larvae have nine pairs there is only one pair of spiracles on the eleventh segment. This pair are at the apex of terminal conical processes surrounded by a fringe of fine, branched setae which enable them to breathe even when the rest of the body is submerged in fluid. The most striking feature of these young larvae is the presence on nine of the body segments of transverse rows of four 'spines'. Each 'spine' consists of a star-like process raised on a short stalk surrounded by six curved hairs. These complicated structures trap soil and organic particles so that the larvae may become camouflaged. In older larvae these complicated 'spines' are replaced by simple setae. The adults emerge from the pupa within a week. Pairing occurs within half an hour and oviposition follows within 2 or 3 days. All the stages of scatopsids are frequently parasitized by an entomogenous nematode *Scatonema wulkeri* Bovien.

As these insects are found only where the compost is putrefying, or otherwise unsatisfactory, no control is necessary.

Chapter 10

Other Arthropod Pests of Mushrooms

Order ACARINA: Mites

Family TYROGLYPHIDAE

Mites of this family have soft, colourless bodies with long flexible hairs which are either quite smooth or bear minute pectinations. There are 4–7 pairs of these hairs forming a 'train' at the end of the body. The slender legs have no lateral spines and project well beyond the abdomen. Both sexes are of similar appearance, the genital openings being concealed between the third and fourth pair of legs.

Under certain conditions a hypopal, or deutonymphal, stage may occur between the proto- and tritonymph in tyroglyphid mites (Fig. 10.1*c*). This hypopa is specially adapted, both to facilitate dispersal and to permit survival under adverse conditions. This form is so different from the other developmental stages that, for many years, hypopae were regarded as distinct species. The hypopa is markedly flattened, with a hard integument, the weak legs are very short, only the front two pairs, which are held straight forward, being easily visible. At the posterior end of the ventral surface is a sucker plate, or adhesive apparatus, provided with several circular suckers by which hypopae attach themselves to moving objects. Flies are sometimes so covered with hypopae that they are incapable of flight.

Under modern commercial conditions tyroglyphid mites are rarely as common as they used to be in pre-war days, probably because of pasteurization of the compost. This practice has decreased secondary bacterial decomposition of the compost before colonization by mushroom mycelium. These mites are saprophagous and are normally found only in foul compost into which mushroom mycelium cannot grow. However, once established in a bed, they may occur on sporophores in large numbers.

Four species commonly found on mushrooms in this country may be distinguished by the following key:

KEY TO TYROPHAGUS ADULTS FOUND ON MUSHROOMS (Fig. 10.1).

1. On tarsi I to III the distal seta is thickened to form a spine.......... *T. fungivorus* Oudemans

   On tarsi I to III seta is unthickened.......................... 2

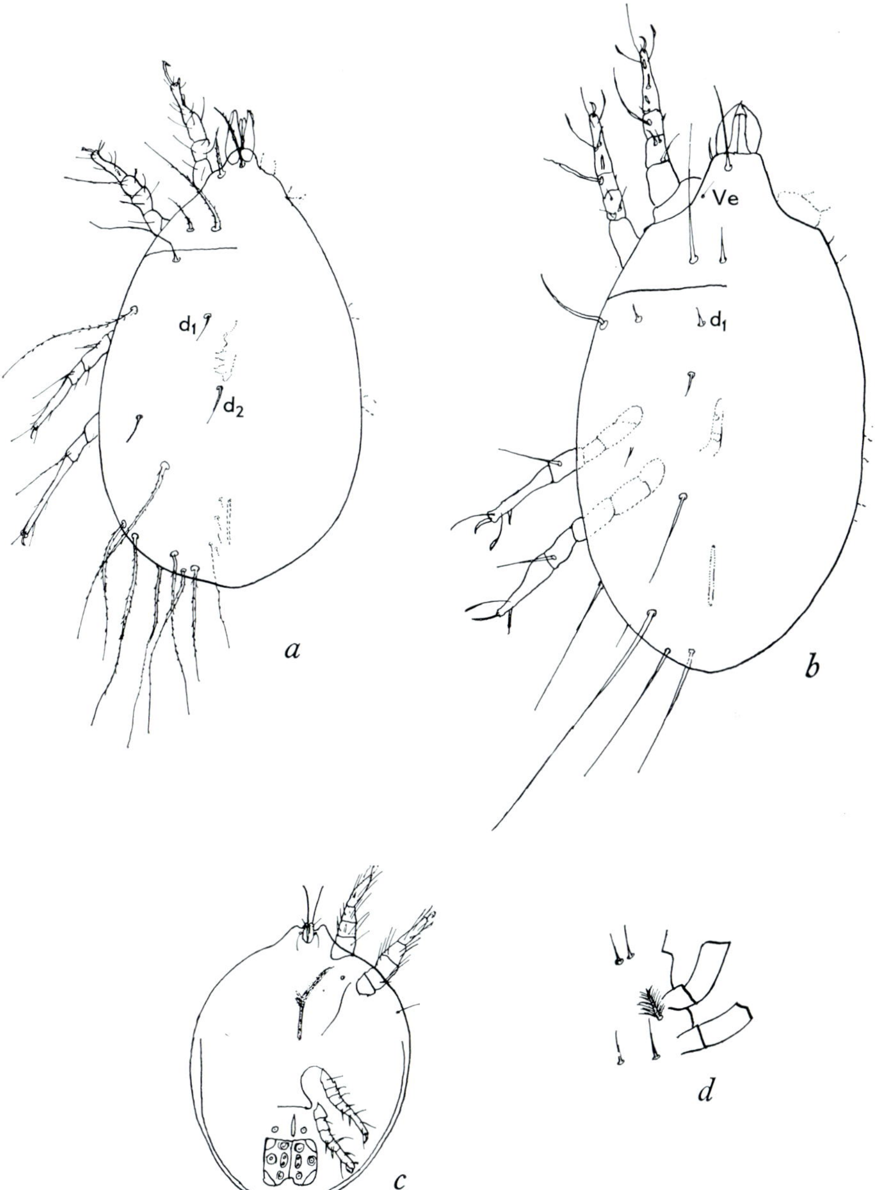

FIG. 10.1. (*a*) Female, *Tyrophagus longior* (dorsal); (*b*) female, *Caloglyphus* (dorsal); (*c*) hypopus of *Caloglyphus* (ventral); and (*d*) supracoxal setae. (For explanation of lettering see key p. 338 and p. 413.)

2. Supra-coxal seta (Fig. 10.1*d*) straight. In dorso-ventral view of male, arms of penis support turned outwards; d_2 (Fig. 10.1*a*) 2·5–3 times longer than d_1 .... *T. putrescentiae* Schrank
In dorso-ventral view of male, arms of penis support turned inwards.. 3
3. Supra-coxal seta curved. Medio-dorsal seta (d_2) short, less than half length of mouthparts and about equal to d_1.. *T. similis* Volgin
Medio-dorsal seta (d_2) longer, more than half length of mouthparts, 1·3–2 times as long as d_1.................. *T. longior* Gervais

Tyrophagus putrescentiae (Schrank) (=*lintneri* (Osborne)=*castellanii* (Hirst)). The adult female, usually translucent pinkish or yellowish with a large median, brownish and two smaller lateral spots, has almost colourless limbs and mouthparts. The body is more slender than in related species, and the long setae do not project so stiffly from the body. The females live for almost a year during which they are mated several times, the first time shortly after maturation. Two to three days later the first eggs are laid and oviposition proceeds steadily, each female producing up to 15 eggs per day, until the total complement of over 500 eggs is produced. Before hatching, the white, elongate-oval eggs ($0{\cdot}12\times0{\cdot}07$ mm) show irregular, longitudinal markings which are alternately lighter and darker. At 75 °F (24 °C) eggs hatch in only 6 days but at 60 °F (16 °C), the normal cropping temperatures, the larvae appear after 11 days. The white, semi-transparent larvae have only six legs and a wet, sticky appearance. After 8 days, at 60 °F (16 °C), the first moult occurs, producing the characteristically diamond-shaped, eight-legged protonymph; this is nearly colourless at first but later becomes light amber-yellow. After a period of activity the protonymph becomes inert and 6 days later emerges as a tritonymph. This closely resembles the adult but possesses only a vestigial genital opening. At 60 °F (16 °C) the tritonymph matures to the adult in 11 days. Under cropping conditions the developmental cycle is therefore completed in 36 days but in only 13 days during spawn-running. Development is retarded by low humidities. At 75 °F (24 °C) the life cycle takes 13 days at 100% R.H. but 24 days at 70% R.H.

Since *T. putrescentiae* adults are able to withstand a wide temperature range, and even dryness, they are able to reach new habitats without recourse to a hypopal stage and this is therefore rarely seen.

In mushroom beds the mites are constantly on the move and since eggs are dropped wherever the female happens to be, infestations tend to be spread evenly throughout the house. The mites feed on the mycelium and the sphorophores. Where numbers are low there is almost no damage and even a fairly high population may not cause obvious symptoms until the failure of the spawn run reveals that the mycelium had been attacked soon after planting. Damage to the sporophores consists of small, irregular pits eaten in the stalks and caps (see Fig. 10.4)—buttons attacked in this way may be completely hollowed out. Close examination shows that these feeding pits are lined with smaller cavities made slimy by bacterial decomposition.

Tyrophagus similis Volgin (=*oudemansi* Robertson=*dimidiatus* (Hermann)) may be recognized by its very long body hairs; some of these, at the end of the abdomen, are almost as long as the whole body. The legs are darker than in the previous species. Like *T. putrescentiae* this species rarely forms hypopae in mushroom beds. At 70 °F (21 °C) the egg, which is covered with a distinct sculpturing, hatches in 5 to 6 days. The larva, protonymph and tritonymphal stages take 4 to 5, 3 to 5 and 5 to 8 days respectively.

In this species, as in other *Tyroglyphidae*, there may be two types of male. One, the homeomorphic form, resembles the female except that it is smaller, though the legs are longer and more slender while the body setae also tend to be longer than in the female. The other form, called the heteromorphic, is darker than its fellows with the third pair of legs markedly thickened and the terminal claws replaced by a blunt process.

Tyrophagus longior (Gervais) (Fig. 10.1*a*) is a large species with dark legs and mouthparts, it has a similar arrangement of setae to *T. putrescentiae*. The eggs are covered with a distinct sculpturing arranged in interrupted, concentric bands. No hypopae are known. *T. longior* is especially common in straw and abounds during the early phases of composting. A fourth species known to attack mushrooms, *T. fungivorus* (Oudemans), is also common in stacks of hay and straw.

Caloglyphus spp. Another genus affecting decaying mushrooms is *Caloglyphus*, which may be distinguished from *Tyrophagus* by the external vertical setae *Ve* (Fig. 10.1*b*) which are long and pectinate in the latter but only very short and fine, or even absent, in *Caloglyphus*. The three species involved are *C. berlesei* (Michael), *C. mycophagus* (Megnin), and *C. krameri* (Berlese) which may be distinguished by the form of the supracoxal seta. This seta is found posterio-laterally to the base of the coxa of the front leg (Fig. 10.1*d*).

KEY TO THE *Caloglyphus* SP. FOUND ON MUSHROOMS (Fig. 10.1*b*).

1. Supracoxal seta distinctly pectinate, expanded towards the base...... *C. krameri*

— Supracoxal seta almost smooth................................ 2

2. Supracoxal seta almost inconspicuous.............. *C. mycophagus*

— Supracoxal seta well developed..................... *C. berlesei*

All species prefer to breed in very moist situations and are usually found on putrefying sporophores but they also eat large cavities in the caps of otherwise normal mushrooms. Unlike *Tyrophagus* sp. they readily form hypopae.

Caloglyphus berlesei (Michael) has a smooth, shining but colourless body though the legs are pale brown. Where the temperature and humidity are high breeding is very rapid. Individual females live up to 3 weeks and may lay several hundred eggs—there is one record of a female producing over 1000 eggs in 39 days, over 200 being produced in one 24-hr period. The smooth, blunt and oval eggs (0·15 × 0·11 mm) hatch in less than 3 days at 70 °F (21 °C). The larvae are active for about 3 days before entering a resting stage for a further day. The protonymph and deutonymph are completed in 2–4 days, with the resting stage lasting about 24 hr in each case. The entire life cycle may therefore be completed in about 10 days.

Control of Tyroglyphid Mites

Many tyroglyphid species are common in hay, straw and other organic materials including chicken manure so that some may be introduced into mushroom beds at filling. However, so long as the peak-heat is efficient these should be eliminated. The ability of many species to form hypopae which are transported on insects breeding in decomposing organic matter ensures that, should bacterial decomposition proceed in mushroom compost, mites which abound in these conditions will readily be spread by these insects as they seek fresh habitats. Since suitable breeding sites exist wherever organic debris, compost or other rubbish is allowed to accumulate in, or about, mushroom buildings it is most important to institute an efficient and routine sanitation treatment on the farm. Floors should be kept clean and washed down with suitable disinfectants especially after manure or compost has been stacked or turned thereon. During composting the outer 6 in. or so of the compost heap is relatively cool and dry and the next inmost layer cool, wet and largely anaerobic. In these conditions, tyroglyphids can breed freely and rapidly and a vast population may develop unless, at each turn of the stack, the outside layers are turned into the centre so that the whole mass, at some time, heats to about 130 °F (55 °C).

Fungal and bacterial contaminants, on which tyroglyphids breed readily, commonly infect the compost and sporophores at the end of the average mushroom crop. Safe disposal of spent compost is therefore essential. Before emptying, the house should either be cooked-out with live steam, so reaching temperatures above 130 °F (55 °C), or fumigated with formalin or methyl bromide. The spent compost should then be removed completely away from the farm site, taking every care to avoid spillage in areas from which contamination of new crops may take place.

REFERENCES

DAVIS, A. C. (1944). The mushroom mite (*Tyrophagus lintneri*) as a pest of cultivated mushrooms. *Tech. Bull. U.S. Dep. Agric.*, **879**, 26 pp.

HUGHES, A. M. (1961). The mites of stored food. *Tech. Bull. Minist. Agric. Fish. Fd, Lond.*, No. **9**, 287 pp.

JARY, S. G. (1938). Investigations on the insect and allied pests of cultivated mushrooms XII. Two more Tyroglyphid mites. *J. S.E. Agric. Coll. Wye*, **42**, 66–81.

JARY, S. G. & STAPLEY, J. H. (1937). Investigations on the insect and allied pests of cultivated mushrooms IX. *Tyroglyphus dimidiatus* Herm. (*longior*) Gerv. *J.S.E. Agric. Coll. Wye*, **40**, 119–33.

RIVARD, I. (1961). Influence of temperature and humidity on mortality and rate of development of immature stages of the mite *Tyrophagus putrescentiae* (Schrank) (Acarina: Acaridae) reared on mold cultures. *Can. J. Zool.*, **39**, 869–76.

ROBERTSON, P. L. (1959). A revision of the genus *Tyrophagus* with a discussion on its taxonomic position in the Acarina. *Aust. J. Zool.*, **7**, 146–81.

SAMSINAK, K. (1962). Beitrage zur Kenntnis der Gattung *Tyrophagus* Oudemans. *Acta Soc. ent. Czechoslov.*, **59**, 266–80.

ZAKHVATKIN, A. A. (1941). Arachnoidea. Vol. 6. Tyroglyphoidea. Fauna U.S.S.R. *Inst. Zool. Acad. Sci. Moscow*, **28**, 1–475.

Family ANOETIDAE

Histiostoma feroniarum (Dufour) (=*rostro-serratum* Megnin) (Fig. 10.2) has highly modified mouthparts adapted for feeding on minute particles in fluid media. The mandibles are a pair of serrated blades while the palps terminate in a flattened bilobed plate bearing bristle-like hairs. This bilobed plate is highly mobile and sweeps fluid food to the mouth. With alternate, forward and backward movements, the mandibles rake in solid particles. *H. feroniarum* may be distinguished from the other three British species as the hairs, at the apex of the papillae, are of approximately equal length. The female has a quadrangular, whitish body containing large quantities of chalky guanin. The cuticle also bears projections to which small particles of debris adhere and lend colour to the mite. On the dorsal surface there are three or four large flat humps. The body of the female measures 0·28–0·5 mm × 0·13–0·25 mm

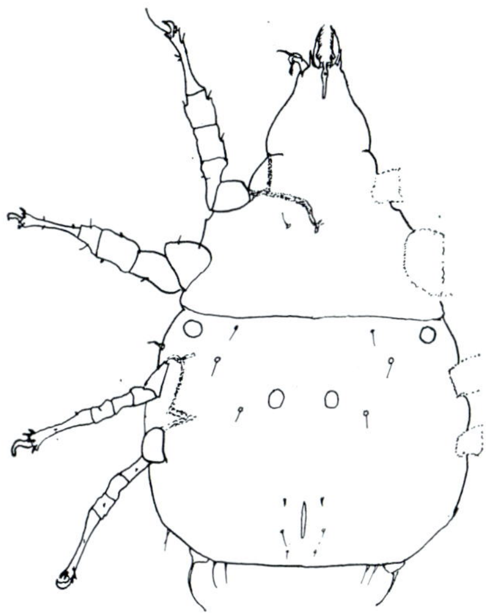

FIG. 10.2. Female, *Histiostoma* sp. (dorsal).

with the posterior margin evenly rounded but the smaller male (0·26–0·40 × 0·25 mm) has a body which is deeply indented in the middle, narrowing abruptly towards the posterior margin.

H. feroniarum is entirely oviparous but may appear, at first sight, to be viviparous because the eggs are difficult to detect on the wet surface on which the mites feed. The eggs (0·16 mm × 0·1 mm) are smooth, whitish at first but later translucent. At 70 °F (21 °C) an active larva hatches after 2 days and itself moults to the protonymph 1 day later. This protonymph moults to the tritonymph within 2 days while the adult matures only 1 day later. Hypopae, commonly produced by protonymphs in this species, are shield-shaped and heavily chitinized. They may be so numerous that flies, or other insects active in mushroom compost, may be completely covered. So long as the carrier insects move the hypopae to suitable conditions they moult within 2 days to a tritonymph and resume the life cycle. *H. feroniarum* is parthenogenetic, there being two types of females, one giving rise to females the other to haploid males.

It is a saprophagous species feeding only on decomposing sporophores which have been killed or damaged by some other agency.

REFERENCES

COOREMAN, J. (1944). Notes et observations sur les Acariens. *Bull. Mus. Hist. nat. Belg.*, **20**, 1–16.

HUGHES, R. D. & JACKSON, C. G. (1958). A review of the Anoetidae (Acari). *Virginia J. Sci.*, **9**, 1–198.

JARY, S. G. & STAPLEY, J. H. (1936). Investigations on the insect and allied pests of cultivated mushrooms VI. Observations on the Tyroglyphid mite *Histiostoma rostro-serratum*. *J. S.E. Agric. Coll. Wye*, **38**, 67–74.

Family EUPODIDAE

Linopodes antennaepes Banks (the long-legged mushroom mite) has long been regarded as a pest of mushrooms though it seems to occur only where the beds are outdoors or on the ground in glasshouses. It is easily recognized by the long front legs which are more than twice the length of the body (0·65 mm × 0·4 mm). The body is light yellowish-brown with a lighter T-shaped mark dorsally. This mark serves to distinguish *L. antennaepes* from reddish, active, predatory gamasid mites which sometimes appear in vast numbers when other mite populations have developed sufficiently to provide abundant prey.

L. antennaepes is an extremely active mite. It runs rapidly about the compost continuously oscillating the elongate forelegs in front of the head, possibly in a sensory capacity. The older literature unanimously concluded that the mites fed at the base of mushroom sporophores in all stages of development. The mycelial attachments were eaten away, causing larger mushrooms to be unstable, and the base of the stipe to become bright rusty-red. Recent observations throw doubt on these beliefs for *L. antennaepes* has also been recorded as a predator of *Caloglyphus krameri* and its general activity strongly suggests that it is *not* a mycophagous species. Further, attempts in America to rear the larvae on mycelium were unsuccessful. Extensive observations of mushrooms with rusty basal marks have revealed that the only mite present was *Tarsonemus myceliophagus* Hussey. It seems most probable that *L. antennaepes* is a predator on the latter species.

Little is known about the life history. Small round eggs, laid in groups in the casing, hatch after 8–10 days into minute, white larvae. The first moult, which occurs after 6–8 days, produces nymphs which have considerably elongated forelegs. No control is necessary.

REFERENCES

AUSTIN, M. D. (1937). The long-legged Mushroom mite *Linopodes antennaepes* Banks. *J. S.E. Agric. Coll. Wye*, **40**, 115–8.

GAHM, O. E. (1930). The Mite, *Linopodes antennaepes* Banks, as a pest of cultivated mushrooms, with preliminary tests toward control. *J. econ. Ent.*, **23**, 744–7.

RIPPER, W. (1931). Eine unbeobachtete Champignonmilbe, *Linopodes motatorius* L. *Pfl. Krankh.*, **41**, 127–31.

Superfamily: TARSONEMINI

Family TARSONEMIDAE

Tarsonemus myceliophagus Hussey. Four *Tarsonemus* spp. are known to be pests of cultivated mushrooms in America but only one, described as *T. myceliophagus* in 1963, has been found in Britain.

These minute mites, only 180–190 μ long, are occasionally found swarming in vast numbers on the sporophores but, although present in almost every crop, they can usually only be found by microscopic examination of the compost.

The body is oval with a pale-brown, shining, smooth cuticle; under a hand-lens a whitish transverse band is visible across the body. The female (Fig. 10.3*a*) has four pairs of legs, each with five free segments, but the tarsus of the fourth leg terminates in two long setae, one much longer than the other. The males (Fig. 10.3*b*), only 140 μ long, have enlarged hindlegs with a bent femur and a long curved terminal claw. These modified legs are used to carry

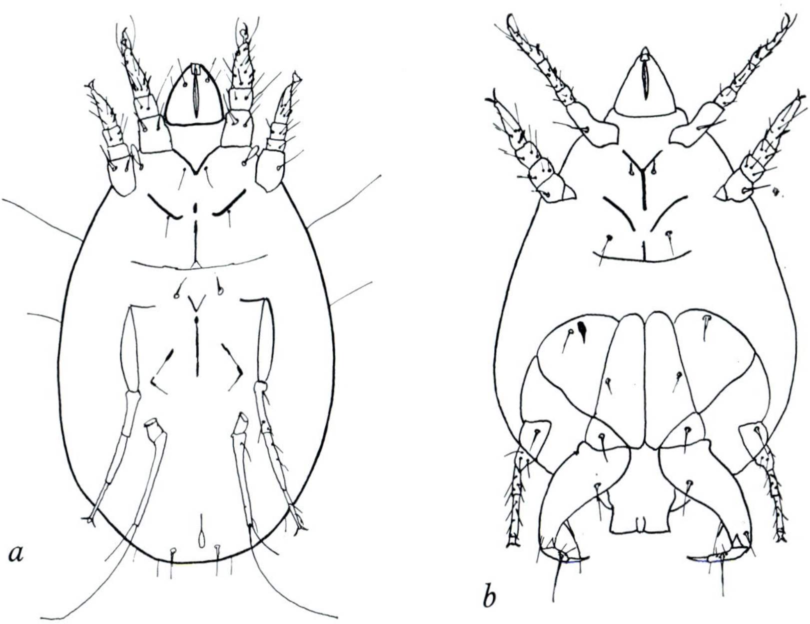

FIG. 10.3. *Tarsonemus myceliophagus*: (*a*) female, (*b*) male (ventral).

the female nymphal resting stage to ensure copulation immediately she matures. Males, carrying female resting stages on their backs, are common in cultures of this mite.

Fertilization may occur successfully at any time after maturation of the female but, if copulation is delayed, the female can produce viable eggs which all develop into males. After mating both sexes are produced. At 75 °F (24 °C) she lays about 14 large, oval and translucent eggs which are smooth on one surface and finely tuberculate on the other. After 2 days the eggs hatch into active six-legged larvae which moult to an eight-legged nymphal stage, itself maturing to the adult only 9 days later. At the normal cropping temperature of 60 °F (16 °C) the total developmental period increases to 12 days. At this lower temperature females live up to 2 months, during which time they each produce an average of 22 eggs. The average length of life of the female at 75 °F (24 °C) is 15 days though a few may live more than 1 month.

T. myceliophagus feeds on the hyphae of both mushroom and weed moulds. The mites cause a bright reddish-brown discolouration at the base of the stipe and, occasionally, actually cut through the basal hyphal attachments so that the sporophores become unstable (Fig. 10.5). However, observations in Britain have revealed cases where several thousand mites on a single mushroom (Fig. 10.6) caused no basal discolouration and yet others where only a few mites produced the typical damage. There is some doubt, therefore, as to whether the discolouration is caused by secondary decomposition of the tissues which becomes visible only under certain environmental conditions. Artificial

FIG. 10.4. Pits caused by tyroglyphid mites. (Courtesy of Glasshouse Crops Research Institute.)

infestation of experimental plots has shown that damage to the earliest flushes occurs only when the mites are introduced at, or soon after, spawning.

It is not possible to incriminate this common species as a primary pest but laboratory experiments have suggested that it may, occasionally, be a vector of the virus disease known as 'die-back'. In the cropping house the mite disperses very slowly, for no spread from artificially infested trays to adjoining, uninfested, trays occurs for some weeks until the population in the former approaches a maximum. In infested beds the mite population tends to be highest in the casing layer and, if this is allowed to dry out, the females come to the surface where, by standing erect on the long terminal hairs of the hindlegs, they can jump several inches.

The most serious aspect of their potential for virus spread is the relative ease with which they appear to survive 'cook-out'. Although laboratory tests have shown that exposure to 120 °F (49 °C) for only 20 min is lethal, even in a saturated atmosphere, living mites could readily be recovered from compost which, according to plunge thermometers, had been above 140 °F (60 °C) for several hours. Evidently, there are many niches within which the temperature does not become lethal during the average 'cook-out' process. Fumigation with methyl bromide is therefore preferable.

In America another species, *T. floricolus* C. and F., is said to eat tiny pits in the stems of sporophores and even inhibit the growth of smaller mushrooms.

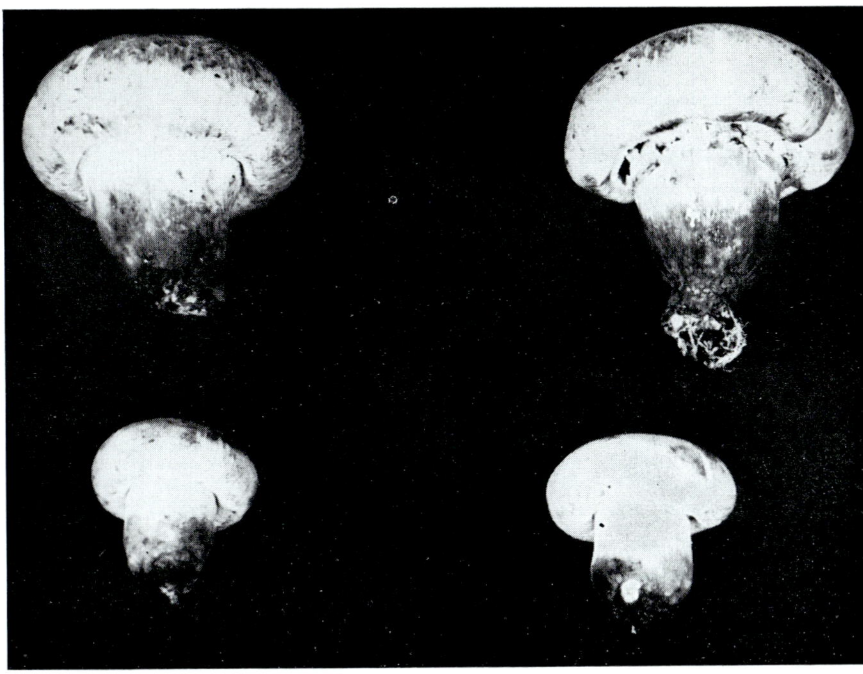

FIG. 10.5. Basal browning of the stipe caused by *Tarsonemus* mites. (Courtesy of Glasshouse Crops Research Institute.)

Control

While current knowledge suggests that control is usually unnecessary it has been shown that the number of mites infesting the sporophores can be drastically reduced by incorporating gamma-BHC in the casing at 50 p.p.m. (1 pint 10% e.c. per ton of casing).

REFERENCES

HUSSEY, N. W. (1963). A new species of *Tarsonemus* (Acarina: Tarsonemidae) from cultivated mushrooms. *Acarologia*, **5**, 540–4.

HUSSEY, N. W. & GURNEY, B. (1967). Bionomics and control of *Tarsonemus myceliophagus* Hussey in mushroom composts. *Ent. exp. & appl.*, **10**, 287–94.

SCHAARSCHMIDT, L. (1959). Systematik und Ökologie der Tarsonemiden. In H. J. Stammer, *Beiträge zur Systematik und Ökologie Mitteleuropäischer Acarina* **I**, Teil **2**, (5). 715–823.

Family PYEMOTIDAE

Pygmephorus spp. (the red pepper mites) are minute, flattened, wedge-shaped mites, about 250 μ long, which have a characteristically blunt end to the body (Fig. 10.7). They are yellowish-brown with a central whitish internal band. Not infrequently, these mites swarm in enormous numbers on the casing surface and the sporophores. These swarms consist of a jostling mass, often

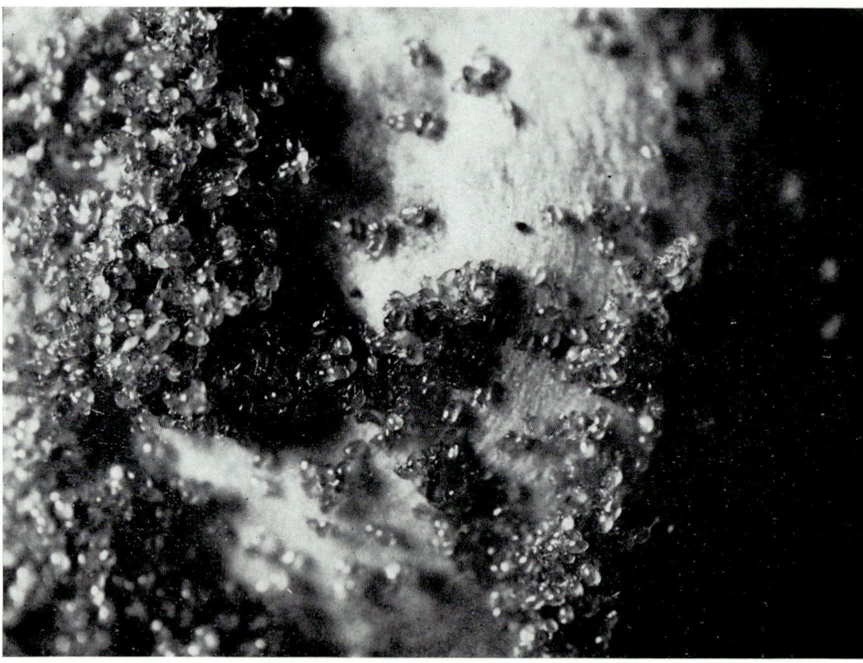

FIG. 10.6. *Tarsonemus myceliophagus* swarming on mushroom. (Courtesy of Glasshouse Crops Research Institute.)

piled several individuals deep, and visible, as a rusty-brown patch, at a distance of several feet. Three species have been recorded from mushroom crops in Britain. *P. mesembrinae* Canestrini (=*americanus* Banks) and *P. sellnicki* Krczal are found frequently but *P. tarsalis* Hirst only rarely. Differentiation from *P. sellnicki* (Fig. 10.7) may be made conveniently as follows: *P. mesembrinae* has three pairs of bristles on the propodosoma (the portion of the body immediately behind the head) and *P. tarsalis*, unlike the other two species, has the epimera of the third and fourth legs widely separated from the coxae.

It is widely believed that *Pygmephorus* spp. are associated with the olive-green weed mould *Chaetomium* which tends to develop in composts which have risen, during peak-heat, to temperatures in excess of 150 °F (65 °C),

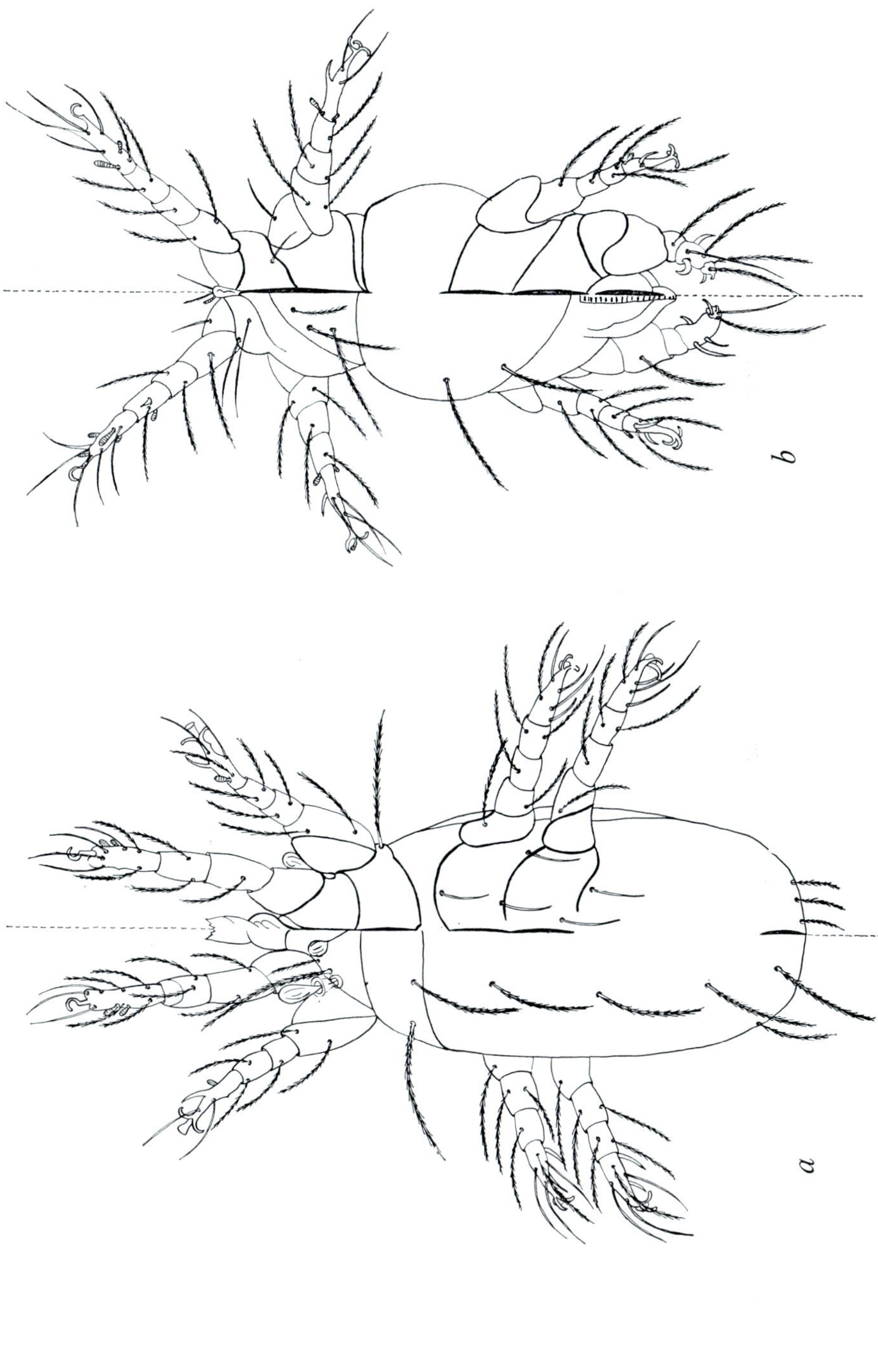

FIG. 10.7. *Pygmephorus sellnicki*: (*a*) female, (*b*) male. (Dorsal to the *left*, ventral to the *right*.)

often with a deficiency of oxygen. American evidence has suggested that other weed moulds including *Trichoderma* and *Penicillium* may also encourage *Pygmephorus*. However, our own observations have shown that, while the life cycle of *P. sellnicki* can be successfully completed on *Trichoderma*, *Monilia* or *Humicola*, *P. mesembrinae* breeds only on *Trichoderma*. Neither species can develop on *Chaetomium* and the association with this mould is, therefore, an indication that the compost is a suitable substrate for both of the fungi on which these mites subsist.

The biology of mites in this large genus is almost entirely unknown and, until recently, nearly all species were regarded either as associated with small rodents or as parasites of various beetles and flies. Should the fungal diet, shown to be favoured by the mushroom species, be more widespread, their frequent discovery in rodent nests is probably due to the presence of these fungi.

Once provided with an abundance of food the abdomen of the females gradually distends to a whitish, bubble-like sphere resembling a droplet of liquid. When fully distended the female is incapable of locomotion, though the legs remain mobile. Eggs are produced every few minutes and within a couple of days a heap of up to 150 is seen beside each dead female. These hatch after 2 days to active, six-legged larvae. A day later eight-legged nymphs abound in the culture and, following an inert resting stage, these mature to active females, so completing the entire life cycle in a maximum of 7 days. Males, which have almost undeveloped heads and no mouthparts (Fig. 10.7*b*) appear to mature directly from six-legged larvae. The high fecundity, coupled with rapid development, provides an enormous reproductive potential, so accounting for the teeming thousands of individuals which may swarm even a few days after casing. The active females move about rapidly using the front legs, which constantly vibrate, as sensory organs rather then walking appendages.

Although it is claimed that some mycelial damage may result if large populations of *Pygmephorus* occur within a few days of spawning, there is no such evidence under British conditions. Indeed, if no control measures are adopted, the swarms of mites will disappear within a few days without damage to the sporophores or reduction in future cropping. Any abnormality in the pattern, or quantity, of sporophore production seems to be associated with unfavourable composts caused by unsuitable peak-heat conditions.

Occasionally, workers may be irritated by these mites and allergies have been reported. However, there is no advantage in attempting chemical control because acaricidal applications can only affect those mites on, or in, the casing and the mites will continue to swarm only while the food supply dwindles.

Pepper mite populations are undoubtedly reduced by large, long-legged, fast-moving red gamasid mites which feed both on mites and eelworms.

REFERENCES

HUSSEY, N. W. & GURNEY, B. (1967). The *Pygmephorus* spp. (Acarina: Pyemotidae) associated with cultivated mushrooms, *Acarologia*, **9**, 353–8.

KRCZAL, H. (1959). Systematik und Ökologie der Pymotiden. In H. J. Stammer, *Beiträge zur Systematik und Ökologie Mitteleuropäischer Acarina* **I**, Teil **2**, (3), 385–625.

MORETON, B. D. (1956). The riddle of the Red Pepper Mite. *Grower*, **46**, 853–5.

Class DIPLOPODA: Millepedes

Several species of millepedes, including *Blaniulus guttulatus*, *Choneiulus palmatus* and *Cylindroiulus britanicus*, eat holes in the base of the stipes of mushrooms. Such damage is uncommon in modern tray culture but may occur when beds are made on the floor of glasshouses.

The life history and control of millepedes is outlined on p. 183.

PHYLUM MOLLUSCA

Class GASTEROPODA: Slugs

Slugs are another group of potential mushroom pests which have virtually disappeared under modern cultural systems but, where beds are made on the floors of unsuitable buildings, they may cause trouble by eating large cavities in the caps and stalks. In one unusual case, slugs sought refuge behind the insulated lining of a modern cropping house which had become loosened by successive ‘cook-outs’. At night they crawled out and damaged mushrooms on the lowest boxes. Suitable control measures are outlined on p. 232.

Chapter 11

Mushroom Nematodes

General

Nematodes are invariably associated with mushroom growing. The substrate for mycelium growth—a compost of some kind—and the 'casing' soil, are favourable environments for saprobic* nematodes and their (eelworm) predators; the mycelium offers abundant food for those saprobes which are specifically fungal-feeders.

A wide range of organic and inorganic materials are used either as raw material for compost or as compost additives. None of these substances is sterile; all are subject to contamination from many sources. For these reasons the range of nematode genera to be found, or expected, in a compost is remarkably wide for, in addition to entirely saprobic forms, many plant-parasitic and animal-parasitic species may be encountered. The same holds true, but to a less extent, of 'casing' soil, for the soil mixtures, or peat and chalk, used for this must inevitably contain nematodes if dug from shallow beds or subjected to flooding or contamination.

Because of its diverse origins the eelworm population in each compost stack is unique. Different composting procedures may change the eelworm populations; as composting proceeds there may be a change of dominance in the nematode species that are present. The composting process involves microbial breakdown of the raw material and the processes involved in this breakdown generate much heat. The temperatures reached in the centre of a compost stack (160–170 °F; 71–77 °C) are lethal to all eelworms, but on the outside of the stack, where it is cooler, many eelworms persist. During composting each stack is an active biological mass in which chemical and physical conditions are continuously changing. It is therefore difficult to assess the part that eelworms play in the composting process, but evidence suggests that they are beneficial because they provide food substances for microorganisms which in turn produce the changes needed to convert raw material to a good medium for mushroom growth. There is no evidence that eelworms are harmful in raw compost; if they were, their specific control would be difficult, and reinfestation would quickly occur for larvae of many species are carried by compost-inhabiting arthropods such as flies and mites.

The presence of eelworms in a 'finished' compost is of little account if composting is completed by a pasteurization technique which ensures that all

* Those that feed on, or are associated with, rotting organic matter.

the compost (and its containers!) reach lethal temperatures (140 °F; 65·5 °C). In practice, however, pasteurization may not be possible, or it may be imperfect, and then eelworm survivors will be present when the compost is spawned. Mushroom beds may become infested with eelworms after a perfect pasteur-

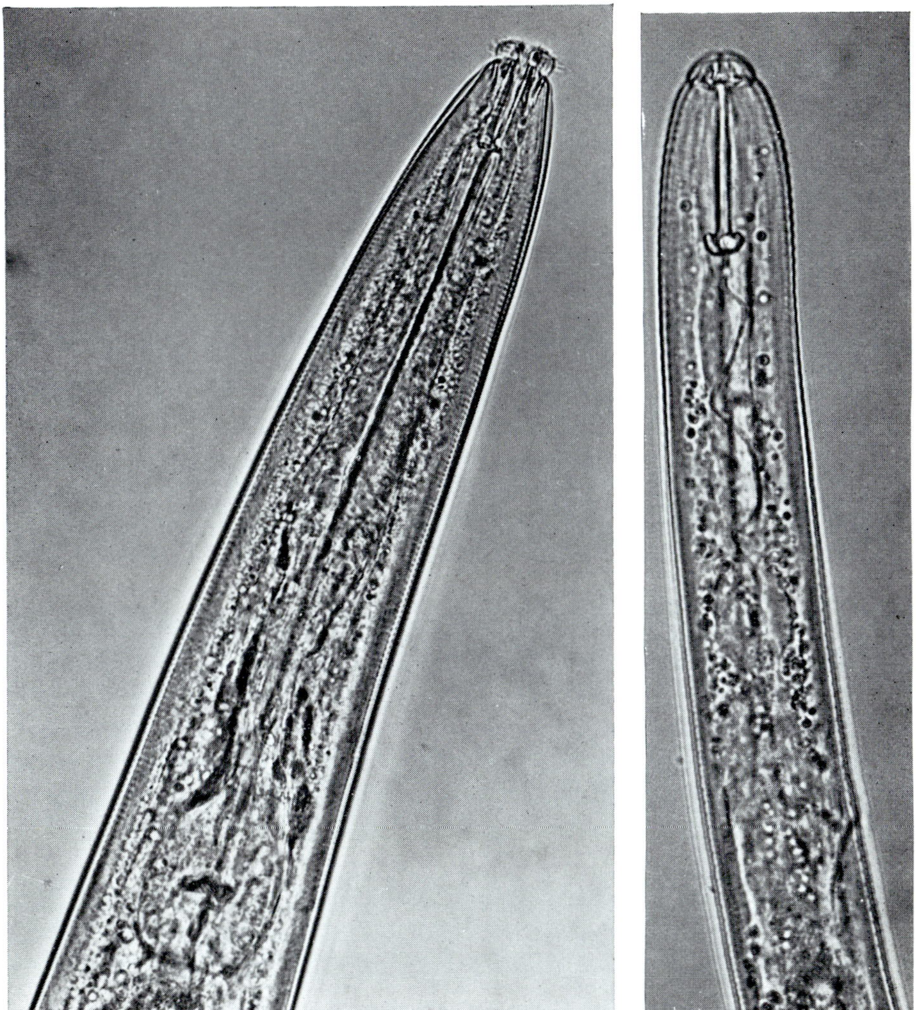

FIG. 11.1. (*a*) Head of rhabditid nematode (saprophagous) (× 600); (*b*) head of a spear-bearing nematode similar to that of fungal-feeding species (× 750). (J. Hesling.)

ization; for example, as a result of careless hygiene, or by introduction on a vector such as a fly, mite or other arthropod, or even on mice which occasionally steal grain-spawn. In practice, eelworm contamination usually occurs before cropping ends, for nematodes are ubiquitous and many have evolved efficient methods of self-dissemination. The effects of the eelworms on mycelial growth and cropping are dependent on their species and numbers, on their immediate, or subsequent, environment in the mushroom bed, and, perhaps above all, on their feeding habits. By their feeding habits, eelworms in mushroom beds can be classed into three main groups. Most common are members of the *saprophagous* (*microphagous*) *group* (Fig. 11.1*a*), which suck up liquid and particulate organic matter including bacteria; most of these belong to the order *Rhabditida*. With them is often found the *predacious group* comprising nematodes from families that are widely separated taxonomically. The third group is the most important because its members are tylenchid fungal-feeders and pathogenic to mushroom (Figs. 11.6, 11.7, 11.8, and 11.9).

Other nematodes, which cannot be placed into any of the above groups, are sometimes found, e.g. *Rotylenchus* spp., *Paratylenchus* spp., *Tylenchorhynchus* spp., *Tylenchus* spp., and some insect-parasitic species; usually all these occur in small numbers and can be dismissed as unimportant and harmless contaminants (Fig. 11.2).

The *predacious group* of mushroom nematodes are of no economic importance; there is no evidence that they usefully control other, pathogenic, forms. In the predacious group Diplogasterids and Aphelenchoidids (*Seinura*, Figs. 11.3 and 11.10) are commonly encountered; Monochids and Dorylaimoids occur less frequently, the last two taxa are probably introduced in unsterilized 'casing' (peat).

SAPROPHAGOUS (MICROPHAGOUS) NEMATODES

These, in mushroom beds, comprise many species and so form the largest group; their importance is difficult to ascertain. Insufficient is known about their biology, and their pathogenicity to mushroom cannot be satisfactorily proven because this requires the sterile culture of mushroom and eelworm together. This cannot be easily achieved because the larvae of some microphagous eelworms need living bacteria as food in order to mature; also, it is claimed that the production of mushrooms (sporophores) requires the presence of bacteria in the casing.

To *saprophagous* eelworms (Fig. 11.1) mushroom beds offer two different environments—compost and casing. Compost at spawning contains much moisture, its pH is approximately 7·5, and its temperature about 75 °F (24 °C). At this stage it is an ideal environment for the breeding of any saprophagous eelworms that have survived the pasteurization process, or have been introduced as contaminants. After pasteurization, compost cannot normally be maintained at its pasteurization-level of 'sterility', and any eelworms in it will feed, and multiply rapidly. However, as the mycelium colonizes the compost, the environment changes, for much free moisture is removed by the mycelium which also creates a bacteriostatic environment.

The 'dryness' of a compost colonized by vigorous mycelium growth probably limits eelworm activity, for all eelworms are dependent on free moisture (e.g. as water films) for efficient locomotion (and hence for breeding); insufficiency of bacteria might prevent eelworm maturation and so restrict increase. Thus, eelworm populations in freshly spawned compost may rise rapidly at first and

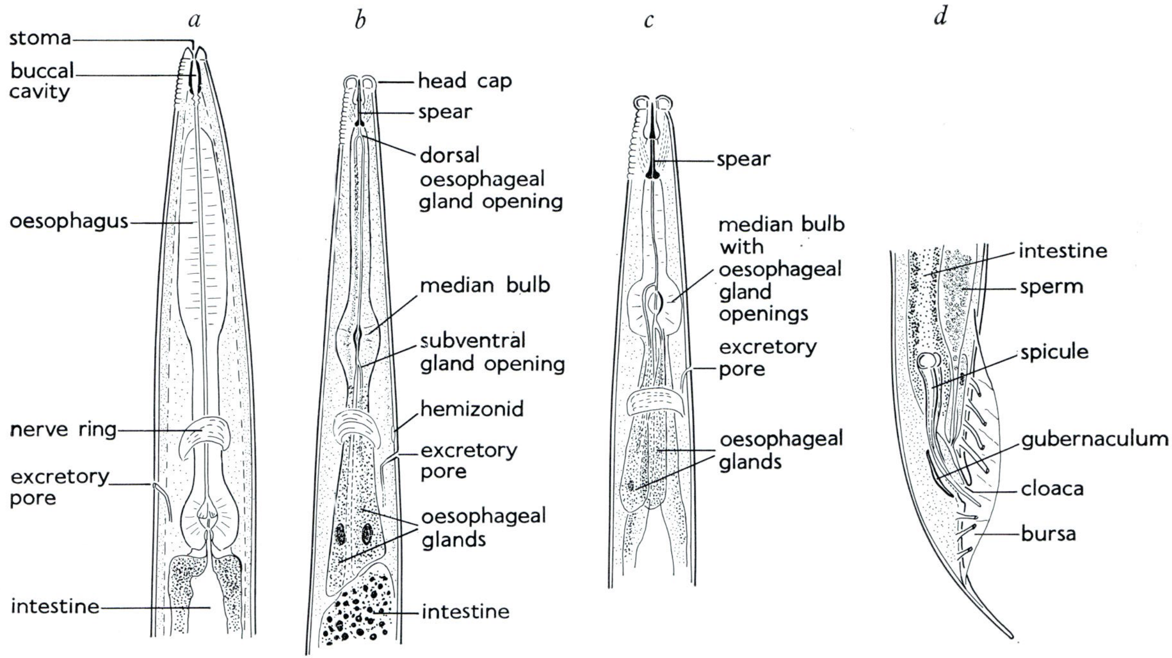

FIG. 11.2. The head and oesophageal regions (diagrammatic) of: (*a*) panagrolaimid—free-living microphagous form; (*b*) tylenchoid; and (*c*) aphelenchoid. (*b*) and (*c*) are both plant- or fungal-feeding forms. (*d*) Anatomy of the tail of a male Rhabditid nematode (hypothetical).

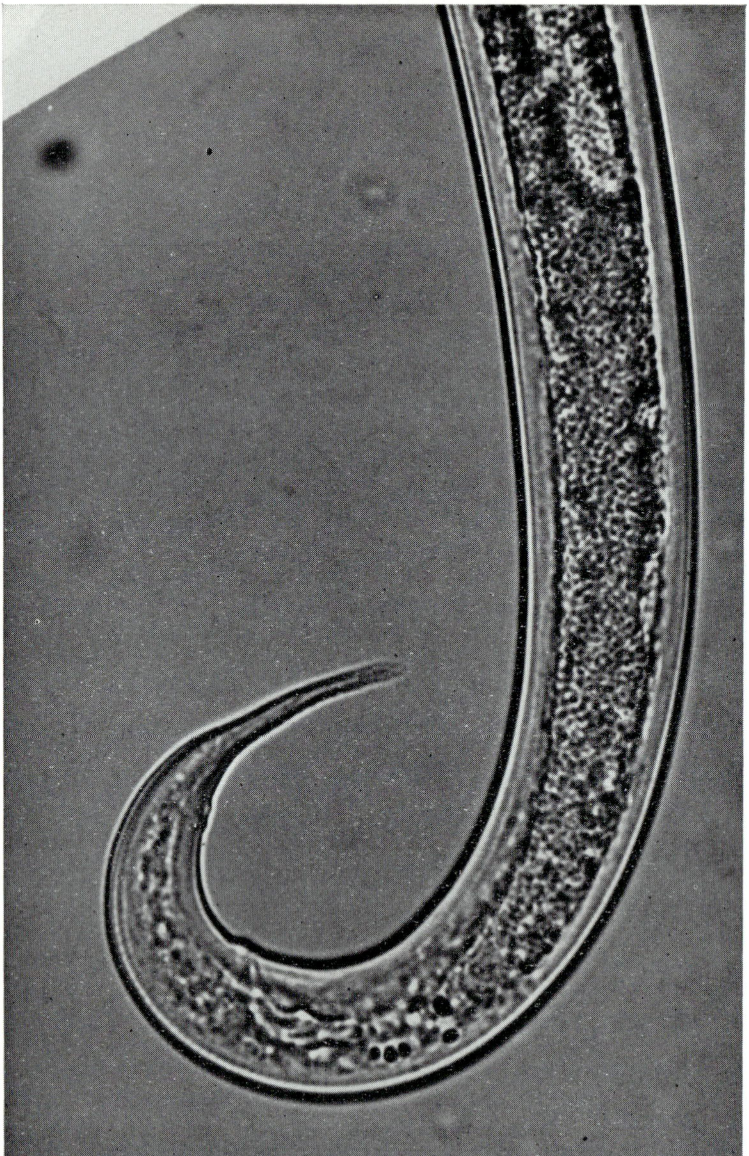

FIG. 11.3. Tail of male *Seinura* sp., a predatory nematode often found in compost (× 1100). (J. Hesling.) (Cf. Fig. 11.6.)

then decline to very low levels as the compost is colonized by mycelium. However, if the mycelium does *not* grow, as when mushroom competitors, diseases, and other pests are present, the eelworm populations may become very large.

These facts suggest that saprophagous eelworms are not primary pathogens of mushroom, but are indicators of bad growing conditions. In experiments, mycelium has successfully competed with and 'controlled' eelworm populations at a density of 1500 saprophagous forms per g compost. Although high, this figure is only a tenth of peak populations sometimes achieved, then mushrooms can hardly be expected to thrive.

Casing (from peat and chalk) seems a more favourable environment for saprophagous eelworms than 'healthy' compost. Ideally, it is moist, approximately neutral in pH, and not well colonized by mycelium. The temperature for cropping—about 62 °F (16·7 °C)—is still favourable for eelworm increase, and the moist conditions and absence of mycelium (and hence of bacteriostasis) favour eelworm movement and multiplication. Enormous populations of saprobic eelworms are often found in casing and some effect on cropping may become noticeable. Experiments have indicated that large populations of eelworms are associated with a more regular appearance of mushrooms—i.e. the 'flushes' are less clearly demarcated in time. Some mycelial decline may be evident after 4–5 weeks cropping but at this stage it is impossible to solely incriminate eelworms because of the co-existence of other micro-organisms. Nevertheless, eelworms possibly play a part in mycelial decline by encouraging bacterial activity unfavourable to mushrooms and everything should be done to discourage their entry to mushroom beds. One species, *Pelodera lambdiensis* is believed to transport the bacterium (*B. tolaasi*), the cause of 'bacterial blotch'; doubtless *any* nematodes in mushroom beds can act as vectors of micro-organisms (Steiner, 1933).

Under favourable conditions saprophagous nematodes increase rapidly, increase rates of 50- to 100-fold per week are commonplace. Some species are parthenogenetic, others hermaphroditic, so that, theoretically, a single larva introduced to a mushroom bed could initiate a population of millions in 3–4 weeks.

Saprophagous eelworms are ubiquitous and easily spread. They contaminate most moist organic matter, in many species there is adaptation to survive adverse conditions (such as drying) and for transport by larger organisms such as flies and mites. Larvae swarm on top of compost or casing and adhere readily to anything that touches them, each larva stands on its tail and waves the rest of its body in the air. Arthropods, human hands and implements can transfer these eelworms from one place to another. The number of swarming larvae is so immense that some successful transfer is assured. It is clearly advisable to try to exclude from mushroom houses any vectors such as flies, to exercise the greatest care with hygiene, and to 'sterilize' implements such as cutting knives at every opportunity. A dip in near-boiling water will kill all eelworms.

FUNGAL-FEEDING NEMATODES

These form the most important group of mushroom nematodes, for in some circumstances (general chronic infestation) they may render mushroom growing completely uneconomic. But in hygienic growing conditions, and with a cropping period of 6–8 weeks, these eelworms rarely cause serious trouble.

Many eelworm species are fungivorous and might, by chance, appear in mushroom beds. However, the various practices associated with mushroom culture, and the materials used for compost, seem to limit the number of species that is normally found in mushroom beds and these are described (pp. 361–6); of these *Ditylenchus myceliophagus* and *Aphelenchoides composticola* seem to occur most commonly.

Fungal-feeding eelworms are introduced with compost and 'casing'; they may persist from crop to crop in cellars and caves or in structures that cannot be efficiently sterilized. Once established they may be carried on contaminated implements, human hands and, in some cases, by arthropods. Although fungal-feeding eelworms seem not to have the behavioral adaptations for spread like some of the rhabditids (see p. 357), their numbers when swarming from compost are so astronomical that some spread by arthropods must occur. Some can remain viable in dry conditions and so can be dispersed by wind; *A. composticola* and *D. myceliophagus* can survive in an active state for at least 6 weeks without food, when temperatures are low.

Fungal-feeding eelworms harm their host by sucking out the contents of hyphal cells. Different species feed in different ways. *A. composticola* spends only seconds feeding at a chosen cell; *D. myceliophagus* may feed for over an hour. It first uses its mouth stylet to puncture the cell wall and inject digestive juices, later the products of this extra-corporeal digestion are drawn through the stylet, and ingested. In this way *D. myceliophagus* can affect several cells other than the one which has been punctured (Doncaster, 1966).

Whatever the mode of feeding, the mycelium is damaged, its cytoplasm weeps from punctures and it eventually dies. As the eelworms increase, the destruction progresses, and subsequent bacterial activity renders the compost soggy, foul, and smelly.

Most of the fungivorous eelworms that have been found harmful to mushrooms have very high rates of increase. Some (e.g. *Aphelenchus avenae*) are parthenogenetic. Increase rates of 100,000-fold in 4 weeks are commonplace. Increase is enhanced if compost moisture is high, and the eelworms seem to have a burst of activity after compost has been watered.

The effects of fungivorous eelworms on cropping is dependent both on the stage of growing at which the beds become infested and on the initial number of eelworms. It may range from failure of the 'spawn-run' on one hand (when compost is infested at spawning)—to slight loss of yield (when eelworms are introduced at the end of cropping). Failure of 'spawn-run' is a catastrophe likely to occur in contaminated caves or cellars where disinfection is impossible, or when peak-heating is not practised; slight loss of yield may occur on the most hygienic farms and may pass, unnoticed. The earliness of eelworm diagnosis is dependent on the detail with which the course of a crop is followed. The eelworm first destroys the fine mycelial threads but at this stage there is little of diagnostic value for the grower to see and few would wish to interfere with the beds without reason. Cropping records might later reveal an inexplicable decline and be the first indication of eelworm trouble. Suspect beds should then be examined—taking care not to spread eelworm in the process. Compost samples can be processed on the Baermann funnel (Fig. 11.4) and the collected eelworms identified under a microscope. It is not too difficult to identify the fungal-feeding genera but the identification of species requires specialist knowledge—see p. 363.

In infested compost the fine mycelial weft declines or disappears, and the mycelium seems 'stringy' because only coarse strands remain. If cropping has ceased, little mycelium is left and ultimately the compost becomes soggy,

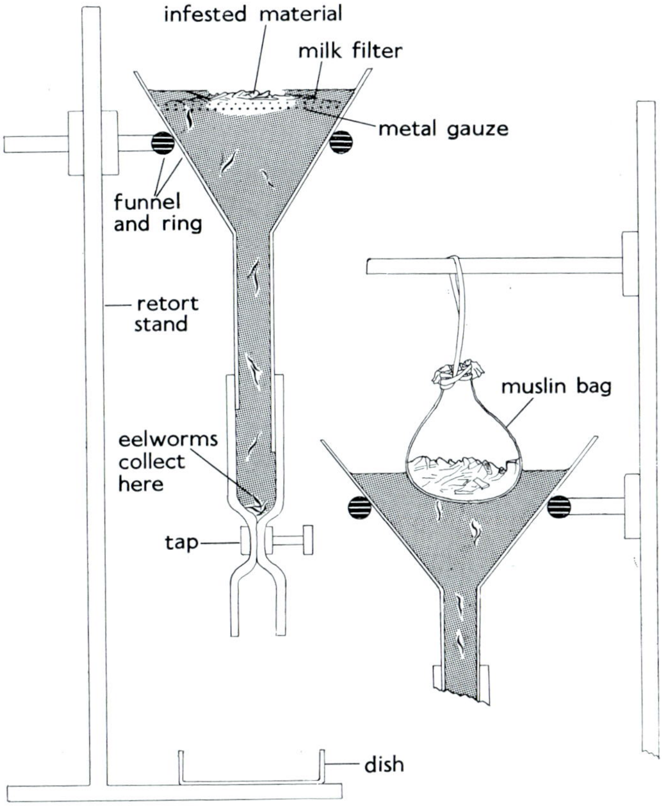

FIG. 11.4. The Baermann funnel. A simple method of extracting eelworms from infested plants, soil, or compost. Infested material is cut up or crumbled, and spread thinly on a cotton wool filter ('milk filter') supported on a 1-mm. (approx.) mesh gauze in a funnel of fresh water. Eelworms migrate from the infested material and fall to the bottom of the funnel. After several hours the tap is opened briefly and the eelworms are run out into the dish for examination. An alternative method is to suspend infested material in a coarse muslin bag dipping into the water of the funnel.

with a characteristic odour. Affected compost sinks in patches and a fine greyish mould (which is an eelworm-trapping fungus) often develops on its surface. At this stage the eelworms tend to leave their fouled environment and millions migrate from it, often congregating in 'curds' or masses (Fig. 11.5), each comprising thousands of eelworms in all stages (including eggs).

These curds may be found on compost, casing and boxes. In drying conditions, eelworms in such curds curl up like watchsprings. If there is ample moisture the eelworms may hang from boxes as inch-long 'stalactites', which break to drop eelworms to lower boxes and to the floor. This is the stage at which spread

FIG. 11.5. *Ditylenchus myceliophagus* swarming on mushroom bed ($\times 200$). (J. Hesling.)

by arthropods and human hands is inevitable, as well as that by wind or in draining water.

Some details of the biology of some fungal-feeding eelworms are given below.

Aphelenchoides bicaudatus (Imamura) Fil. and Sch. Stek. This eelworm, recently redescribed by Siddiqui and Taylor (1967) is a widely distributed soil nematode which can be distinguished by its small size (0·5 mm long) and by

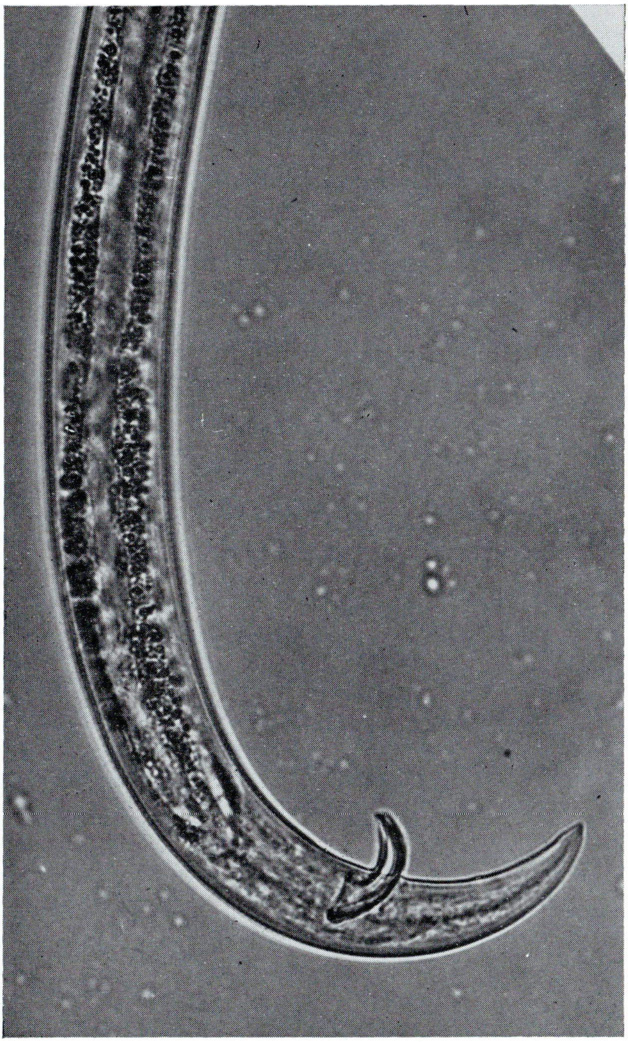

FIG. 11.6. Tail of male *Aphelenchoides composticola* (× 1100). (J. Hesling.) N.B. Thorn-shaped spicule.

the bifurcate tail of the female. It is a parasite of mushroom and can reduce yield, but is considered unlikely to be a serious pest because of its slow rate of reproduction in mushroom beds (McLeod, 1967).

Aphelenchoides composticola Franklin, 1957 (Figs. 11.6 and 11.7). This species has been reported from most countries with a mushroom industry. In addition to mushroom it feeds and breeds on many common fungi; it survives freezing and desiccation, and starvation for at least 6 weeks at about 60 °F and

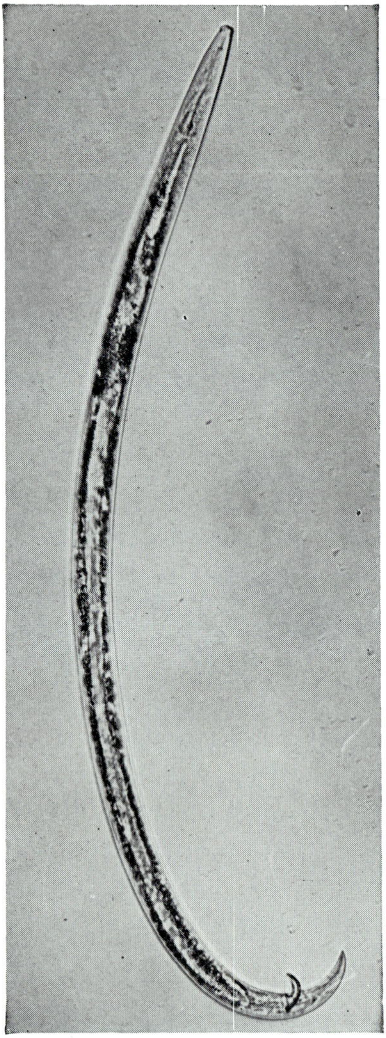

FIG. 11.7. *Aphelenchoides composticola*—adult male (× 250). (J. Hesling.)

longer at cooler temperatures. On mushroom it breeds rapidly, increase rates of 100,000-fold in 6 weeks are common, and the eelworms become so numerous that they cluster together in curds. In water *A. composticola* tend to adhere to one another and after an hour or so form whitish clumps of thousands of eelworms. This behaviour probably ensures that new infestations are started by at least several eelworms—this is essential because *A. composticola* is a bisexual species and does not reproduce by parthenogenesis. The life cycle from egg to adult occupies 10 days at 64 °F (18 °C), and 8 days at 73 °F (23 °C) (Cayrol, 1967a).

A. composticola is small (about 0·5 mm long); it varies in size and shape according to its food supply; peak populations that have destroyed mycelium comprise smaller individuals and care is needed to avoid misidentification.

Aphelenchoides cyrtus Paesler, 1957. This species was described from mushroom beds in Germany. Taxonomically it is very close to *A. composticola*, a valid difference seems to be in the length of the post-vulval sac which, in *A. cyrtus*, extends less than one-quarter of the distance between vulva and anus; in *A. composticola* the post-vulval sac extends one half to two-thirds of the distance to the anus. Little is known of its biology.

Aphelenchoides dactylocercus, and ***A. sacchari*** both, Hooper 1958. The type habitat of *A. dactylocercus* was described as 'a pot of soil growing young Sitka spruce trees'; that of *A. sacchari* as 'a rotting root of sugar cane from Jamaica'. Investigations showed that both species fed and multiplied on, and destroyed, mushroom mycelium. Both of these eelworms represent examples of a world-wide and possibly large number of undescribed *Aphelenchoides* species that are pathogenic to mushroom and which may, by chance, find their way into a mushroom bed. *A. dactylocercus* seems to be a parthenogenetic form; *A. sacchari* is dioecious but males occur in the ratio of 1:50 females. *A. sacchari* has been reported from mushroom beds in Poland.

Aphelenchoides limberi Steiner, 1936 is characterized by the absence of males and by the obtuse-to-truncate female tail; there is, however, great variation in the form of the tail-tip. Hooper (1962) showed that this species reproduced rapidly on mushroom, and on *Botrytis cinerea*. In Europe and the U.S.A. the species has been found in plant material in contact with soil; it is probably entirely mycophagous. Reproduction is parthenogenetic; increase rates of 1500-fold in 4 weeks have been reported on mushroom cultured on agar, but in mushroom beds its occurrence seems rare and then its numbers are often low. Cayrol (1967b) showed that the optimum temperature for reproduction was 64 °F; no reproduction occurred at 77 °F.

Aphelenchoides saprophilus Franklin, 1957, is frequently found in rotting plant tissues that have been in contact with soil and so it is likely to find its way into mushroom beds. It is a dioecious species that feeds on other common fungi; little else is known about it.

Aphelenchoides spinosus Paesler, 1957. This meagrely described species was discovered in association with mushroom growing in Germany. It is a small, dioecious species; a characteristic feature is the short broad male tail which has a sharply pointed, offset, prominent mucro. Nothing seems known of its biology; its pathogenicity to mushroom was only inferred when it was described.

The taxonomy and identification of *Aphelenchoides* is difficult. Many of the species listed above are included in the '*parietinus*' group within *Aphelenchoides*, and this group alone (which includes species at one time possibly identified as *A. parietinus* (Bastian, 1865), Steiner, 1932—a lichenicolous form re-described by Franklin in 1955) comprises perhaps fifteen species which need specialist identification. Inclusion of the detail needed to identify *Aphelenchoides* from mushroom beds is outside the scope of this book. Reference should be made to Sanwal (1961; 1965); Franklin (1955; 1965), Siddiqi & Franklin (1967) and Siddiqui & Taylor (1967). The few publications listing nematodes found in mushroom beds each record the occurrence of one or more unidentified species of *Aphelenchoides*. These occurrences must be expected. There is always some risk that a persistent infestation of *Aphelenchoides* may arise on a mushroom farm and this is another reason for strict hygiene, efficient pasteurization and 'cooking-out' after cropping.

Ditylenchus destructor Thorne, 1945. Before the description of *D. myceliophagus* by Goodey in 1958 (see below), *Ditylenchus* nematodes in mushroom beds were usually identified as *D. destructor*. *D. destructor sensu stricto* is thought of today as a pest of higher plants; it is dealt with on p. 297. Some isolates of *D. destructor* thrive only on higher plants, but others can be cultured on mushroom, which they destroy. Since 1958 most reports of *Ditylenchus* damage in mushroom beds cite *D. myceliophagus* as the pathogen, but the possibility that *D. destructor* may be responsible must not be overlooked. Correct identification is needed, not only to clarify the taxonomy of isolates of *D. destructor*, but also because this species does not survive drying, as does *D. myceliophagus*. Both species are about 1 mm long and very similar; features useful in differentiating these two species are: the relative sizes and shapes of the tails of both sexes, the shape of the spicules, and the length of the post-vulval sac (Figs. 8.16 and 8.17). For further details, reference should be made to Goodey (1958) and Thorne (1945).

Ditylenchus myceliophagus Goodey, 1958 (the mushroom spawn nematode). This species (Fig. 11.8) thrives on many common fungi; it is a worldwide pest of mushrooms on which it breeds rapidly at normal growing temperatures. Below 45 °F (7 °C), and above 78 °F (26 °C), its rate of increase declines rapidly; at 65 °F (18 °C), the life cycle is completed in about 3 weeks, and increase rates of 50,000- to 100,000-fold in 8 weeks are commonplace. The species is dioecious; parthenogenetic reproduction is unknown. *D. myceliophagus* is able to withstand slow drying, late-stage larvae and adults being most resistant; the eelworm has been revived by re-wetting infested compost after 3 years. *D. myceliophagus* is killed by rapid drying; on a mushroom holding this is likely to occur only on the surface of a bed and in a dry atmosphere. Because dry nematodes seem more resistant to heat and to chemicals, dry infested materials should be damped down before control treatments are applied.

Paraphelenchus myceliophthorus Goodey, 1958 (Fig. 11.9). *P. myceliophthorus* was discovered in mushroom compost; experiments showed that as few as eight eelworms per 100 g compost could destroy the mycelium in 8 weeks with an eelworm increase of over 50,000-fold in this time (Goodey, 1960). Fortunately this species seems uncommon. Little else is known about its biology, though it is known to feed on the fungi *Alternaria tenuis* and *Botrytis cinerea*, and it can be assumed that it normally exists in rotting organic matter where it feeds on fungi.

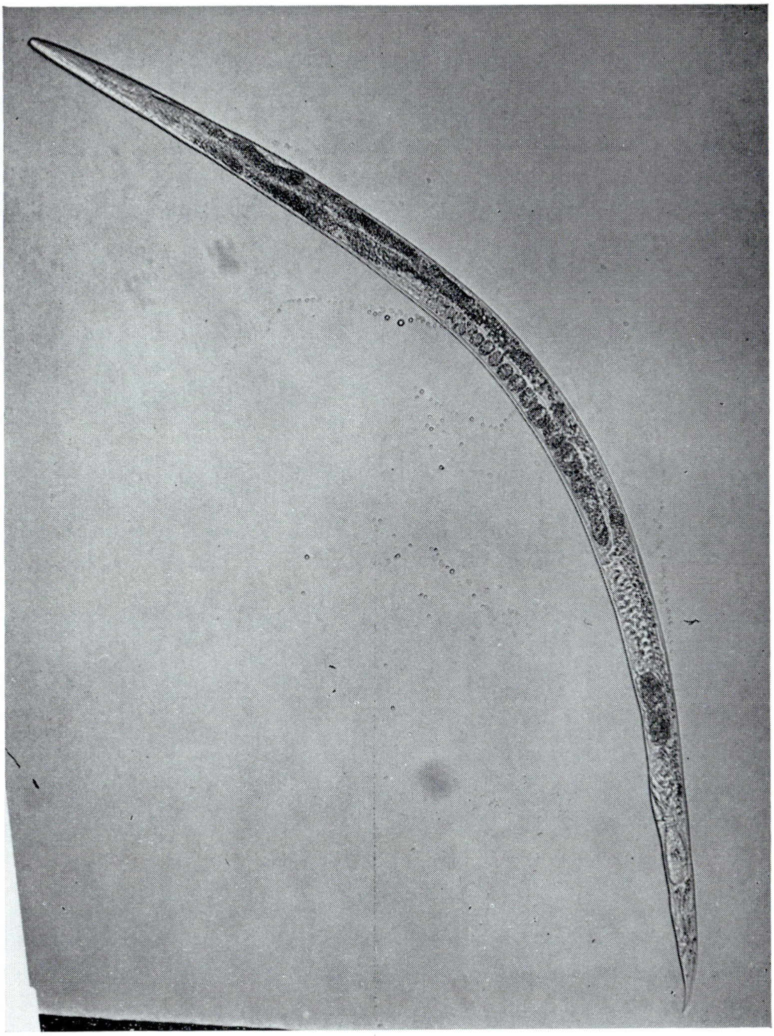

FIG. 11.8. *Ditylenchus myceliophagus*—adult female (×170). (J. Hesling.)

Aphelenchus avenae Bastian, 1865. This is a common cosmopolitan species that has been shown to destroy mycelium in mushroom beds as well as in laboratory cultures. However, on mushroom farms it seems to occur in damaging numbers only infrequently. *A. avenae* is about 0·75 mm long; characteristic features are its bluntly rounded tail and a marked narrowing of the body immediately behind the vulva. Different isolates show variation in size, internal morphology, and in sex ratio; males are usually extremely rare (1:80,000 females). The species is parthenogenetic and its reproductive rate on mushroom is 70,000- to 90,000-fold in 3 weeks. At 65 °F (18 °C), the life cycle is completed in about 2 weeks. *A. avenae* can be cultured on many soil and saprophytic fungi and seems on occasion to live on higher plants. These factors doubtless account for its cosmopolitan nature. It is a fungal-feeding

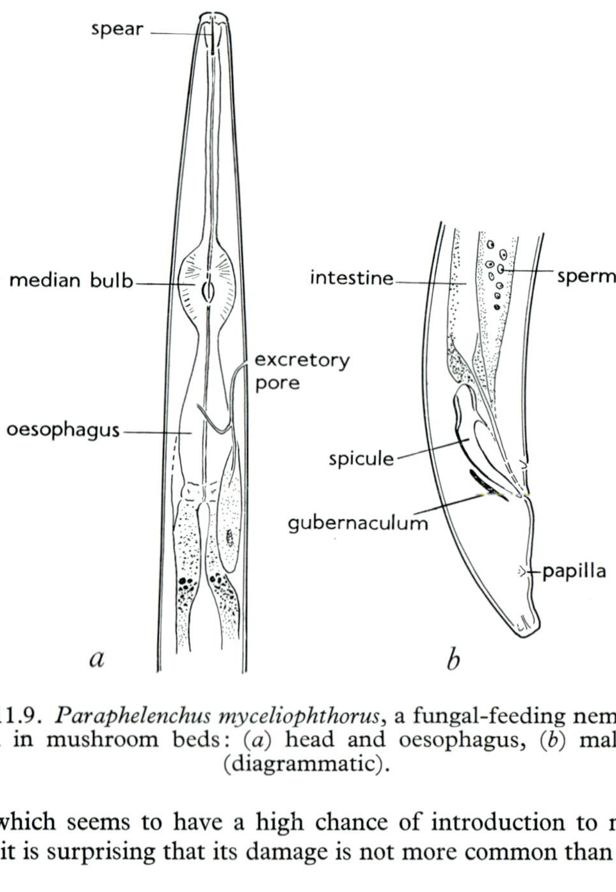

FIG. 11.9. *Paraphelenchus myceliophthorus*, a fungal-feeding nematode found in mushroom beds: (*a*) head and oesophagus, (*b*) male tail (diagrammatic).

eelworm which seems to have a high chance of introduction to mushroom beds, and it is surprising that its damage is not more common than seems the case.

Control of mushroom eelworms

There is nothing to be gained by controlling eelworms during composting before pasteurization (peak-heat), because recontamination would be inevitable in commercial practice. On most farms the chance of contamination after pasteurization is much less; it is vital that pasteurization is effective and uniform so that the crop has a pest-free start. Ensuring thorough peak-heat is an exercise in heat transfer; this exercise must vary from farm to farm according to the equipment and heating facilities available, as well as to the type of compost. Generally distributed rather than local sources of heat, ample circulation of heated air, and open packing of boxes and of compost, help to ensure adequate and uniform temperatures. Cold spots near badly fitting doors, or similar sources of draught, will take much longer to heat, and usually convection causes cooler conditions near the floor. It is unwise to place the lowest boxes directly on the floor unless this is heated. Attainment (by the eelworms!) of the usual peak-heat temperatures of 135–140 °F (57–60 °C) for a few minutes is lethal.

If peak heating is not possible—as when mushrooms are grown as a catch crop in glasshouses—'finished' compost can be fumigated with methyl bromide before spawning. This should only be done by contractor's personnel, equipped with effective respirators, necessary protective clothing, gas-detection equipment and warning notices, for methyl bromide is a very potent and insidious poison. In some countries (e.g. the U.K.) only approved operators should use this fumigant. A proven practical method is to seal the compost stack with polythene sheeting beneath which canisters of gas are placed on special spiked stands. The sides of the sheet must be well sealed, e.g. by burying them in a trench, or by weighting them down with loose, moist sand, or earth. There should be no leaks. When all is sealed, the canisters are pressed on to the spikes by hand and the methyl bromide is released. Fumigation then proceeds for a time dependent on the quantity of gas used, rate of any leakage and amount of sorption, until the C.T.P.* exceeds a value of 600 mg/hr/l. This dose is sufficient to kill arthropods and eelworm *pests* (but it may not control fungal diseases, for methyl bromide is not an efficient fungicide at this C.T.P. unless the temperature is raised to 75° F.) (see p. 34).

After fumigation the polythene sheet is removed and the stack ventilated until gas has dispersed. The compost can then be boxed, or made into beds, and spawned in the usual way. Methyl bromide is a penetrating gas and seems effective throughout compost stacks.

The calculation of the volume of a covered and amorphous compost stack, so that an economical quantity of methyl bromide is used is only one of the practical problems associated with this fumigant. The weight of methyl bromide required depends on the volume of the enclosed compost stack and on the losses due to fumigant sorption and any leakage. The amount of the latter is greater with long duration treatment and so the duration of treatment must be taken into account when estimating the amount of methyl bromide required. It is important to monitor the concentration at intervals so that any loss of fumigant may be compensated.

If the vapour concentration in the stack can be measured at intervals it is possible to decide not only how long fumigation should take but also whether or not 'topping-up' is needed, so that a C.T.P. of at least 600 mg h/l. is achieved.

* C.T.P. = concentration time product. *Within limits*, a brief exposure to a large dose of fumigant has the same effect as a long exposure to a small dose. Thus, a vapour concentration of 1 g/l. for 1 hr has the same effect as 0·1 g/l. for 10 hr. C.T.P. is usually expressed as milligrams (0·001 g)/hr/l., or oz/hr/1000 ft^3, these two products being equivalent for all practical purposes. In the above example the C.T.P. would be

$$\left.\begin{matrix}(1000 \text{ mg}/1 \text{ hr/l.})\\(100 \text{ mg}/10 \text{ hr/l.})\end{matrix}\right\} 1000 \text{ mg/hr/l.}$$

If the gas concentration cannot be measured, the use of 1 lb of methyl bromide per $3\frac{1}{2}$ yd^3 of closely sheeted compost for 24 hr will provide adequate control provided that the stack height does not exceed 5 ft, and that the compost is not wet and soggy.

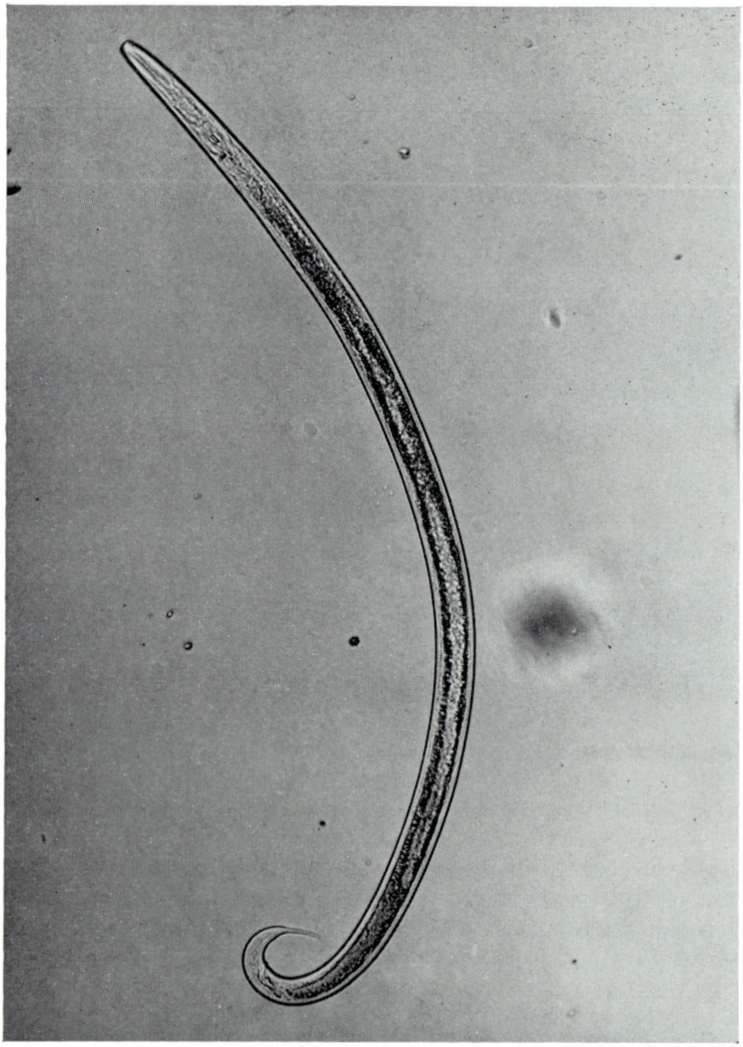

FIG. 11.10. *Seinura* sp.—adult male ($\times 250$). (N.B. long thin tail, cf. *Aphelenchoides*.) (J. Hesling.)

If mushroom beds become infested with destructive eelworms during cropping there is little that can be done because there are no liquid nematicides available that are safe to use—all are fungitoxic, hazardous to man, or unsuitable in some other way. Infested trays or compost should be removed and thoroughly heat-sterilized, taking great care not to spill and thereby spread eelworms. In most cases it is extremely likely that local eelworm spread will have occurred before action is taken, and every precaution should be observed to avoid spillage of any sort, and so prevent eelworm transfer (as contamination) to other houses. At the end of cropping it is wise to regard the infestation as universal in the infested house, and to 'cook-out', or fumigate, every bit of compost, each tray, structure, and, in fact, everything humanly possible. Experiments have shown that thionazin at 20 p.p.m. incorporated in compost controls fungal-feeding eelworms satisfactorily and this technique may be developed commercially.

'Cooking-out' at the end of cropping is the best way to control eelworm pests. As with peak-heating, success in pest control by 'cook-out' is achieved only by ensuring that the eelworms in infested compost and structures reach a lethal temperature; the general principles of effecting efficient heat transfer, outlined on p. 367, similarly apply. Eelworm control during 'cook-out' has a greater chance of success than during peak-heat for the temperatures reached can be, and are, usually higher and of ample duration.

Some mushroom houses cannot be 'cooked-out'; if they are infested, the trays, shelf boards and other structures should be removed and sterilized by steaming or immersion in boiling or near-boiling water. Cresylic-acid-type disinfectants may be added to the water to kill heat-resistant diseases. Exposure to heat must be long enough (not less than 10 min) to ensure effective treatment. A quick dip is not enough. Structures which cannot be removed for treatment can be heated with a jet of steam, or washed down liberally with a solution of formaldehyde (formalin). A 2% solution of commercial formalin may be effective, but if the house to be treated is wet, the concentration should be increased to, say, 5% to offset the diluting effect of the free water in the structures. Walls and floors should be sterilized at the same time. Washing-down, or 'flash' steaming cannot kill eelworms protected in crevices in wood, and these survivors can be expected to attack the next crop. Intercrop treatments become essential. Rubber gloves, boots, respirator, hood and waterproof overall or oilskins should be worn when washing down with formalin.

Eelworms may be introduced in casing material, but this may be pasteurized before use. Steam can be used for this purpose, the casing being heated throughout to, and maintained at, at least 140 °F (60 °C) for 10 min. Great care must be taken to prevent its recontamination, e.g. by transport in dirty barrows, or by piling it on 'dirty' ground. Eelworms may breed in wet peat so it is essential to store it in a dry state. It cannot be over-stressed that the grower's most effective defences against eelworms are: insistence on hygiene; exclusion of eelworm vectors such as flies; and routine and thorough 'cooking-out' at the end of cropping.

REFERENCES

BRZESKI, M. W. & JANKOWSKA, J. (1966). [Nematode survey in cultivated-mushroom beds]. *Roczn. Nauk Roln.*, Ser. A. Roslina, **91**, 245–54.

CAYROL, J. C. (1964). Études preliminaires sur le cycle evolutif de *Ditylenchus myceliophagus* Goodey. *Nematologica*, **10**, 361–8.

— (1967a). Étude du cycle evolutif d'*Aphelenchoides composticola*. *Nematologica*, **13**, 23–32.

CAYROL, J. C. (1967b). Les conditions de culture et de milieu et le développement des principaux nématodes nuisibles aux champignons de couche. *Mushr. Sci.* **VI**. 475–84.

DONCASTER, C. C. (1966). Nematode feeding mechanisms. 2. Observations on *Ditylenchus destructor* and *D. myceliophagus* feeding on *Botrytis cinerea. Nematologica,* **12**, 417–27.

FRANKLIN, M. T. (1955). A redescription of *Aphelenchoides parietinus* (Bastian, 1865) Steiner, 1932. *J. Helminth.*, **29**, 65–76.

— (1965). *Aphelenchoides. Tech. Bull. Minist. Agric. Fish. Fd,* **7**, 2nd ed. 131–41.

GOODEY, J. B. (1958a). *Ditylenchus myceliophagus* n.sp. (*Nematoda: Tylenchidae*). *Nematologica,* **3**, 91–6.

— (1958b). *Paraphelenchus myceliophthorus* n.sp. (*Nematoda: Aphelenchidae*). *Nematologica,* **3**, 1–5.

— (1960). Observations on the effects of the parasitic nematodes *Ditylenchus myceliophagus, Aphelenchoides composticola,* and *Paraphelenchus myceliophthorus* on the growth and cropping of mushrooms. *Ann. appl. Biol.*, **48**, (3), 655–64.

GOODEY, J. B. & HOOPER, D. J. (1965). A neotype of *Aphelenchus avenae* Bastian, 1865 and the rejection of *Metaphelenchus* Steiner, 1943. *Nematologica,* **11**, 55–65.

HECHLER, H. C. (1962). The development of *Aphelenchus avenae* Bastian, 1865 in fungus culture. *Proc. helminth. Soc. Wash.*, **29**, 162–7.

HESLING, J. J. (1963). Eelworms and mushroom-growing. *Mushr. Sci.*, **V**. 497–502.

— (1966). The effect of some microphagous saprobic nematodes on mushroom culture. *Ann. appl. Biol.*, **58**, 477–86.

LENGWEILER-REY, V. (1966). Über die Bekämpfung von Nematoden in Champignonbetrieben mit Kistensystem. *Z. Pflkrankh. Pfl. Path. PflSchutz,* **73**, 46–56.

MCLEOD, R. W. (1967). *Aphelenchoides bicaudatus,* a parasite of cultivated mushroom. *Nature, Lond.*, **214**, 1163–4.

MORETON, B. D. (1966). Mite and Nematode Pests. *M.G.A. Bull.*, **204**, 691–701.

MORETON, B. D. & JOHN, M. E. (1962). *Peak heat:* Is yours what you think it is? *M.G.A. Bull.*, **147**, 91–6.

MORETON, B. D., JOHN, M. E., & GOODEY, J. B. (1957). *Aphelenchoides* sp. destroying mushroom mycelium. *Nature, Lond.*, **177**, 795.

PAESLER, F. (1957a). Beitrag zur kenntnis Nematodenfauna in Champignonkulturen. *Nachr Bl. dt. PflSchutzdienst, Berl.*, **7**, 129–36.

— (1957b). Beschreibung einiger Nematoden aus Champignonbeeten. *Nematologica,* **2**, 314–28.

SANWAL, K. C. (1961). A key to the species of the nematode genus *Aphelenchoides* Fischer 1894. *Can. J. Zool.*, **39**, 143–8.

— (1965). Appraisal of taxonomic characters of '*Parietinus* Group' of species of the genus *Aphelenchoides* Fischer, 1894 (*Nematoda: Aphelenchoididae*). *Can. J. Zool.*, **43**, 987–95.

SIDDIQI, M. R. & FRANKLIN, M. T. (1967) *Aphelenchoides goodeyi* n.sp. (*Nematoda: Aphelenchoidea*) a mycophagous nematode from South India. *Nematologica,* **13**, 125–30.

SIDDIQUI, I. A. & TAYLOR, D. P. (1967). A redescription of *Aphelenchoides bicaudatus* (Imamora, 1931) Filipjev & Schuurmans Stekhoven, 1941 (*Nematoda: Aphelenchoididae*), with a description of the previously undescribed male. *Nematologica,* **13**, 581–5.

STEINER, G. (1933). *Rhabditis lambdiensis,* a nematode possibly acting as a disease agent in mushroom beds. *J. agric. Res.*, **46**, 427–35.

THORNE, G. (1945). *Ditylenchus destructor,* n.sp., the potato rot nematode, and *Ditylenchus dipsaci* (Kuhn, 1857) Filipjev 1936, the teasel nematode (*Nematoda: Tylenchidae*). *Proc. helminth. Soc. Wash.*, **12**, (2), 27–34.

TOWNSHEND, J. L. (1964). Fungus hosts of *Aphelenchus avenae* Bastian, 1865 and *Bursaphelenchus fungivorus* Franklin & Hooper, 1962 and their attractiveness to these nematode species. *Can. J. Microbiol.*, **10**, 727–37.

Appendix I

Information on Chemicals Used to Control Glasshouse Pests

Scheduled chemicals are those to which the compulsory requirements of the Agriculture (Poisonous Substances) Regulations apply. It should not be assumed, however, that unscheduled chemicals are harmless.

'*Part*' indicates the group (part) into which the scheduled chemical has been placed in the appendix to the above regulations. See p. 38 for protective clothing required.

L.D.50 figures (see p. 36) give some indication of the relative dangers of the chemicals. Other factors, however, such as the method of application, duration of exposure and previous exposure to other chemicals, greatly influence the risks of poisoning.

Acceptable uses are those agreed, under the Pesticides Safety Precaution Scheme, as not constituting a hazard to consumers and wild life. They do not imply that the chemical is not damaging to plants. A non-edible crop is one which is not used for human or animal consumption, either wholly or in part.

Always read the directions accompanying an insecticide before using it. The information presented here is believed to be correct at the time of writing but the reader should remember that amendments may be issued from time to time.

ALDRIN

Not scheduled.
Properties—Acute oral L.D.50 = 40–60 mg/kg: dermal L.D.50 > 200 mg/kg. A very persistent organochlorine insecticide—both in the environment and in animal tissues.

Acceptable glasshouse uses in Great Britain

1. As a dust in potting composts used for ornamental plants at rates up to 2 oz of $1\frac{1}{4}\%$ dust/bushel of compost for controlling vine weevil.
2. As a narcissus bulb dip, at a rate up to 6 oz a.i./20 gal (0·187%), for the control of narcissus fly.

AZINPHOS-METHYL

Scheduled—Part III.
Properties—Acute oral L.D.50 = 7–13 mg/kg: dermal L.D.50 = 280 mg/kg. A non-systemic organophosphorus insecticide and acaricide.

Acceptable glasshouse uses in Great Britain

1. May be used on any non-edible crop.

2. As an aerosol on cucumbers. Up to 5 applications per crop—each not exceeding 1·5 g/1000 ft^3 of glasshouse space. There must be an interval of at least 48 hr between the last application and picking.
3. As an aerosol on tomatoes. Up to 3 applications per crop—each not exceeding 1·5 g/1000 ft^3. There must be an interval of at least 48 hr between the last application and picking tomatoes.

AZOBENZENE

Not scheduled.
Properties—no data available on oral and dermal L.D.50's.
A specific acaricide—effective against eggs and immature mites. Volatile and non-persistent. Superseded in commercial glasshouses by more potent acaricides, but smoke formulations are convenient for amateur use. Available to commercial growers, admixed with parathion, in smoke canisters.

Acceptable glasshouse uses in Great Britain

May be used as an aerosol or smoke on any crop. The regulations applying to parathion must be observed when using azobenzene-parathion mixtures.

BHC

BHC is available either as the 99% pure gamma-BHC (lindane) or as technical BHC containing only a proportion of the insecticidally active gamma-BHC. Gamma-BHC should be used in preference to technical BHC, although the latter may be used in pre-planting treatments to control certain soil pests. The data below relates to gamma-BHC (lindane).
Not scheduled.
Properties—Acute oral L.D.50 = 200 mg/kg: dermal L.D.50 = 500 mg/kg.
An organochlorine insecticide effective against a wide range of insects. Somewhat volatile and of only moderate persistence on foliage, although more persistent in the soil. Gamma-BHC is liable to taint strawberries and grapes so it should not be applied after the fruits have formed.

Acceptable glasshouse uses in Great Britain

BHC may be used on any edible, or non-edible, crop provided the following intervals elapse between last application and harvesting:

Sprays and dusts—14 days
Aerosols and smokes (including BHC with DDT)—48 hr
Thermal vaporizers—no officially recommended interval, but 48 hr is advisable if the rate of vaporization was high.
Soil drenches, where there is no contamination of the aerial parts of plants, require no interval.

With thermal vaporizers (electric fumigators) the amount of BHC (or of a BHC-DDT mixture) should not exceed 0·5 g/1000 ft^3 in any period of 7 days. Glasshouses, or other buildings, should not be occupied if gamma-BHC (or the BHC-DDT mixture) is being vaporized at a rate exceeding 5 mg/hr 1000 ft^3. If this rate is exceeded the building must be aired before occupation.

BINAPACRYL

Not scheduled.
Properties—Acute oral L.D.50 = 140–185 mg/kg: dermal L.D.50 = 1350 mg/kg (rabbit).
A dinitro compound of high acaricidal activity but phytotoxic. Has shown promise for the control of red spider on cut-back or rested roses. Fungicidal to some powdery mildews.

Acceptable glasshouse uses in Great Britain

May be used only on non-edible crops.

CHLORFENSON

Not scheduled.
Properties—Acute oral L.D.50 = 2000 mg/kg.
A specific acaricide, somewhat phytotoxic to cucumbers. Now largely replaced by more potent acaricides. Normally applied as an aerosol. Moderately persistent but slow in action since it kills only young stages and eggs.

Acceptable glasshouse uses in Great Britain

May be used on any crop: no specified interval between application and harvesting.

CHLORFENVINPHOS

Scheduled—Part III substance.
Properties—Acute oral L.D. 50 = 10–39 mg/kg: dermal LD 50 = 31–108 mg/kg.
An organophosphorus compound, very effective against dipterous larvae, including sciarid larvae.

Acceptable glasshouse uses in Great Britain

May be used on non-edible crops.
May be admixed with mushroom composts at the time of preparation of the compost at a rate not exceeding 4 oz active ingredient per ton of compost.

CRESYLIC ACID—see TAR ACIDS

DAZOMET

Not scheduled.
Properties—Acute oral L.D.50 = 500 mg/kg.
A soil fumigating powder or 'prill', technical products usually containing 85% active ingredient, effective against some nematodes and soil fungi. Functions by conversion to methyl isothiocyanate (q.v.) in moist soil.

Acceptable glasshouse uses in Great Britain

May be used as a pre-planting soil treatment for any crop.

DD

Not scheduled.
Properties—Volatile liquid consisting mainly of dichloropropane and dichloropropene, used to control certain nematodes. Irritating by contact or inhalation of vapour. Ventilate glasshouses fully during application and during the first subsequent cultivation.

Acceptable glasshouse uses in Great Britain

May be used on any crop as a pre-planting soil-injection treatment.
Note Residues in soil, or vapour in the atmosphere, are injurious to growing plants.

DDT

Not scheduled.
Properties—Acute oral L.D.50 = 300 mg/kg: dermal L.D.50 = 2500 mg/kg.
A persistent organochlorine insecticide effective against many insects but not against mites or eelworms. Very injurious when applied to the foliage of cucurbits. Tends to accumulate in animal tissues and its use may be restricted on this account.

Acceptable glasshouse uses in Great Britain

DDT may be used on any edible or non-edible crops, with the following restrictions:

1. Intervals which must elapse between the last application and harvesting an edible crop

   Sprays and dusts—14 days

   Aerosols and smokes (including DDT with BHC)—48 hr

   Thermal vaporizers—no officially recommended interval, but 48 hr is desirable if the vaporization rate was high.

   Soil drenches require no interval where there is no contamination of the aerial parts of plants.

2. The maximum amount of DDT vaporized should not exceed 0·5 g/1000 ft^3 in any period of 7 days. Glasshouses should not be occupied if DDT is vaporized at a rate exceeding 5 mg/1000 ft^3/hr and should be aired before reoccupation. If mixtures of DDT and gamma BHC are vaporized the rates of vaporization given above apply to the mixture of the two insecticides.

DEMETON-S-METHYL

Scheduled—Part III—Statutory Regulations apply.
Properties—Acute oral L.D.50 = 40 mg/kg: dermal L.D.50 = 85 mg/kg.
A systemic organophosphorus insecticide and acaricide, having medium persistence in plants. Low solubility in water, and absorbed more readily by waxy-surfaced leaves than oxydemeton-methyl. Readily translocated following root uptake. Odour, which may persist several days in flowers, more unpleasant than that of oxydemeton-methyl.

Acceptable glasshouse uses in Great Britain

May be used on any non-edible crop or on any edible crop provided there is an interval of at least 3 weeks between the last application—including application to the soil—and harvesting.

DIAZINON

Not scheduled.
Properties—Acute oral L.D.50 = 300–600 mg/kg: dermal L.D.50 = 500–>1200 mg/kg.
A non-systemic organophosphorus insecticide and acaricide effective against a wide range of pests including leafminers and scale insects.

Acceptable glasshouse uses in Great Britain

May be used on any non-edible crop and on edible crops, except as direct sprays on mushrooms, provided the following intervals elapse between the last application and harvesting

Sprays—2 weeks.

Aerosols, smokes or dusts on mushrooms, cucumbers and tomatoes—48 hr.

DICHLORVOS

Scheduled—Part III substance. Resin strips are not included in the Regulations.
Properties—Acute oral L.D.50 = 25–30 mg/kg: dermal L.D.50 = 75–900 mg/kg.
A volatile, non-systemic, organophosphorus insecticide and acaricide of moderate phytotoxicity. Very short persistence due to its rapid decomposition by moisture. Can be applied as a fumigant by slow vaporization from impregnated resin strips. The vapour is highly effective against mushroom flies and whitefly adults.

Acceptable uses in glasshouses in Great Britain

May be used without restriction on any non-edible crop and in mushroom houses during spawn-running. Its use on cropping mushrooms at a rate not exceeding

5 g a.i./1000 ft^3 is acceptable provided there is an interval of at least 24 hr between its application and picking.
It may be used on any other edible crop provided the dosage does not exceed 2 g/1000 ft^3 and there is an interval of at least 24 hr between the last application and harvesting.
Impregnated resin strips (measuring approx. 10 in. × $2\frac{1}{2}$ in. × $\frac{1}{4}$ in. and containing not more than 20% by weight of dichlorvos) may be used in glasshouses and mushroom houses at a dosage rate not exceeding 1 strip/750 ft^3.

DICOFOL

Not scheduled.
Properties—Acute oral L.D.50 = >500 mg/kg.
A relatively persistent compound closely related, chemically, to DDT, but specifically acaricidal. Has translaminar activity.

Acceptable glasshouse uses in Great Britain

May be used on any crop, non-edible or edible, provided that with indoor edible crops there is a minimum interval of 48 hr between the last application and harvesting.

DIELDRIN

Not scheduled.
Properties—Acute oral L.D.50 = 40 mg/kg: dermal L.D.50 = >100 mg/kg.
A persistent organochlorine compound. Tends to accumulate in animal tissues and to persist outside in the soil. Its use is therefore restricted.

Acceptable glasshouse uses in Great Britain

1. As a dust in potting composts used for ornamental plants, at a rate up to 2 oz of $1\frac{1}{4}$% dust/bushel of compost, for controlling vine weevil.
2. As a dip for narcissus bulbs at a rate up to 6 oz a.i./gal (0·187%) to control narcissus fly.

DIMETHOATE

Not scheduled. Preparations containing not more than 16% dimethoate are acceptable for home garden use.
Properties—Acute oral L.D.50 = 200–300 mg/kg: dermal L.D.50 = 750–1150 mg/kg.
A systemic organophosphorus insecticide and acaricide of medium persistence. Readily translocated following root uptake. Slightly soluble in water. Odour less persistent than that of some other systemic insecticides.

Acceptable glasshouse uses in Great Britain

May be used on any crop provided that, on an edible crop, there is an interval of at least 7 clear days between the last application and harvesting.

DINOBUTON

Not scheduled.
Properties—Acute oral L.D.50 = 140 mg/kg.
A somewhat phytotoxic, dinitro compound having acaricidal and fungicidal activity.

Acceptable glasshouse uses in Great Britain

May be used on any non-edible crop. May be applied to cucumbers at the following rates and frequencies per crop.
1. High volume sprays of suspensions—up to 5 applications each at not more than 8 oz/100 gal.

2. As an aerosol—up to 5 applications each at 4 g/1000 ft^3.
   In both cases there should be a minimum of 3 days between the last application and picking cucumbers.

ENDOSULFAN

Scheduled—Part II substance.
Properties—Acute oral L.D.50 = 35 mg/kg: dermal L.D.50 = 74–130 mg/kg.
A persistent organochlorine insecticide and acaricide effective against tarsonemid mites.

Acceptable glasshouse uses in Great Britain

1. On narcissus to control bulb scale mite.
2. On ornamentals to control cyclamen mite.

ENDRIN

Scheduled—Part II substance.
Properties—Acute oral L.D.50 = 3–6 mg/kg: dermal L.D.50 = 60–120 mg/kg.
A persistent organochlorine compound with restricted uses.

Acceptable glasshouse uses in Great Britain

1. On narcissus to control bulb scale mite.
2. On ornamentals to control tarsonemid mite.

FENSON

Not scheduled.
Properties—A specific acaricide of low mammalian toxicity (L.D.50 values not available) and very similar to chlorfenson. Largely superseded by more potent and less phytotoxic acaricides. Has pronounced translaminar activity and is usually applied as an aerosol.

Acceptable glasshouse uses in Great Britain

May be used on any edible or non-edible crop. No interval specified between application and harvesting edible crops.

FORMALDEHYDE

Not scheduled.
Properties—A highly irritant and phytotoxic vapour sold as a 38–40% solution in water (formalin). Solutions containing a stabilizer to prevent polymerization normally used in horticulture. Rubber boots, rubber gloves, overall, face shield, hood and respirator should be worn when washing or spraying houses with solutions. To fumigate with formaldehyde add 15 fl. oz. of the 40% solution to ½ lb potassium permanganate crystals in a glass or earthenware container using 1 container/15,000 ft^3. A respirator may be necessary where a block of houses is treated.

Acceptable glasshouse uses in Great Britain

May be used before or after any crop.

FORMOTHION

Not scheduled.
Properties—Acute oral L.D.50 = 400 mg/kg: dermal L.D.50 = 400–600 mg/kg.
A systemic organophosphorus insecticide and acaricide.

Acceptable glasshouse uses in Great Britain

May be used on any non-edible crop, and on any edible crop provided there is an interval of at least 7 days between the last application and harvesting.

LINDANE—see under BHC

MALATHION

Not scheduled.
Properties—Acute oral L.D.50 = 1400–1900 mg/kg: dermal L.D.50 = >4000 mg/kg.
A non-systemic insecticide with some acaricidal activity and of slight-moderate persistence.

Acceptable glasshouse uses in Great Britain

May be used on any non-edible or edible crop. The required minimum interval between the last application and harvesting an edible crop is 1 day, but an interval of 4 days is advisable to avoid possible taint.

METALDEHYDE

Not scheduled.
Properties—A molluscicide which apparently anaesthetizes slugs and snails, causing them to die of desiccation—consequently most effective under drying conditions. Used in baits or aqueous suspension sprays.

Acceptable glasshouse uses in Great Britain

May be used on any non-edible crop. May be used on any edible crop provided that, where sprays are used, there is an interval of 10 days between the last application and harvesting. Baits should be applied only to the soil.

METHAM-SODIUM

Not scheduled.
Properties—Acute oral L.D.50 = 285 mg/kg.
A nematicidal and fungicidal soil fumigant, commercially available as aqueous solutions containing about 33% a.i. Functions by rapid decomposition in the soil to produce methyl isothiocyanate.

Acceptable glasshouse uses in Great Britain

May be used as a pre-planting soil or compost treatment for any edible or non-edible crop.

METHIOCARB

Not scheduled.
Properties—Acute oral L.D.50 = 100 mg/kg: dermal L.D.50 = 1000 mg/kg.
A carbamate compound used as a molluscicide in dry baits.

Acceptable glasshouse uses in Great Britain

May be used on any non-edible crop, and on any edible crop provided there is an interval of at least 7 days between application and harvesting.

METHYL BROMIDE

Not scheduled.
Official recommendations apply only to the use of methyl bromide (containing 2% chloropicrin as a warning agent) as a fumigant for soil under sheeting and supplied in 1 lb sealed containers.
Properties—Gaseous at temperatures above 39° F (4 °C).
Vapour very poisonous to mammals. Used as a soil fumigant to control eelworms and certain soil fungi.

Acceptable glasshouse uses in Great Britain

Only to be used for soil or compost fumigation by contract servicing companies who are responsible for its distribution and use. Can be used prior to planting out any edible or non-edible crop. Tests for dangerous concentrations of methyl bromide should be made before entering treated houses.

METHYL ISOTHIOCYANATE

Not scheduled.
Properties—A volatile liquid highly irritating to the skin. The vapour is irritating to the nose and eyes. Usually sold diluted with other organic liquids. Used for the control of certain eelworms and soil fungi. Residues in the soil or vapour in the atmosphere are extremely injurious to plants.

Acceptable glasshouse uses in Great Britain

May be used as a pre-planting soil or compost treatment for any edible or non-edible crop.

MORPHOTHION

Not scheduled.

Properties—Acute oral L.D.50 = 200 mg/kg : dermal L.D.50 = 283 mg/kg. A systemic organophosphorus insecticide of low acaricidal activity.

Acceptable glasshouse uses in Great Britain

May be used on any non-edible crop. Not cleared for use on edible glasshouse crops.

NALED (DIBROM)

Not scheduled.
Properties—Acute oral L.D.50 = 430 mg/kg : dermal L.D.50 = 1100 mg/kg. A volatile organophosphorus compound usually applied by volatilization from heating pipes. Of short persistence.

Acceptable glasshouse uses in Great Britain

May be used on any non-edible crop. May be used on tomatoes and cucumbers (and in mushroom houses) at a rate not exceeding 3 g a.i./1000 ft^3 provided there is an interval of at least 24 hr between the last application and harvesting.

NICOTINE

Scheduled—Part III substance if the concentrate as bought contains more than 7·5% by weight of nicotine.
Properties—Acute oral L.D.50 = 70 mg/kg : dermal L.D.50 = 140 mg/kg. A naturally occurring insecticide, soluble in water, sufficiently volatile to be used as a fumigant. Of very short persistence. Controls many sap-sucking insects but not mites. Most effective at temperatures above 65 °F (16 °C).
Nicotine is very rapidly absorbed through the skin—wash off splashes immediately. The vapour may be hazardous when spraying in hot, unventilated houses. L.V. spraying requires the use of a respirator which is compulsory for applying aerosols even where there is less than 7·5% nicotine in the concentrate.

Acceptable glasshouse uses in Great Britain

May be used on any non-edible crop, and on any edible crop provided the following minimum intervals elapse between the last application and harvesting

Sprays, dusts or aerosols	48 hr
Smokes	24 hr

OXYDEMETON-METHYL

Scheduled—Part III substance. The regulations apply when low-pressure aerosol packs containing 0·25% oxydemeton-methyl are used in glasshouses.
Properties—Acute oral L.D.50 = 57 mg/kg : dermal L.D.50 = 100 mg/kg.

A water-soluble systemic organophosphorus insecticide and acaricide. Absorbed by foliage and roots, but may be less readily translocated than demeton-S-methyl. Moderately persistent. Odour not so pronounced as demeton-S-methyl, hence preferable to that compound for use on some flower crops.

Acceptable glasshouse uses in Great Britain

May be used on any non-edible crop, and on edible crops provided there is an interval of at least 3 weeks between the last application and harvesting.

PARATHION

Scheduled—Part II substance.
Properties—Acute oral L.D.50 = 3–6 mg/kg: dermal L.D.50 = 4–200 mg/kg.
A very dangerous non-systemic organophosphorus compound of medium persistence. Effective against a wide range of nematodes, insects and mites (active stages).

Acceptable glasshouse uses in Great Britain

May be used on any non-edible crop. (It is inadvisable to handle plants sprayed with parathion until 4 days have elapsed or until 1 day after using smokes or aerosols.) It may be used on tomatoes and cucumbers under the following conditions:

1. As a spray—Up to 3 applications per crop at up to 2 oz/100 gal (0·0125%) with a minimum interval of 4 weeks between the last application and harvesting.
2. As a liquid soil application—(a) Tomatoes—1 application at 6 oz/100 gal (0·0375%) at planting time *or* up to 3 applications each at $2\frac{1}{2}$ oz/100 gal (0·015%), (b) Cucumbers—1 application at 1 oz/100 gal (0·0063%) with a minimum interval of at least 4 weeks between the last application and harvesting.
3. As an aerosol—Up to 5 applications each not exceeding 1·4 g/1000 ft^3, with a minimum interval of 48 hr between the last application and harvesting.
4. As a smoke—Up to 5 applications each not exceeding 4·0 g/1000 ft^3, with a minimum interval of 24 hr between the last application and harvesting.

PARIS GREEN

Not scheduled.
Properties—No data on L.D.50 values.
A highly poisonous arsenical compound used in baits to control woodlice, leatherjackets and molluscs.

Acceptable glasshouse uses in Great Britain

Baits, applied only to the soil, staging, etc., may be used for any crop. Edible crops must not be contaminated.

PETROLEUM OIL

Not scheduled.
Properties—A very safe acaricide, also useful against scales. Spray formulations consist of a highly refined, emulsified petroleum.
Summer stock Emulsions for Glasshouse Use are covered by a M.A.F.F. Specification. Effective as H.V. sprays against all strains of red spider mite. Not persistent in action.

Acceptable glasshouse uses in Great Britain

May be used on all crops. No interval required between application and harvesting.

PHOSPHAMIDON

Scheduled—Part III substance.
Properties—Acute oral L.D.50 = 15 mg/kg: dermal L.D.50 = 125 mg/kg.
A systemic, organophosphorus insecticide and acaricide of relatively short persistence. Sometimes used to control red spider mite on ornamentals.

Acceptable glasshouse uses in Great Britain

May be used on any non-edible crop. Not accepted for use on edible crops.

PROPOXUR (ARPROCARB)

Not scheduled.
Properties—Acute oral L.D.50 = 100 mg/kg: dermal L.D.50 = > 1000 mg/kg.
A carbamate compound used in insecticidal sprays and in ant baits (1%).

Acceptable glasshouse uses in Great Britain

May be used on any non-edible crop. Not yet approved as a spray on glasshouse edible crops. (Outdoor edible crops require a minimum interval of 7 days between application and harvesting.) Edible crops must not be contaminated with baits.

PYRETHRUM (PYRETHRINS)

Not scheduled.
Properties (Pyrethrins)—Acute oral L.D.50 = 200 mg/kg: dermal L.D.50 = > 1800 mg/kg.
Pyrethrum insecticides may be prepared either from the powdered flowers of *Chrysanthemum cinerariaefolium* or, more usually, from the extracted, insecticidally active pyrethrins. The latter often contain a synergist, e.g. piperonyl butoxide.

Acceptable glasshouse uses in Great Britain

Pyrethrum and pyrethrin insecticides may be used on any crop.

QUINOMETHIONATE (OXYTHIOQUINOX)

Not scheduled.
Properties—Acute oral L.D.50 = 2000 mg/kg: dermal L.D.50 = 3000 mg/kg.
An acaricide and fungicide. Controls strains of red spider mite resistant to most other acaricides, and also powdery mildews.

Acceptable glasshouse uses in Great Britain

May be used on any non-edible crop, and on cucumbers provided there is an interval of 48 hr between the last application and harvesting.

SCHRADAN

Scheduled—Part II substance.
Properties—Acute oral L.D.50 = 5 mg/kg: dermal L.D.50 = 50–100 mg/kg.
A very poisonous, persistent, systemic organophosphorus insecticide and acaricide. Largely superseded by less poisonous and cheaper products but used occasionally on young main-crop cucumber plants and certain ornamentals.

Acceptable glasshouse uses in Great Britain

May be used on any non-edible crop. May be used on main-crop cucumbers—1 application only at 16 fl.oz a.i./acre with an interval of at least 4 weeks if applied from April to July inclusive and 6 weeks if applied between the beginning of August and mid-September. It should not be used on edible crops after 15th September.

SULFOTEP

Scheduled—Part II substance.
Properties—Acute oral L.D.50 = 1–5 mg/kg: oral L.D.50 = 20 mg/kg (rabbit).
A non-systemic organophosphorus compound of short persistence. Insecticidal and acaricidal (active stages only). Available only in smoke generators.

Acceptable glasshouse uses in Great Britain

May be used on any non-edible crop. May be used on any edible crop, at a rate not exceeding 1 g a.i./1000 ft^3, provided there is a minimum interval of 24 hr between the last application and harvesting.

TAR ACIDS

Not scheduled.
Properties—Insecticidal and fungicidal phenolic acids derived from tar distillation and varying in composition. Poisonous if swallowed, and strongly irritant to the skin—particularly cresylic acid. Household disinfectants containing tar oils should not be used in glasshouses. 'Solubilized xylenols' are miscible xylenol concentrates containing a high proportion of wetting agent to promote greater soil penetration.

Acceptable glasshouse uses in Great Britain

May be used as preplanting treatment for any crop.

TDE

Not scheduled.
Properties—Acute oral L.D.50 = 900 mg/kg: dermal L.D.50 = 10,000 mg/kg.
A persistent organochlorine compound resembling DDT and somewhat more effective against certain tortrix larvae.

Acceptable glasshouse uses in Great Britain

May be used on any non-edible crop. Insufficient information for recommendations on edible glasshouse crops.

TETRADIFON

Not scheduled.
Properties—Acute oral L.D.50 = >5000 mg/kg.
A specific acaricide capable of translaminar action. Acaricidal activity mainly results from its ovicidal effect. Slow control but persistent.

Acceptable glasshouse uses in Great Britain

May be used on any crop. No interval stipulated between application to edible crops and harvesting.

THIONAZIN

Scheduled—Part II substance.
Properties—Acute oral L.D.50 = 8–16 mg/kg: dermal L.D.50 = 10–35 mg/kg.
A systemic organophosphorus insecticide, acaricide and nematicide. Uses restricted by its high mammalian toxicity and persistence.

Acceptable glasshouse uses in Great Britain

May be used on any non-edible crop, and on tomatoes provided there is a minimum of 11 weeks between application to the soil and harvesting. (Note: it may be desirable to ventilate glasshouses in which the soil has been treated, before working therein during the first few days after its application.)

Disposal of liquid from bulb dipping baths

The manufacturer's directions should be followed.

XYLENOLS—see TAR ACIDS

Appendix II

Tolerance of Plants to Pesticides

The phytotoxicity of a pesticide is influenced by several factors including conditions of cultivation (e.g. CO_2 supplementation, soil composition and fertilizer treatment), climatic conditions, age of plant and injury by pests or disease. The response of different cultivars of a single species may differ significantly. There is some evidence that variegated-leaved varieties may be more susceptible to injury than their plain-leaved relatives. Furthermore, the nature of other materials than the pesticide in proprietary products—solvents, emulsifiers and wetting agents—may also influence phytotoxicity.

The safety of pesticides on ornamental plants which have not been subjected to 'phytotoxicity' tests is usually doubtful. With certain crops (e.g. chrysanthemums) the large number of cultivars makes it impracticable to consider all those in commercial use. If a grower has no experience of the response of a pesticide listed as being tolerated by 'most varieties', he is advised to make a preliminary test or to seek information from the N.A.A.S., the manufacturer, or the supplier of the cuttings or plants.

The appended list applies only to materials used at the concentrations mentioned in the text. 'Sprays' refers to H.V. spray application.

N.B. Remember that young seedlings and recently transplanted plants are abnormally susceptible and that, if avoidable, open flowers should not be sprayed.

Abutilon TOLERANT to aldrin (in compost), azobenzene (smoke), gamma-BHC (in summer), DDT, demeton-S-methyl, diazinon (in summer), dicofol, dieldrin (in compost) formothion, malathion, nicotine, oxydemeton-methyl, tetradifon.

Acalypha hispida, *A. wilkesiana* TOLERANT to gamma-BHC, malathion, oxydemeton-methyl, parathion (smoke): SUSCEPTIBLE to petroleum.

Achimenes longiflora TOLERANT to aldrin (in compost), DDT, diazinon, endrin, malathion, parathion (smoke), nicotine (smoke): SUSCEPTIBLE to azobenzene, gamma BHC, demeton-S-methyl, formothion, petroleum. Flowers injured by diazinon and malathion aerosols and most sprays. Do not treat in bright sunlight or in cold, humid conditions.

Adiantum cuneatum (maidenhair fern) TOLERANT to BHC (smoke), nicotine smoke: SUSCEPTIBLE to azobenzene, gamma-BHC e.c. sprays, demeton-S-methyl, diazinon, formothion, malathion, petroleum.

Aechmea rhodocyanea TOLERANT to azinphos-methyl (aerosol), gamma-BHC (smoke), DDT (smoke, aerosol and dust), nicotine: SUSCEPTIBLE to most organophosphorus sprays, petroleum. *Note*. Drain plants after spraying.

Aeschyanthus lobbianus TOLERANT to aldrin (in compost), DDT, diazinon, endrin, malathion, parathion (smoke), nicotine (smoke): SUSCEPTIBLE to azobenzene, gamma-BHC, demeton-S-methyl, formothion, petroleum. Flowers susceptible to most sprays.

Amaryllis see *Hippeastrum*

Ampelopsis (*Parthenocissus*) *heterophylla* TOLERANT to gamma-BHC (except *A. h* var. Elegance), DDT, malathion: SUSCEPTIBLE to demeton-S-methyl, diazinon, oxydemeton-methyl, parathion.

Anthurium scherzerianum (Guatemala types), *A. andreanum* TOLERANT to azinphos-methyl, gamma-BHC (smoke), DDT (aerosol or dust), nicotine (smoke): SUSCEPTIBLE to most insecticides—especially diazinon, demeton-S-methyl, dimethoate, formothion, malathion, oxydemeton-methyl, parathion, petroleum. Leaf-yellowing reaction to insecticides is slow in developing.

Antirrhinum TOLERANT to aldrin (in compost), gamma-BHC, DDT, demeton-S-methyl, diazinon (most vars.), dicofol, endrin, malathion (most vars.), naled, nicotine (most vars.), oxydemeton-methyl, parathion (smoke): SUSCEPTIBLE to azobenzene, nicotine shortly following sulphur injures flowers.

Aphelandra squarrosa TOLERANT to aldrin (in compost), azinphos-methyl (aerosol), demeton-S-methyl (soil drench), formothion, endosulfan, endrin, oxydemeton-methyl (soil drench), petroleum: SUSCEPTIBLE to dimethoate, quinomethionate.

Aralia spp. TOLERANT to gamma BHC (not frequent applications), DDT, petroleum (not applied in bright sunshine): SUSCEPTIBLE to demeton-S-methyl, dimethoate, formothion, malathion, oxydemeton-methyl, parathion, quinomethionate.

Arum lily see *Zantedeschia*

Asparagus plumosus, *A. sprengeri*. TOLERANT to aldrin (in compost) gamma-BHC (smoke), DDT, demeton-S-methyl, diazinon, dicofol, dimethoate, malathion (not in hot weather), naled, nicotine, oxydemeton-methyl, tetradifon: SUSCEPTIBLE to azinphos-methyl, petroleum. Do not apply insecticides in bright sunshine, at low temperatures or when plants are dry at the roots.

Azalea indica, Kurume hybrids TOLERANT to aldrin (in compost), gamma-BHC, DDT, demeton-S-methyl, diazinon, dicofol, dieldrin (in compost), malathion, nicotine, oxydemeton-methyl, parathion (most vars.), petroleum, quinomethionate (some vars.) sulfotep, tetradifon.

Bean French and Climbing French TOLERANT to gamma-BHC, DDT, demeton-S-methyl (not seedlings), diazinon, dicofol (not seedlings), formothion (not seedlings), malathion, nicotine, oxydemeton-methyl (not seedlings), sulfotep, tetradifon (not seedlings): SUSCEPTIBLE to azobenzene (moderately safe at temps. 60°–80 °F), fenson, petroleum.

Begonia TOLERANT to (see note below) aldrin (in compost), azobenzene, gamma-BHC, DDT (care with e.c. sprays), diazinon, dichlorvos, dicofol, dieldrin (in compost), endosulfan, endrin, fenson, malathion, naled, nicotine, oxydemeton-methyl (most vars.), parathion (incl. soil drench), quinomethionate (some vars.), sulfotep, tetradifon: SUSCEPTIBLE to dimethoate (*B. maculata*), formothion (some vars.). *Note*. Flowers may be damaged by many insecticidal sprays, particularly by organophosphorus compounds. Do not saturate compost when applying soil drenches of systemic compounds.

Bouvardia TOLERANT to malathion, parathion (except in sunny weather), petroleum: SUSCEPTIBLE to azobenzene, gamma-BHC.

Cacti TOLERANT to gamma-BHC (most spp.) including soil drenches, DDT (aerosols, dusts and w.p. sprays), demeton-S-methyl (H.V.), dicofol, dimethoate, nicotine, oxydemeton-methyl (H.V.), parathion (including soil application), tetradifon: SUSCEPTIBLE to diazinon (some spp.), malathion (some spp.), petroleum (temporary removal of 'bloom'). Epiphyllums are sensitive to dicofol and most O.P. compounds. Do not treat when dry at roots; apply sprays in cool, cloudy weather.

Calceolaria TOLERANT to azobenzene, gamma-BHC (not flowers), DDT, demeton-S-methyl (soil drench), malathion (care required), nicotine, oxydemeton-methyl (soil drench), parathion, tetradifon: SUSCEPTIBLE to dimethoate.

Calla lily see *Zantedeschia*

Camellia TOLERANT to aldrin (in compost), gamma-BHC (most vars.), demeton-S-methyl, dicofol, dieldrin (in compost), malathion, nicotine, oxydemeton-methyl, petroleum: SUSCEPTIBLE to DDT. Avoid treating open blooms.

Carnation TOLERANT to aldrin (in compost), azinphos-methyl (aerosol), azobenzene, gamma-BHC (except Ashington vars. in autumn and winter), DDT, demeton-S-methyl, diazinon, dichlorvos, dicofol, dimethoate, fenson, formothion, malathion, morphothion, naled, nicotine (not frequent applics.), oxydemeton-methyl, parathion, schradan, TDE, sulfotep, tetradifon: SUSCEPTIBLE to petroleum.

Chrysanthemum TOLERANT to aldrin (in soil), azobenzene, gamma-BHC (some e.c. sprays suspected of inducing blindness in young plants of some vars. such as Fred Shoesmith—do not apply to young plants or stock plants shortly before taking cuttings), DDT, demeton-S-methyl (most vars.—soil drenches safest), diazinon (most vars. except Mermaid, Fred Shoesmith), dichlorvos (most vars.), dicofol (not Bluechip, Momento or Iceberg), dieldrin (in soil), endrin, endosulfan, malathion (not on open buds), morphothion (most vars.), naled (most vars.), nicotine (not Iceberg—foliage should be dry when fumigated), oxydemeton-methyl (most vars.), parathion (smokes may injure flowers at low temp.), petroleum and petroleum/DDT (not Sweetheart vars., tend to encourage grey mould under humid conditions), schradan (most vars.), quinomethionate (most vars.—some risk to flowers), TDE, tetradifon (Favourite vars. may be susceptible), thionazin (rooted cuttings and young plants): SUSCEPTIBLE to binapacryl, carbaryl, demeton-S-methyl (some vars. including Balcombe Perfection, Christmas Greetings, Friendly Rival, Galaxy, Golden Balcombe Perfection, Harlequin, Hi-Fi, the Minstrels, Moonbeam, Princess Anne, Revelry, Silversmith, Taffeta, Yellow Fred Shoesmith), dichlorvos (some vars. including Calvert's Gold, Chairman, Columbia, Ducon Star, Galaxy, Golden Charm, Hurricane, Iceberg, Parade, Shasta, Taffeta, Vanguard), dimethoate, dinobuton, fenson, formothion, naled (as for dichlorvos), phosphamidon, petroleum/BHC (some vars.), schradan.

Cissus TOLERANT to gamma-BHC, nicotine, oxydemeton-methyl: SUSCEPTIBLE to diazinon, dimethoate, petroleum oil, tetradifon.

Codiaeum (Croton) TOLERANT to gamma-BHC, demeton-S-methyl, formothion, malathion, nicotine, oxydemeton-methyl, sulfotep.

Coleus TOLERANT to gamma-BHC, DDT, demeton-S-methyl malathion, nicotine, parathion: SUSCEPTIBLE to formothion, dimethoate. Do not apply sprays during bright sunshine.

Crassula see *Rochea*.

Cucumber See pp. 199–200 regarding azinphos-methyl, demeton-S-methyl, diazinon, dicofol, dimethoate, dinobuton, formothion, oxydemeton-methyl, parathion, petroleum, quinomethionate, schradan, sulfotep and tetradifon. TOLERANT to gamma-BHC (infrequent smokes on established plants after May), DDT (SOIL APPLICATION ONLY), malathion, (aerosol or spray after mid-May), nicotine; SUSCEPTIBLE to azobenzene (aerosols and, to a lesser extent smokes), binapacryl, gamma-BHC (aerosols and sprays or frequent smokes), DDT, dichlorvos, fenson, TDE.

Note. The susceptibility of cucumbers is greatly influenced during the winter and early spring by light and cultural conditions.

Cyclamen TOLERANT to aldrin (in compost), gamma-BHC (not aerosols or sprays on flowers—soil drenches safe), DDT, demeton-S-methyl (soil drenches), diazinon, dicofol, dieldrin (in compost), fenson, formothion, endosulfan, endrin, malathion, nicotine (not on flowers), oxydemeton-methyl, parathion (soil drenches only), sulfotep: SUSCEPTIBLE to binapacryl, dimethoate, dinobuton, parathion (spray or smoke).
Note. Try to avoid the necessity to treat flowering plants.

Dahlia (established rooted cuttings) TOLERANT to aldrin (in compost), azobenzene, gamma-BHC, DDT, demeton-S-methyl, diazinon, dicofol, dieldrin (in compost), dimethoate, endosulfan, endrin, fenson, formothion, malathion, nicotine, oxydemeton-methyl, petroleum, phosphamidon, TDE, tetradifon (most vars.).

Dieffenbachia TOLERANT to diazinon, dicofol, demeton-S-methyl, malathion, nicotine, parathion, petroleum.

Dipladenia splendens TOLERANT to gamma-BHC, DDT, malathion, nicotine, oxydemeton-methyl, parathion: SUSCEPTIBLE to petroleum.

Easter lily see Lilies.

Echeveria elegans, *E. harmisii* TOLERANT to aldrin in soil, gamma-BHC, demeton-S-methyl (soil drench only, applied when compost is moist), diazinon, dicofol, dieldrin (in compost), nicotine, parathion (smoke): SUSCEPTIBLE to demeton-S-methyl (spray), malathion, petroleum.

Epiphyllum see Cacti.

Erica TOLERANT to gamma BHC, DDT, dichlorvos, dieldrin (in compost), malathion, oxydemetonmethyl (most vars.), parathion (not seedlings), TDE.

Euphorbia pulcherrima, *E. splendens* TOLERANT (see note below) to aldrin (in compost), gamma-BHC, DDT, dichlorvos, endosulfan, malathion, naled, oxydemeton-methyl (some vars. *E. pulcherrima*), nicotine, parathion (smoke), sulfotep.
Note. Do not use insecticides after bracts show colour: SUSCEPTIBLE to azobenzene, demeton-S-methyl, diazinon, petroleum.

Ferns TOLERANT to gamma-BHC (dust or smoke), DDT (dust or smoke), nicotine (care on *Adiantum*), schradan (*Asplenium*): SUSCEPTIBLE to azobenzene, DDT (*Adiantum*), demeton-S-methyl (*Adiantum*), diazinon (especially *Adiantum*), dimethoate, formothion, malathion (0·05% safe most vars.), metaldehyde (on fronds or crowns), morphothion, petroleum.

Ficus TOLERANT to azinphos-methyl (aerosols), gamma-BHC, DDT (most vars.), diazinon (most vars.), malathion, nicotine, oxydemeton-methyl, parathion (most vars.), tetradifon: SUSCEPTIBLE to demeton-S-methyl (transient), dimethoate, formothion.

Freesia TOLERANT to aldrin (in soil), azobenzene, gamma-BHC, DDT, demeton-S-methyl, dichlorvos, dicofol, dimethoate, formothion, malathion, naled, nicotine, oxydemeton-methyl, parathion (smoke—not seedlings), sulfotep, tetradifon. Avoid spraying flowers, especially with malodorous insecticides.

Fuchsia TOLERANT to gamma-BHC, DDT, demeton-S-methyl (soil drench—most vars.), diazinon, dicofol, malathion, naled, nicotine, oxydemeton-methyl, TDE, sulfotep: SUSCEPTIBLE to azobenzene (young foliage), demeton-S-methyl (some vars.—transient), dimethoate, petroleum.

Gardenia TOLERANT to aldrin (in soil), gamma-BHC, DDT (use dust or smoke on flowers), dicofol, dieldrin (in soil), endosulfan, malathion, oxydemeton-methyl, nicotine, parathion, petroleum, quinomethionate: SUSCEPTIBLE to diazinon.

Geranium see Pelargonium.

Gerbera jamesonii TOLERANT to gamma-BHC (may redden some open flowers), DDT (as BHC), demeton-S-methyl (soil drench), diazinon, dicofol, naled, nicotine, parathion: SUSCEPTIBLE to azobenzene, demeton-S-methyl (spray), dimethoate (spray), formothion, malathion (reported injurious but has been applied at 10-day intervals without damage in tests).

Gladioli TOLERANT to aldrin (in soil), azobenzene, gamma-BHC, DDT (incl. corm treatment), demeton-S-methyl, diazinon (incl. soil drench), dicofol, dimethoate, fenson, formothion, malathion, nicotine, oxydemeton-methyl, parathion (incl. soil drench), TDE, tetradifon. Some e.c. sprays cause flower spotting—smokes or dusts preferable on open flowers.

Gloxinia TOLERANT to aldrin (in compost), DDT (dust safest on flowers), diazinon (care required), endosulfan, endrin, malathion (care required), nicotine, parathion: SUSCEPTIBLE to azobenzene, gamma BHC, formothion, petroleum. Flowers are sensitive to sprays. Do not treat in bright sunshine or under cold, humid conditions.

Grape TOLERANT to azobenzene (May onwards), gamma-BHC, DDT, diazinon, dicofol (May onwards), malathion, nicotine, petroleum (before flowering), TDE, tetradifon (May onwards): SUSCEPTIBLE to fenson.
Note. Spray before berries swell to avoid marring the 'bloom'. Smokes or aerosols preferable after this stage. Do not treat within 10 days of picking.

Hedera TOLERANT to aldrin (in compost), gamma-BHC (smokes—not variegated vars. in winter), DDT, diazinon, dicofol, endosulfan, endrin, malathion (not variegated vars.), nicotine: SUSCEPTIBLE to demeton-S-methyl (variegated vars.).

Hippeastrum (Amaryllis) TOLERANT to aldrin (in compost), azobenzene, gamma-BHC, DDT, demeton-S-methyl, diazinon, dimethoate, malathion, nicotine, oxydemeton-methyl, schradan, sulfotep. Avoid spraying flowers.

Hyacinth TOLERANT to aldrin (bulb dip), gamma-BHC (incl. bulb dip), DDT, diazinon, endosulfan, malathion, nicotine.

Hydrangea TOLERANT to DDT, demeton-S-methyl, diazinon, dicofol (not on soft young growth), malathion, naled, nicotine, petroleum: SUSCEPTIBLE to gamma-BHC, dimethoate, quinomethionate, thionazin.
Note. Do not spray in bright sunshine or if plants are dry at the roots.

Iris Tolerant to aldrin (bulb dip), gamma-BHC (incl. bulb dip and soil treatment), DDT, demeton-S-methyl, diazinon, dicofol, dieldrin (bulb dip), dimethoate, malathion, nicotine, parathion (incl. soil treatment).

Kalanchoe TOLERANT to demeton-S-methyl (soil drench), dicofol, diazinon, oxydemeton-methyl (soil drench), nicotine: SUSCEPTIBLE to DDT, endrin (sprays delay flowering some vars.), gamma-BHC, malathion (some vars.), parathion (some vars.).

Lettuce TOLERANT to aldrin (in soil), gamma-BHC (incl. soil treatment), DDT, demeton-S-methyl, diazinon (incl. soil treatment), dimethoate, formothion, malathion (incl. transplant dip), metaldehyde (spray), nicotine (incl. transpl. dip), oxydemeton-methyl, parathion (incl. soil treatment).

Lilies TOLERANT to aldrin (in soil), gamma-BHC, DDT, demeton-S-methyl (not Croft var. of Easter lily), diazinon, dichlorvos, dicofol, endosulfan, malathion, nicotine, oxydemeton-methyl (most vars.), schradan (most vars.): SUSCEPTIBLE to parathion (var. Fine King).

Matthiola see Stock.

Mint TOLERANT to gamma-BHC, DDT, diazinon, malathion.

Mushrooms (Sporophores) TOLERANT to gamma-BHC (smoke, aerosol and dust), DDT (as for BHC), diazinon (aerosol and dust), dichlorvos (aerosol and vapour), malathion (aerosol and dust and e.c. drench), pyrethrum (aerosol and dust), sulfotep (smoke): SUSCEPTIBLE to most sprays (caps marked). Compost and casing treatments, see pp. 315, 324.

Narcissus TOLERANT to aldrin (cold bulb dip before roots start), gamma-BHC (including bulb dip—as aldrin), DDT, diazinon, dieldrin (as aldrin), endosulfan, endrin, malathion, nicotine, thionazin (bulb dip).

Nectarine see Peach.

Orchids TOLERANT to gamma-BHC, DDT, diazinon (incl. compost treatment), dicofol, dimethoate, fenson (most vars.), malathion (most vars.), metaldehyde (spray), nicotine, naled, oxydemeton-methyl, sulfotep (smoke), tetradifon: SUSCEPTIBLE to demeton-S-methyl (*Colanthe*, *Phragmopedilum* and *Stanhopaea* spp.), petroleum (some vars.—particularly cattleyas).
Note. Spray under drying conditions but not in bright sunshine, drain sheath-leaved plants after spraying.

Palms TOLERANT to gamma-BHC (incl. soil drench), DDT, diazinon, malathion: SUSCEPTIBLE to petroleum.

Peach, Nectarine TOLERANT to azobenzene, gamma-BHC (not within several weeks of ripening as may taint), DDT, demeton-S-methyl, diazinon, dichlorvos, dicofol, dimethoate, fenson (as BHC), malathion (as BHC), nicotine, oxydemeton-methyl, petroleum (after picking), tetradifon.

Pelargonium, Geranium TOLERANT to azobenzene, gamma BHC, DDT, demeton-S-methyl (not sprays on zonals), diazinon, dicofol, dimethoate (not sprays on zonals), endosulfan, endrin, formothion, malathion, nicotine, oxydemeton-methyl, sulfotep, TDE, tetradifon: SUSCEPTIBLE to parathion (some vars. in hot weather), petroleum.

Peperomia TOLERANT to aldrin (in compost), gamma-BHC, DDT, diazinon (most vars.), dicofol, endosulfan, malathion, naled, nicotine, oxydemeton-methyl: SUSCEPTIBLE to parathion, petroleum.
Note. Spray under drying conditions but not in sunshine, compost must be moist.

Petunia TOLERANT to gamma-BHC, DDT, diazinon, dicofol (not seedlings), dimethoate (not seedlings), endrin, malathion (not aerosol on seedlings), oxydemeton-methyl (not seedlings), tetradifon.

Philodendron TOLERANT to gamma-BHC, demeton-S-methyl (soil drench), diazinon, dicofol, endosulfan, malathion, nicotine.

Pilea TOLERANT to gamma-BHC, demeton-S-methyl (soil drench), oxydemeton-methyl: SUSCEPTIBLE to azobenzene, diazinon, malathion.

Primula TOLERANT to gamma-BHC (flowers sensitive), DDT (smoke, dust or spray), demeton-S-methyl (care with *P. malacoides*), diazinon, dicofol, dimethoate (care required), formothion (care with *P. malacoides*), malathion, naled, oxydemeton-methyl, parathion (incl. soil drench), TDE: SUSCEPTIBLE to dieldrin (in compost), phosphamidon.
Note. Spray under drying conditions but not during sunshine: do not apply insecticides when plants are dry at roots.

Rhoicissus rhomboidea TOLERANT to diazinon, malathion, naled, nicotine, oxydemeton-methyl, parathion, petroleum: SUSCEPTIBLE to gamma-BHC, DDT.

Rochea (*Crassula*) TOLERANT to azinphos-methyl, gamma-BHC, dicofol, diazinon, nicotine: SUSCEPTIBLE to DDT, demeton-S-methyl (some vars.), dimethoate (some vars.), malathion (some vars.), parathion, petroleum, phosphamidon, tetradifon.

Rose TOLERANT to BHC (smoke), DDT (not e.c. sprays when in bud), demeton-S-methyl (most vars.—not Roselandia or Baccara), diazinon, dicofol, dimethoate, endosulfan, malathion, naled (most vars.—not Baccara, Dr. Verhage, Golden Rapture, Lady Sylvia or White Butterfly when in bud), nicotine (not frequent applications), oxydemeton-methyl, petroleum (not within 10 days of sulphur), TDE, tetradifon (most vars.—not on yellow vars. and White Butterfly and Cinderella): SUSCEPTIBLE to azobenzene, binapacryl (except when 'cut back'), gamma-BHC (sprays), dichlorvos, dinobuton (as binapacryl), parathion (when in bud), quinomethionate (except when cut back).

Note. The soft growth and flowers of glasshouse roses are very sensitive.

Saintpaulia TOLERANT to gamma-BHC (not flowers), DDT, demeton-S-methyl (soil drench), diazinon (not flowers), endosulfan, endrin, malathion, naled, schradan, tetradifon: SUSCEPTIBLE to nicotine.

Note. Saintpaulia is susceptible to most insecticides applied in sunshine, under slow drying conditions or if the compost is dry. Use tepid water for sprays and place sprayed plants on their sides to drain.

Salvia splendens TOLERANT to gamma-BHC, DDT, demeton-S-methyl, dimethoate, diazinon, formothion, malathion, nicotine, oxydemeton-methyl, phosphamidon: SUSCEPTIBLE to petroleum.

Smilax as for Asparagus fern.

Solanum capsicastrum TOLERANT to azobenzene, gamma-BHC (not seedlings), DDT, demeton-S-methyl, diazinon, dicofol, dimethoate, formothion, malathion, nicotine, oxydemeton-methyl, tetradifon: SUSCEPTIBLE to petroleum (when berries formed).

Stock TOLERANT to gamma-BHC, DDT, demeton-S-methyl, diazinon, dicofol, dimethoate, formothion, malathion, nicotine, oxydemeton-methyl: SUSCEPTIBLE to petroleum.

Strawberry TOLERANT to gamma-BHC (before flower set to avoid taint), DDT, demeton-S-methyl, diazinon, dicofol, dieldrin (in soil), dimethoate, endosulfan, endrin, formothion, malathion (not after berries begin to colour), nicotine, oxydemeton-methyl, parathion, TDE, tetradifon.

Sweet Pea TOLERANT to gamma-BHC, DDT, demeton-S-methyl (soil drench), diazinon, malathion (most vars.): SUSCEPTIBLE to azobenzene, dicofol, petroleum, quinomethionate.

Tomato TOLERANT to azinphos-methyl, azobenzene (established plants April onwards), gamma-BHC (not young plants—particularly e.c. sprays), DDT (not seedlings or e.c. sprays on young plants), demeton-S-methyl, diazinon (incl. soil drench), dichlorvos (fruiting plants only), dicofol (see p. 200), dimethoate, fenson (see p. 201), formothion, malathion, nicotine, oxydemeton-methyl, parathion, petroleum (fruiting plants), sulfotep, TDE, tetradifon (see p. 201): Young plants SUSCEPTIBLE to gamma-BHC, dichlorvos, petroleum (oedema under some cultural conditions), traces of methyl isothiocyanate (see p. 264).

Tulip TOLERANT to gamma-BHC (incl. bulb dip), DDT, demeton-S-methyl, diazinon, dimethoate, formothion, malathion, naled, nicotine, oxydemeton-methyl, thionazin.

Note. Avoid applying sprays or aerosols to flowers.

Zantedeschia (Arum or Calla-Lily) TOLERANT to azobenzene, gamma-BHC, DDT, demeton-S-methyl, diazinon, dichlorvos, dicofol, dimethoate, fenson, formothion, malathion, naled, nicotine, oxydemeton-methyl, petroleum, schradan, sulfotep, tetradifon.

Zinnia TOLERANT to gamma-BHC, DDT, demeton-S-methyl, (soil drench), diazinon, dicofol, malathion (not young plants—care on older plants), nicotine: SUSCEPTIBLE to demeton-S-methyl (sprays), dimethoate, formothion, petroleum.

ABBREVIATIONS

ac	acre(s)	l.	litre
a.i.	active ingredient	kg	kilogrammes
°C	degree (Celsius) (Centigrade)	lb	pound(s)
cc	cubic centimetre (equivalent to ml.)	m	metre(s)
cwt	hundredweight(s)	m^2	square metre(s)
°F	degree Fahrenheit	m^3	cubic metre(s)
fl.	fluid	min	minute
ft	foot (feet)	ml.	millilitre(s)
ft^2	square foot	oz	ounce(s)
ft^3	cubic foot	st	standard
g	gramme(s)	temp.	temperature
gal	gallon(s)	p.p.m.	parts per million
ha	hectare(s)	yd	yard(s)
hr	hour	yd^2	square yard(s)
in.	inch(es)	w/v	weight in volume
		μ	micro (10^{-6} or one millionth)

EQUIVALENTS

Volume

1 fl. oz = 28·41 ml. = 0·962 fl. oz (U.S.) = 8 standard teaspoonsful = 2 st. tablespoonsful
1 pint = 568 ml. = 20 fl. oz = 1·2 pint (U.S.)
1 gal = 4·55 l. = 160 fl. oz = 1·2 gal (U.S.)
1 bushel = 36·37 l. = 8 gal = 1·25 ft^3 = 70–80 lb of potting soil
1 ft^3 = 0·0283 m^3 = 28·3 l. = 6·23 gal
1 yd^3 = 0·765 m^3 = 21 bushels = 14–16 cwt of potting soil = 7½ cwt (approx) prepared mushroom compost
1 ml./1000 ft^3 = 3·52 ml./100 m^3
8 fl. oz/100 gal = 0·5 ml./l. = 2½ ml./gal = 0·05%
1 fl. oz/1000 ft^3 = 100 ml./100 m^3 = 28·4 ml./1000 ft^3
1 fl. oz/ft^2 = 306 ml./m^2 = 9 fl. oz/yd^2
1 pint/ft^2 = 6·12 l./m^2
1 gal/yd^2 = 4·98 l./m^2 = 1·25 gal (U.S.)/yd^2
1 gal/100 ft^2 = 49·5 l./100 m^2 = 416 gal/ac

Weight

1 oz = 28·35 g = wt. of 1 fl. oz water
1 lb = 453·6 g = 16 oz
1 cwt = 50·8 kg = 112 lb
1 ton (long) = 1016 kg = 1·016 tonne = 2240 lb
1 oz/yd^2 = 33·9 g/m^2 = 303 lb/ac
1 oz/gal = 6·25 g/l. = 0·625% w/v
1 lb/100 gal = 100 g/100 l. = 0·1% w/v
1 lb/ac = 1·12 kg/ha
1 g/1000 ft^3 = 3·52 g/100 m^3 = 35 μ g/l.
1 oz/1000 ft^3 = 100 g/100 m^3
1 oz/yd^3 = 37g/m^3 = 40 p.p.m. in potting compost = approx. 80 p.p.m. in prepared mushroom compost

Area

1 ft^2 = 929 cm^2 = 0·093 m^2 = 144 in^2
1 yd^2 = 0·836 m^2 = 9 ft^2
100 ft^2 = 9·3 m^2 = 11·1 yd^2
1 ac = 0·405 ha = 4840 yd^2 = 2½ vergees (Guernsey) = 2¼ vergees (Jersey)

Length

1 in. = 2·54 cm
1 ft = 30·48 cm = 12 in.
1 yd = 0·914 m = 3 ft

Concentration × Time Product (C.T.P.) in fumigation mg hr/l. = oz. hr/1000 ft^3

Capacity (cubic) of glasshouses and mushroom houses

A. Even span, lean-to and flat-roofed houses.
Calculate the average height if necessary:
Average height (ft) = height to ridge (ft) + height to eaves (ft) ÷ 2
Capacity in cubic feet (ft^3) = average height (ft) × length (ft) × width (ft)

B. Uneven span or three-quarter span roofed houses.
Treat each side of the ridge as a separate lean-to house as at A and add the results.

Capacity of rectangular tanks in gallons

Height (in.) × length (in.) × width (in.) ÷ 277
or Height (ft) × length (ft) × width (ft) × 6·23

Capacity of cylindrical tanks in gallons

Diameter (in.) × diameter (in.) × height (in.) ÷ 353
or Diameter (ft) × diameter (ft) × height (ft) × 4·9

Capacity of rectangular tanks in litres

Length (cm) × width (cm) × height (cm) ÷ 1000

Capacity of cylindrical tanks in litres

Diameter (cm) × diameter (cm) × height (cm) ÷ 1274

DILUTION TABLES

A. Quantities of liquid concentrates required to prepare 10 gal of spray

Percentage concentration required on dilution	Percentage of active ingredient in concentrate													
	5	**10**	**15**	**18**	**20**	**25**	**30**	**40**	**50**	**60**	**70**	**80**	**100**	
0·002	18	9	6	5	4½	3½	3	2¼	1¾	1½	1¼	1	1	
0·004	1¼	¾	15	10	9	7	6	4½	3½	3	2½	2	1¾	
0·005	1½	¾	15	17	12	9	8	5½	4½	4	3½	2¾	2	
0·01	3¼	1½	1	25	23	18	16	11	9	7½	6½	5½	4¼	Millilitres
0·02	6½	3¼	2	1¾	1½	1¼	1	24	18	15	13	11¼	8½	ml. (or cc)
0·025	8	4	2¾	2¼	2	1½	1½	1	23	19	16	14	11	1 teaspoon = 4 ml. approx.
0·03	9½	4¾	3¼	2¾	2¼	2	1¾	1¼	1	26	19	17	13½	
0·04	12¾	6½	4¼	3½	3¼	2½	2	1½	1¼	1	23	22½	18	
0·05	16	8	5¼	4½	4	3¼	2¾	2	1½	1¼	1¼	1	23	
0·08	1¼	12¾	8½	7	6½	5	4¼	3¼	2½	2	1¾	1½	1½	
0·1	1½	16	10¾	8¾	8	6½	5¼	4	3¼	2¾	2¼	2	1¾	Fluid ounces
0·2	3¼	1½	1	17¾	16	13	10½	8	6½	5¼	4½	4	3¼	(fl. oz)
0·5	8	4	2¾	2¼	2	1½	1¼	1	16	13¼	11½	10	8	
1·0	16	8	5¼	4½	4	3¼	2½	2¼	2	1¾	1¼	1	¾	Pints

B. Weights of wettable powder concentrates required to prepare 10 gal of spray

Percentage concentration required on dilution	Percentage of active ingredient in concentrate										
	10	15	20	25	30	40	50	60	70	80	100
0·002	9	6	4½	3½	3	2¼	1¾	1½	1¼	1¼	1
0·004	½	12	9	7	6	4½	3½	3	2½	2	1¾
0·005	¾	½	½	½	¼	5½	4½	4	3½	2¾	2
0·01	1½	1	¾	¾	½	½	¼	¼	6½	5½	4¼
0·02	3¼	2	1½	1¼	1	¾	¾	½	½	11¼	8½
0·025	4	2¾	2	1½	1½	1	¾	¾	½	½	11
0·03	4¾	3¼	2¼	2	1¾	1¼	1	1	¾	½	½
0·04	6½	4¼	3¼	2½	2	1¾	1¼	1	¾	¾	½
0·05	8	5¼	4	3¼	2¾	2	1½	1¼	1¼	1	¾
0·08	12¾	8½	6½	5	4¼	3¼	1	2	1¾	1½	1½
0·1	1	10¾	8	6½	5¼	4	3¼	2¾	2¼	2	1¾
0·2	2	1¼	1	13	10½	8	6½	5¼	4½	4	3¼
0·5	5	3¼	2½	2	1¼	1¼	1	13¼	11½	10	8
1·0	10	6½	5	4	3¼	2½	2	1¾	1½	1¼	1

Grammes (g)
(1 sixpenny bit = 2½ g approx.
1 shilling bit = 5 g approx.)

Ounces (advp.)—(oz)
(1 bronze threepenny bit = ¼ oz approx.)

Pounds (advp.)—(lb)

Index

Index

Bold figures indicate main references. Italic figures indicate pages bearing illustrations.